John Carstarphen
U. of Nevada
Reno, Nev.

JOHN CARSTARPHEN
P. O. Box 5773
Reno, Nevada 89503

$$d \ln x = \frac{dx}{x}$$

$$y = e^x \text{ or } \ln y = x$$

$$dx^3 = 3x^2 \quad \int 3x^2 = \frac{3x^3}{3} = x^3 \checkmark$$

$$\int x\,dy + y\,dx = xy$$

$$d(\cos\theta) = -\sin\theta$$

DIFFERENTIAL EQUATIONS

Differential Equations

LESTER R. FORD, Ph. D.

Professor Emeritus of Mathematics
Illinois Institute of Technology

SECOND EDITION

McGRAW-HILL BOOK COMPANY, INC.

New York   Toronto   London

1955

DIFFERENTIAL EQUATIONS

Copyright © 1955 by the McGraw-Hill Book Company, Inc.

Copyright, 1933, by the McGraw-Hill Book Company, Inc. Printed in the United States of America. All rights reserved. This book, or parts thereof, may not be reproduced in any form without permission of the publishers.

Library of Congress Catalog Card Number 54-12675

IX

21509

THE MAPLE PRESS COMPANY, YORK, PA.

PREFACE

This second edition follows the use of the book as a text for more than twenty years, and this experience has guided the revision. Many parts have been rewritten and rearranged in the interests of a clearer presentation, a smoother and more natural approach, and a more teachable body of material. More exercises have been worked out as a guide to the student. Numerous additions to the lists of problems include many simple exercises as well as those which challenge the student's ability and insight. In response to a persistent demand a set of review exercises has been put at the end of Chap. 2. A complete set of answers has been included in the book.

The strictly new subject matter has, for the most part, been often used by the author as supplementary material: Riccati's equation, elastic vibrations, planetary motion, and, at the end of the book, the simple numerical methods which are used in the approximate solution of Laplace's equation. There have also been added several pages on the Laplace transform, designed to give the student some acquaintance with this popular tool.

LESTER R. FORD

A SHORT COURSE

The following sections are suggested for a one-semester course of four hours. With some variations this material has stood the test of long experience. The basic theory and the chief applications covered are: (1) an introduction to differential equations and various elementary applications; (2) the highly useful linear equations; (3) a classical equation in the complex domain; (4) an existence theorem; (5) the partial differential equation of the vibrating string; (6) a treatment of planetary motion.

For a still shorter course the instructor must sacrifice some of these materials. For a three-hour course the author suggests the omission of some of the starred items.

The book is rich in materials for special assignments to able students and in intriguing problems.

Chap. 1: 1–13, 15–17.
      2: 1, 2*, 3*, 4, 7, 8.
      3: 1–10, 13*.
      4: 1, 2, 5–9.
      5: 1, 2*–5*.
      6: 4, 5, 11*, 12*.
      7: 1, 2, 9.
    11: 1*–4*, 6*.

CONTENTS

Preface v

A Short Course vii

CHAPTER 1
INTRODUCTION TO DIFFERENTIAL EQUATIONS

1-1. Differential Equations 1
1-2. Solutions of Differential Equations 2
1-3. First Method of Solution. Variables Separable 4
1-4. Arbitrary Constants 5
1-5. Solutions Satisfying Specified Conditions 6
1-6. Velocities. 6
1-7. The Direction Field 9
1-8. The Local Behavior of a Solution 10
1-9. Solutions in Series 12
1-10. One-parameter Families of Curves 14
1-11. Clairaut's Equation. 16
1-12. Singular Solutions 17
1-13. n-Parameter Families 19
1-14. Geometrical Applications 20
1-15. Trajectories 22
1-16. Equations of the Second Order Solved by First-order Methods . . 24
1-17. Motion of a Particle 26

CHAPTER 2
SPECIAL METHODS FOR THE EQUATION OF THE FIRST ORDER

2-1. Linear Equations 33
2-2. Bernoulli's Equation 36
2-3. Riccati's Equation 38
2-4. Homogeneous Equations 40
2-5. The Equations $y' = f\left(\dfrac{l_1x + m_1y + q_1}{l_2x + m_2y + q_2}\right)$ and $y' = F(ax + by)$ . . 43
2-6. The Linear Fractional Equation 45
2-7. Exact Equations 49
2-8. Integrating Factors 52
2-9. On the Making of Rules 55

CHAPTER 3
LINEAR EQUATIONS OF THE SECOND ORDER

3-1. Linear Equations 60
3-2. The Reduced Equation 61
3-3. Linear Dependence. Wronskians 62
3-4. The Wronskian of Two Solutions 63
3-5. The Reduced Equation with Constant Coefficients 65

CONTENTS

3-6. Roots of Solutions of the Reduced Equation 69
3-7. The Complete Equation. Method of Undetermined Coefficients . . 70
3-8. The Method of the Variation of Parameters 72
3-9. The Use of a Known Solution of the Reduced Equation 73
3-10. The Euler Equation 76
3-11. Exact Equations 77
3-12. Integrating Factors 78
3-13. Vibrations 79

CHAPTER 4
GENERAL LINEAR EQUATIONS

4-1. The Linear Equation 84
4-2. The Reduced Equation 84
4-3. Wronskians 85
4-4. Gramians 88
4-5. The Reduced Equation with Constant Coefficients 89
4-6. The Complete Equation with Constant Coefficients. Method of Undetermined Coefficients 91
4-7. Symbolic Methods 92
4-8. Method of the Variation of Parameters 96
4-9. Reduction of the Order of the Equation 97

LAPLACE TRANSFORMS

4-10. The Laplace Transform 99
4-11. Laplace Transforms of Differential Equations 101
4-12. Uniqueness of the Inverse Transformation 103
4-13. Further Properties 105

CHAPTER 5
THE METHOD OF SUCCESSIVE APPROXIMATIONS

5-1. The Method 109
5-2. The Successive Approximations 111
5-3. The Lipschitz Condition 112
5-4. The Convergence 113
5-5. Uniqueness of the Solution 115
5-6. Remarks 116
5-7. Alteration of the Function 117
5-8. Change of Initial Conditions 118

CHAPTER 6
SYSTEMS OF ORDINARY EQUATIONS

6-1. Equations of the First Order 120
6-2. Linear Systems 122
6-3. Geometrical Interpretations 124
6-4. Elementary Methods of Solution 127
6-5. Two Applications 131
6-6. The Equation of the nth Order 133
6-7. Systems of Equations of Higher Order 135
6-8. Total Differential Equations 135
6-9. The Integrable Case 137
6-10. Systems of Linear Equations 141
6-11. The Motion of a Particle 144
6-12. Planetary Motion 146

CHAPTER 7
CERTAIN CLASSICAL EQUATIONS

7-1.	Analytic Solutions	150
7-2.	Regular Singular Points	152
7-3.	The Hypergeometric Differential Equation	155
7-4.	The Legendre Differential Equation	158
7-5.	Legendre Polynomials	159
7-6.	Integral Properties of Legendre Polynomials	160
7-7.	Expansion of an Arbitrary Function	163
7-8.	Roots of Legendre Polynomials	163
7-9.	Bessel's Differential Equation	165
7-10.	Roots of Bessel Functions	168
7-11.	Integral Properties. Expansions	171

CHAPTER 8
INTERPOLATION AND NUMERICAL INTEGRATION

8-1.	Interpolation	174
8-2.	Parabolic Interpolation. Lagrange's Formula	175
8-3.	Finite Differences	177
8-4.	Difference Tables	178
8-5.	Newton's Interpolation Formula	179
8-6.	The Error in Parabolic Interpolation	182
8-7.	Linear Interpolation	184
8-8.	Numerical Integration	185
8-9.	Simpson's Rule	186
8-10.	The Error in Simpson's Rule	188
8-11.	An Application of Legendre Polynomials	190
8-12.	Finite Difference Integration Formulas	191
8-13.	An Integration Formula Involving Derivatives	192
8-14.	Numerical Differentiation	194

Symbolic Methods

8-15.	Operators	197
8-16.	The Algebra of Operators	197
8-17.	The Fundamental Equations	198
8-18.	Application to a Difference Table	199
8-19.	The Operators D^{-1} and Δ^{-1}	200
8-20.	Various Formulas	202

CHAPTER 9
THE NUMERICAL SOLUTION OF DIFFERENTIAL EQUATIONS

9-1.	The Start of the Solution	208
9-2.	The Subsequent Process	210
9-3.	Aids to Good Guessing	212
9-4.	Checking	214
9-5.	Practical Hints	215
9-6.	Questions of Convergence	216
9-7.	Equations of Higher Order and Systems of Equations	217
9-8.	The Cauchy-Lipschitz Method	220
9-9.	The Runge-Kutta Formulas	221
9-10.	Finite Difference Methods	223

CHAPTER 10
PARTIAL DIFFERENTIAL EQUATIONS OF THE FIRST ORDER

- 10-1. Partial Differential Equations 225
- 10-2. Derivation of the Equation of the First Order 226
- 10-3. Types of Solutions 228
- 10-4. Geometric Interpretation 230
- 10-5. Envelopes 231
- 10-6. The Complete Integral. 233
- 10-7. Characteristic Strips 235
- 10-8. Differential Equations of the Characteristic Strip 236
- 10-9. The Integral Surface through a Given Curve 239
- 10-10. Complete Solutions. Charpit's Method. 242
- 10-11. Linear Partial Differential Equations 245
- 10-12. Integrating Factors. 247
- 10-13. Equations in Three or More Independent Variables 248

CHAPTER 11
PARTIAL DIFFERENTIAL EQUATIONS OF THE SECOND ORDER

- 11-1. Certain Elementary Cases. 251
- 11-2. Euler's Equation 253
- 11-3. Homogeneous Linear Equations with Constant Coefficients 257
- 11-4. The Completely Linear Equation 259
- 11-5. Linear with Constant Coefficients 262
- 11-6. The Vibrating String 265
- 11-7. The Vibrating Membrane 269
- 11-8. The Conduction of Heat 272
- 11-9. Laplace's Equation. 275
- 11-10. Approximate Solutions. 276

Answers 279

Index 289

DIFFERENTIAL EQUATIONS

CHAPTER 1

INTRODUCTION TO DIFFERENTIAL EQUATIONS

1-1. Differential Equations. An equation

$$f\left(x, y, \frac{dy}{dx}, \frac{d^2y}{dx^2}, \ldots, \frac{d^ny}{dx^n}\right) = 0 \tag{1-1}$$

involving a function $y(x)$ and certain of its derivatives is called a *differential equation*.

By the *order* of a differential equation is meant the order of the highest derivative which appears. Thus the differential equations

$$\frac{dy}{dx} = y - 2x^2 \tag{1-2}$$

$$\frac{d^3y}{dx^3} - \frac{dy}{dx} = 0 \tag{1-3}$$

and

$$\frac{d^2s}{dt^2} - 3\frac{ds}{dt} + 2s = t^2 \tag{1-4}$$

are of the first, third, and second orders, respectively.

The study of differential equations is important because of the frequency with which they arise in the applications of mathematics to scientific problems. The student has already found in his study of the calculus that the derivative appears in a great variety of problems—as the slope of a curve, as a velocity or acceleration in the study of motion, as the rate of change of some function in a great many connections. Now, in the exact sciences a vast number of problems arise in which the quantity whose value is sought is known only through some relation satisfied by its derivative. Thus the velocity or acceleration of a moving body may be known and the distance traveled in a given time required; or the rate at which a quantity is increasing or decreasing may be given and the magnitude of the quantity itself be sought. In such cases the conditions of the problem supply us with a differential equation satisfied by the unknown function, and we are faced with the problem of finding what the function is. The process of finding the function that satisfies a differential equation is called *solving* the equation.

The differential equations at the beginning of this section are of a particular kind. Each contains one independent variable and one dependent variable (or function). More generally, *a differential equation is an equation connecting certain independent variables, certain functions (dependent variables) of these variables, and certain derivatives of these functions with respect to the independent variables.* Differential equations are divided into classes according as there are one or more independent variables.

If there is a single independent variable, so that the derivatives are ordinary derivatives, the equation is called an *ordinary differential equation.*

If there are two or more independent variables, so that the derivatives are partial derivatives, the equation is called a *partial differential equation.* Thus

$$\frac{\partial^2 z}{\partial x^2} + \frac{\partial z}{\partial y} = yz$$

is a partial differential equation. Here z, the dependent variable, is a function of the two independent variables x and y.

In later chapters we shall consider simultaneous equations in which there are two or more dependent variables satisfying two or more differential equations. For example, it might be proposed to find two functions, $x(t)$ and $y(t)$, satisfying the equations

$$\frac{dx}{dt} = 2y + x \qquad \frac{dy}{dt} = 3y + 4x$$

In the early part of our study, however, we shall be concerned with a single equation with one dependent variable. We shall begin with the simplest case, the equation of the first order. This equation, when solved for the derivative, appears in the form

$$\frac{dy}{dx} = f(x,y)$$

1-2. Solutions of Differential Equations. A relation $y = g(x)$ is a *solution* or *integral* of (1-1) if

$$f[x, g(x), g'(x), \ldots, g^{(n)}(x)] \equiv 0$$

that is, y is such a function of x that if y and its derivatives be expressed in terms of x and substituted into the differential equation, the equation is identically satisfied. Thus $y = x^2 + 2x + 2$ is a solution of (1-2), for on making the substitution we have $2x + 2 = x^2 + 2x + 2 - x^2$, which holds for all values of x. Similarly, $y = e^x$ is an integral of (1-3), for on substituting in (1-3) we have the identity $e^x - e^x = 0$.

It is frequently neither convenient nor desirable to express the dependent variable in terms of the independent variable. An implicit relation,

$F(x,y) = 0$, is a solution, if when solved explicitly for y in terms of x, it yields a solution in the way described above. However, the implicit relation can be differentiated and the derivatives found in terms of x and y and tested by substitution in the differential equation without the necessity of solving explicitly. A test can thus be made when it is difficult or altogether impossible to solve for y in terms of x. For example, let us show that
$$x^2 = 2y^2 \log y$$
is a solution of the differential equation
$$\frac{dy}{dx} = \frac{xy}{x^2 + y^2}$$

Differentiating the proposed solution, we have
$$2x = (2y + 4y \log y) \frac{dy}{dx}$$

Solving for dy/dx and replacing $\log y$ by its value, $x^2/2y^2$,
$$\frac{dy}{dx} = \frac{x}{y + 2y \log y} = \frac{x}{y + \frac{x^2}{y}} = \frac{xy}{x^2 + y^2}$$

The differential equation is satisfied.

By differentiating and substituting in the differential equation we can test whether a given relation is a solution of a given differential equation, but we have as yet no clue as to how the solution is found. A considerable part of our further study will consist in devising methods of finding solutions of particular classes of equations.

The student has already had practice in solving differential equations of a particularly simple form. The problem of integration is to find a function whose derivative is a given function of the independent variable
$$\frac{dy}{dx} = f(x)$$
The most general solution of this differential equation is
$$y = \int f(x)\, dx + C$$
where C is an arbitrary constant.

We shall now consider a simple problem in which a method of solution readily occurs to us.

Problem. The function e^x has the property that the derivative is equal to the function. Find the most general function with this property.

Let y be the function; then the required property is expressed by the differential equation
$$\frac{dy}{dx} = y \tag{1-5}$$

One solution of this is obviously $y = 0$. If $y \neq 0$, we can divide by y and put the equation in the form

$$\frac{dy}{y} = dx$$

The first member is the differential of log y or log $(-y)$, according as y is positive or negative; and the second is the differential of x. Since two functions whose differentials are equal differ at most by a constant, we have, integrating,

$$\log(\pm y) = x + C$$

or

$$y = \pm e^{x+C}$$

where C is an arbitrary constant. This value of y, together with $y = 0$, gives all functions with the required property.

We can put the result in a different form by setting

$$\pm e^C = K$$

thus changing the form of the constant. Then

$$y = Ke^x \tag{1-6}$$

1-3. First Method of Solution. Variables Separable. The solution of the preceding problem was effected by writing the equation in two terms, one of which is a function of x alone, the other a function of y alone, from which an integration gave the solution at once. If a differential equation can be written in the form

$$M(x)\,dx + N(y)\,dy = 0 \tag{1-7}$$

where, as the notation indicates, M is a function of x alone and N a function of y alone, the solution is

$$\int M(x)\,dx + \int N(y)\,dy = C \tag{1-8}$$

where C is an arbitrary constant. The problem is then reduced to the problem of evaluating the two integrals in (1-8). In Eq. (1-7) we say that the variables are *separated*.

It is clear that we can separate the variables in only a limited class of differential equations. It happens, however, that in many of the simpler equations met with in the applications of mathematics the variables can be separated, so that the method is one of importance.

Example. Solve the equation

$$\frac{dy}{dx} = \frac{x}{y\sqrt{1-x^2}}$$

Separating the variables,

$$y\,dy - \frac{x\,dx}{\sqrt{1-x^2}} = 0$$

Integrating,
$$\tfrac{1}{2}y^2 + \sqrt{1-x^2} = C$$
or
$$y^2 + 2\sqrt{1-x^2} = C' \qquad C' = 2C$$

1-4. Arbitrary Constants. The number of solutions of Eq. (1-7) is infinite, since C in (1-8) may be given any value. The infinitude of its solutions is characteristic of a differential equation. In the process of solution of a differential equation of the first order there comes a step by which the differentials or derivatives are removed by an integration, and this integration introduces an arbitrary constant.[1] The solution containing this arbitrary constant is called the *general solution* of the equation. A solution which results from giving a particular value to the arbitrary constant is called a *particular solution*. Thus $y = 3e^x$ and $y = -2e^x$ are particular solutions of (1-5).

By saying that a constant is arbitrary we mean that it can be given any value within a certain range of values. Frequently any value whatever may be given the constant; sometimes only a limited range of values will yield real solutions. For example, in $y = Cx$, C may have any value; in $x^2 + y^2 = C$ only positive values of C give y as a real function of x. In simplifying the solution of a differential equation it is advantageous to replace a function of an arbitrary constant by a new constant since the function is itself an arbitrary constant. Thus, in the solution of (1-5) we replaced $\pm e^c$ by the simpler constant K.

In general, a function of one or more arbitrary constants is itself an arbitrary constant. An expression may have more apparent arbitrary constants than essential ones, for we may be able to replace the constants that appear by a smaller number. For example, $y = Ke^{x+c}$ is a solution of (1-5) with two constants, but if we replace Ke^c by the new constant C', we have $y = C'e^x$ in which there is a single constant. Again

$$y = x^2 + A + B$$

is no more general than $y = x^2 + C$, for to give arbitrary values to A and B is equivalent to giving arbitrary values to C. A less obvious case is the equation

$$x^3y^3 + C_1x^2y^2 + C_2xy + C_3 = 0$$

which contains three constants. But this is a cubic equation in xy whose solution is some function of the coefficients

$$xy = f(C_1, C_2, C_3)$$

and this can be written in the equally general form

$$xy = C$$

---

[1] The presence of an arbitrary constant in the solution will be given a rigorous demonstration later.

It will be found that an equation of the nth order has a solution containing n essential arbitrary constants. Such a solution will be called a *general solution*. A solution obtained by giving particular values to the constants is a *particular solution*.

1-5. Solutions Satisfying Specified Conditions. Owing to the presence of the arbitrary constant in the general solution of the differential equation of the first order, we are able to make the solution satisfy one condition by particularizing the constant. The commonest form of the condition is that the dependent variable shall have a specified value for a given value of the independent variable. This condition is satisfied by substituting the given values in the general solution and solving for the constant. Thus the solution of (1-5) such that $y = 1$ when $x = 0$ is found by setting $x = 0$, $y = 1$ in the general solution (1-6),

$$1 = K$$

whence

$$y = e^x$$

is the solution with the required property.

A condition to be satisfied by the solution may appear in various other ways. For example, find a solution of (1-5) such that y has a value 1 greater at $x = 1$ than it has at $x = 0$. This gives, substituting in (1-6),

$$Ke = K + 1 \quad \text{or} \quad K = \frac{1}{e - 1}$$

The solution is

$$y = \frac{e^x}{e - 1}$$

The method of procedure in any case is to express the required condition as an equation in which the arbitrary constant appears. From this equation the value of the constant is determined. Sometimes, of course, no value of the constant will satisfy the equation, in which case there is no solution with the required property. Sometimes, also, several values of the constant are determined, and there are several solutions with the required property.

1-6. Velocities. Velocities, and rates of change generally, are derivatives. Let an object be moving along a path. At time t let its distance from a fixed point O of the path, measured along the path, be s (see Fig. 8). By convention s will be positive on one side of O and negative on the other. At time $t + \Delta t$ let its distance be $s + \Delta s$. The average velocity for this period of time Δt is $\Delta s/\Delta t$. The instantaneous velocity v at time t is the limit of this ratio,

$$v = \frac{ds}{dt}$$

INTRODUCTION TO DIFFERENTIAL EQUATIONS

If v is positive, the object is moving as t increases in the direction of the positive end of the path; if negative, the motion is in the opposite direction.

Suppose that the velocity is given in terms of s, or t, or both. We have then a differential equation whose solution—a relation between s and t—will enable us to locate the object at a given time. An example will make the matter clear.

Problem. I live on a straight road 6 miles due north of school and I leave home going south at a speed of 30 miles an hour. If my velocity is proportional to the square of my distance from school, find my motion. When, if ever, will I reach school?

Let s be my distance from school t hr after I leave home, distances north from school being considered positive. We are given that

$$v = \frac{ds}{dt} = ks^2$$

The factor of proportionality k is determined by the knowledge that $v = +30$ when $s = 6$,

$$+30 = 36k \qquad k = +\frac{5}{6}$$

The differential equation of the motion is then

$$\frac{ds}{dt} = +\frac{5}{6}s^2$$

We solve this by separating variables,

$$\frac{ds}{s^2} = \frac{5}{6}dt$$

whence, integrating,

$$\frac{1}{s} = \frac{5}{6}t + C$$

When $t = 0$, $s = 6$, whence $C = \frac{1}{6}$. We have, finally

$$s = \frac{6}{5t + 1}$$

This gives my precise position at any time.

Since no value of t will give $s = 0$, the school is never reached.

EXERCISES

Solve the problem of the journey to school under the following six hypotheses about the velocity, the other conditions remaining the same.

1. The velocity is proportional to the distance from school.
2. The velocity is proportional to the square root of the distance from school.
3. The velocity is inversely proportional to the distance from school.
4. The velocity is proportional to the distance from a tower 4 miles south of school.
5. The velocity is proportional to the square of the distance from the tower.

6. The velocity is proportional to the square of the distance from a point 8 miles east of school.

7. Find the most general function whose derivative is equal to the square of the function. Find the particular one of these functions which has the value 1 when the independent variable is 0.

8. Find the most general function such that the product of the function and its derivative is a given constant.

9. Sin x has the property that the square of the function plus the square of its derivative is equal to 1. Find the most general function with this property. Show that the constant of integration can be so chosen that sin x and cos x are yielded as particular functions.

10. What function is equal to the cube of its derivative?

11. Solve: $(1 + x)\,dy + (1 + y)\,dx = 0$.

12. Solve
$$(1 - x^2)y'^2 = 1 - y^2$$
Show that the solution can be written in the form
$$x^2 - 2Cxy + y^2 = 1 - C^2$$
and show that this is a family of conics touching the four sides of a square.

13. Show that $y = Ce^x + x^2 + 2x + 2$ is a solution of (1-2).

14. Show that $y = 5e^x$, $y = \frac{1}{2}e^{-x}$, $y = 8$ are all solutions of (1-3).

15. Show that $s = \frac{1}{4}(2t^2 + 6t + 7)$ satisfies (1-4).

16. Show that $x = ae^{5t} - be^{-t}$, $y = 2ae^{5t} + be^{-t}$ are solutions of the simultaneous equations at the end of Sec. 1-1.

17. Show that $y = A \sin(2x + B)$ is a solution of $y'' + 4y = 0$.

18. Show that each of the following equations contains a single essential constant, and write each in its simplest form.

(a) $C_1(x^2 + y^2) + C_2 = 0$.   (b) $x^2 + \log C_1 y + C_2 = 0$.
(c) $C_1 x^2 + C_2 xy + C_3 y^2 = 2xy$.   (d) $y^2 + C_1 \sin 2x = C_2 \sin x \cos x$.

19. Show that $x^2 + y^2 = Cx^2y^2$ is a solution of $y^3 + x^3 y' = 0$.

20. Show that $y = x \tan(x + C)$ is a solution of $xy' = x^2 + y^2 + y$.

21. Show that $x^2 y + 2x = Cy$ is a solution of $xy' = xy^2 + y$.

Solve:

22. $\dfrac{dy}{dx} = \dfrac{xy}{x-1}$.   **23.** $e^{x^2+y}\,dx + \dfrac{y}{x}\,dy = 0$.

24. $xyy' = (x+1)(y+1)$.

Show that in the following the substitution $y = vx$ changes the equation into an equation in v and x in which the variables are separable; hence solve:

25. $\dfrac{dy}{dx} = \dfrac{x^2 + y^2}{2x^2}$.   **26.** $\dfrac{dy}{dx} = \dfrac{x+y}{x-y}$.   **27.** $\dfrac{dy}{dx} - \dfrac{y}{x} = \dfrac{y^2}{x^2}$.

28. Show that the substitution $y = v/x$ leads to an equation in v and x in which the variables are separable, and solve:
$$\frac{dy}{dx} = \frac{y + xy^2}{x + 2x^2 y}$$

29. Show that the equation

$$\frac{dy}{dx} = my + f(x)$$

can be solved by setting $y = e^{mx}v$.

Solve:

30. $y' = 2y + x$. **31.** $y' + 3y = e^x$. **32.** $y' + y = e^{-x}$.

1-7. The Direction Field. The general solution of a differential equation of the first order,

$$\frac{dy}{dx} = f(x,y) \tag{1-9}$$

is an equation connecting x and y and an arbitrary constant. For each value of the constant we have a particular solution. If we let x and y be the rectangular coordinates of a point in a plane, then each particular solution is the equation of a curve which we shall call an *integral curve* of the differential equation. By varying the arbitrary constant we get an infinite number of integral curves. Each integral curve has the property that at each of its points the coordinates x and y and the slope dy/dx satisfy the differential equation. Conversely, any curve such that the coordinates and slope at every point satisfy the differential equation is an integral curve, and the equation of which the curve is the locus is a solution of the differential equation.

Let us see what information regarding the integral curves we can get from graphical considerations, without solving the differential equation. If an integral curve passes through a point (x,y), we can compute from (1-9) its slope at that point. Let us draw through the point a small piece of straight line having the computed slope. We shall call this piece of line a *lineal element*. The curve must have at the point the direction of the lineal element, that is, the integral curve is tangent to the lineal element at the point. We can draw a lineal element for each point for which the function $f(x,y)$ is defined. An array of lineal elements showing the directions of the integral curves will be called a *direction field*.

In Fig. 1 is shown a direction field for the differential equation

$$\frac{dy}{dx} = x - y \tag{1-10}$$

Thus at $(1,1)$, $dy/dx = 0$ and the lineal element is horizontal; at $(1,2)$, $dy/dx = -1$ and the lineal element has the slope -1, and so on.

We shall subsequently find a method of solving Eq. (1-10). However, without a solution, we can get a fair idea of the behavior of the integral curves by an inspection of the direction field. Any curve which is tangent at each of its points to the lineal element there is an integral curve.

10   DIFFERENTIAL EQUATIONS

We can get approximate solutions graphically by drawing curves across the figure which have, as nearly as we can judge, the proper directions at all points. In the figure we have drawn the integral curve which passes through the origin.

From the direction field we can get an intuitive verification of the fact that the general solution involves an arbitrary constant. Consider, for

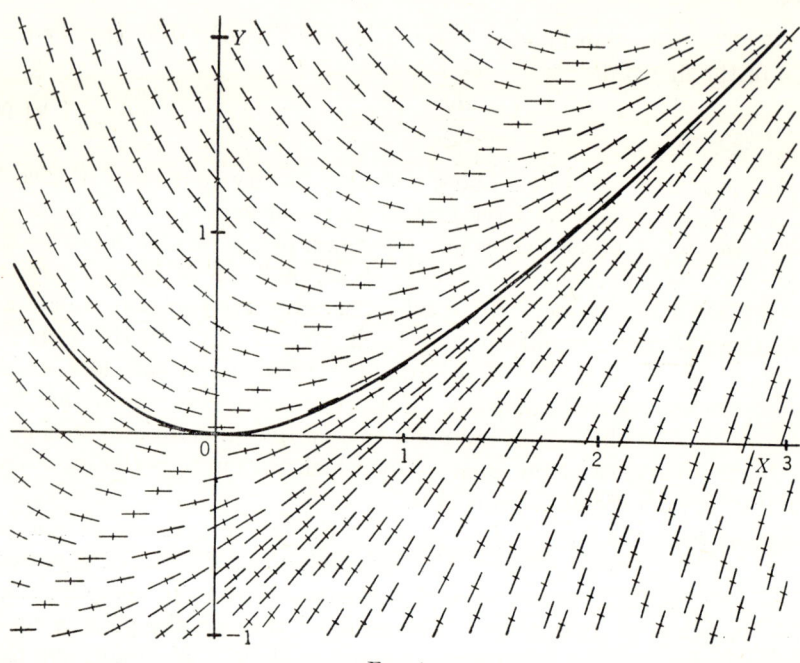

Fig. 1.

example, the integral curves through points on the y axis. Each integral curve meets the y axis in a definite point, and through each point $(0,y_0)$ on the y axis there passes a definite integral curve. By selecting different values of y_0, we get different integral curves. The solution then depends upon y_0, which is an arbitrary constant.

1-8. The Local Behavior of a Solution. Let us consider the integral curve which passes through a point (x,y). By a direct study of the differential equation—without solving it—we are able to discover many things about the behavior of the solution in the neighborhood of the point.

Is the curve concave up or concave down? This depends on the sign of the second derivative, and we can find the second derivative by differentiating,

$$\frac{d^2y}{dx^2} = 1 - \frac{dy}{dx} = 1 - x + y$$

At the origin, for instance, $d^2y/dx^2 = 1$, and the curve is concave up. An integral curve of (1-10) is concave up if $1 - x + y > 0$ and concave down if $1 - x + y < 0$. Hence the curve is concave up at the point (x,y) if the point lies above the line $y = x - 1$ and concave down if the point lies below this line.

Knowing the first and second derivatives, we can find the radius of curvature at the point. Thus, at the origin

$$R = \frac{(1 + y'^2)^{\frac{3}{2}}}{|y''|} = 1$$

By laying off the radius of curvature along a normal to the lineal element at the point we can locate the center of curvature.

We can, if we like, replace the lineal elements of the direction field by circular elements showing at each point both the direction and curvature of the integral curve. This has been done for Eq. (1-10) in Fig. 2. Each arc in the figure is a portion of the osculating circle at the point. This gives a much more accurate idea of the trend of the curves than is furnished by the lineal elements. However, a good deal of labor is involved in the construction of a field of circular elements, and the process is not practical.

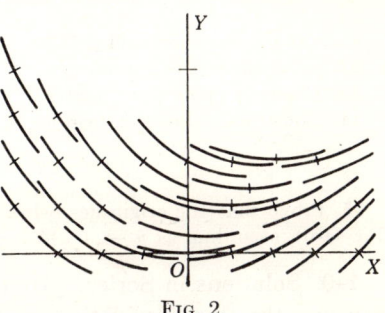

Fig. 2.

EXERCISES

1. By laying a straightedge on Fig. 1, find a rectilinear integral curve. Find its equation and verify that it is a solution.

2. Solve the preceding problem by substituting $y = mx + b$ and determining m and b so that Eq. (1-10) is satisfied.

3. Show that the lineal elements of Eq. (1-10), which are parallel, are at points lying on a straight line.

4. Prove that for Eq. (1-10) the locus of a point whose lineal element is directed toward the origin is a hyperbola.

5. Prove that the points at which the lineal elements of

$$y' = lx + my + q$$

are directed toward a given point lie on a conic.

6. Construct a direction field for the differential equation

$$y' = 1 - x^2 - y^2$$

and sketch the integral curve through the origin.

NOTE: As an aid to plotting find first the locus of a point at which the slope has a constant value so that all lineal elements with the same slope may be readily drawn.

7. Show that the integral curves of the preceding equation and those of Eq. (1-10) touch along a certain circle.

8. Construct a direction field for $dy/dx = y$.

9. Construct a direction field for

$$y = x\frac{dy}{dx} + \left(\frac{dy}{dx}\right)^2$$

Sketch the integral curve through the point (1,2). Can you find the general solution by an inspection of the field?

10. Show that $y = -x^2/4$ is a solution of the preceding equation. Show that $y'' = 0$ for all other solutions.

11. The differential equation of the lines of force due to a short magnet lying along the y axis at the origin is

$$3xyy' = 2y^2 - x^2$$

Construct the direction field. (These are the directions that iron filings would take.)

12. Solve the preceding equation by putting $y = vx$.

13. Change to polar coordinates in Exercise 11 and solve: $x = r \cos \theta$, $y = r \sin \theta$.

14. Construct a direction field for

$$xy' = 1 - x^2 - y^2$$

15. Prove that the origin lies below or above the tangent to an integral curve of Eq. (1-9), according as $y - xf(x,y)$ is positive or negative.

1-9. Solutions in Series. By differentiating the second derivative we can find the third derivative, and we can proceed to find higher derivatives at the point so long as the functions which arise have derivatives.

Let (x_0, y_0) be the point under consideration, and let y_0', y_0'', etc., represent the first, second, etc., derivatives of y with respect to x at (x_0, y_0). Then, assuming that y can be developed in a series of powers of $x - x_0$, we have the solution in a Taylor's series:

$$y = y_0 + y_0'(x - x_0) + \frac{y_0''}{2!}(x - x_0)^2 + \frac{y_0'''}{3!}(x - x_0)^3 + \cdots$$

From this series the solution can be calculated to any desired degree of accuracy.

Let us find the solution of (1-10) with the condition that $y = 0$ when $x = 0$—the solution found graphically in Fig. 1. Here $x_0 = 0$, $y_0 = 0$. We have, differentiating,

$$\begin{aligned} y' &= x - y & y_0' &= 0 \\ y'' &= 1 - y' & y_0'' &= 1 \\ y''' &= -y'' & y_0''' &= -1 \\ y'''' &= -y''' & y_0'''' &= 1 \end{aligned}$$

and clearly thereafter the derivatives are alternately -1 and $+1$. The solution is then

$$y = \frac{x^2}{2!} - \frac{x^3}{3!} + \frac{x^4}{4!} - \frac{x^5}{5!} + \cdots$$

This solution can be put in finite form if we happen to notice the similarity of the preceding series to the series

$$e^{-x} = 1 - x + \frac{x^2}{2!} - \frac{x^3}{3!} + \frac{x^4}{4!} - \cdots$$

We have then

$$y = e^{-x} + x - 1$$

It will usually not be possible to express the series that we get in terms of elementary functions.

An alternative way of proceeding is to assume a solution in a series of the desired form with undetermined coefficients,

$$y = y_0 + C_1(x - x_0) + C_2(x - x_0)^2 + \cdots + C_n(x - x_0)^n + \cdots$$

and substitute it into the differential equation. We can ordinarily determine the coefficients C_n in order by equating like powers of $x - x_0$ in the two members of the equation.

For example, in the preceding problem, put

$$y = C_1 x + C_2 x^2 + \cdots + C_n x^n + \cdots$$

On substituting into Eq. (1-10),

$$C_1 + 2C_2 x + \cdots + nC_n x^{n-1} + \cdots = x - [C_1 x + C_2 x^2 + \cdots + C_n x^n + \cdots]$$

Equating coefficients,

$$C_1 = 0$$
$$2C_2 = 1 - C_1 \qquad C_2 = \tfrac{1}{2}$$
$$3C_3 = -C_2 \qquad C_3 = -1/3!$$
$$\cdots\cdots\cdots\cdots\cdots\cdots\cdots$$
$$nC_n = -C_{n-1} \qquad C_n = (-1)^n/n!$$

We thus get the series found previously.

EXERCISES

Find the series expansions to terms in x^4 of the solutions of the following differential equations with the condition that $y = 0$ when $x = 0$:

1. $y' = 1 - x^2 - y^2$.
2. $y' = x + \sin y$.
3. $y' = 2 + xy + x^3 y^2$.
4. $y' = x^2 + y^2$.
5. $y' = \cos x + \sin y$.
6. $y = x \dfrac{dy}{dx} + \left(\dfrac{dy}{dx}\right)^3 - 1$.
7. $(x + y + 1)y' = x$.

Get five terms of the series solution of $y' = x^2 + y^2$:
8. Given $y = 1$ when $x = 0$.
9. Given $y = 1$ when $x = 1$.

10. Solve to the fourth nonvanishing term

$$y' = 1 + y^2$$

given $y = 0$ when $x = 0$, thus getting the beginning of the series expansion of $\tan x$.

Get the solution in infinite series, finding the general term. If possible put in finite form and check:
11. $y' = y$, where $y = 1$ when $x = 0$.   12. $y' = x - y$, where $y = C$ when $x = 0$.
13. $y' = 2y + e^x$, where $y = 0$ when $x = 0$.

Solve similarly $y'' + y = 0$, where, when $x = 0$:
14. $y = 0, y' = 1$.   15. $y = 1, y' = 0$.   16. $y = a, y' = b$.

1-10. One-parameter Families of Curves. We have found that the general equation of the integral curves of a differential equation of the

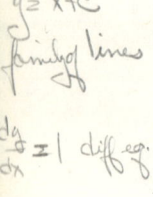

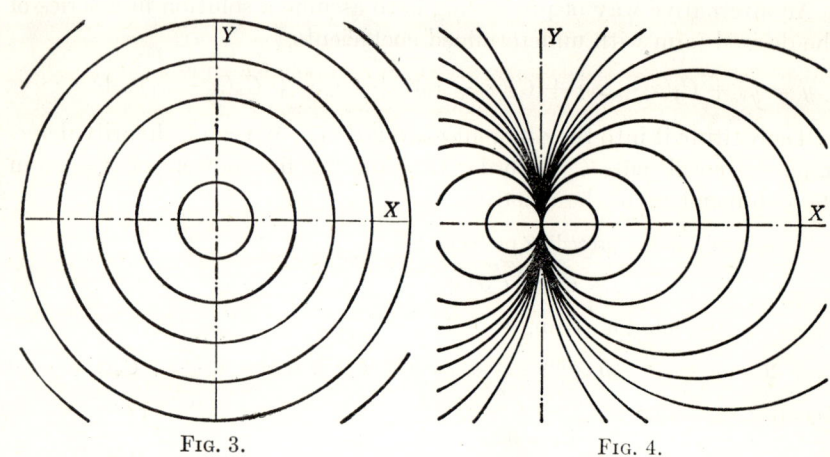

Fig. 3.   Fig. 4.

first order contains an arbitrary constant. An arbitrary constant is also called a *parameter*, and a family of curves depending upon one arbitrary constant is called a *one-parameter family of curves*. For example,

$$x^2 + y^2 = c^2 \qquad (1\text{-}11)$$

is a one-parameter family of circles with centers at the origin (Fig. 3). By giving values to c we get various members of the family. The family

$$x^2 + y^2 + Cx = 0 \qquad (1\text{-}12)$$

is a family of circles with centers on the x axis and passing through the origin (Fig. 4).

We can show that, as we might expect, the curves of a one-parameter family are integral curves of some differential equation of the first order. Let the family be

$$F(x,y,C) = 0 \qquad (1\text{-}13)$$

On differentiating with respect to x we have a relation of the form

$$g\left(x, y, \frac{dy}{dx}, C\right) = 0 \qquad (1\text{-}14)$$

INTRODUCTION TO DIFFERENTIAL EQUATIONS

For a fixed value of C we have a particular curve, and the coordinates and slope at any point on that curve satisfy the two equations (1-13) and (1-14). The coordinates and slope will then satisfy any equation obtained from these two by legitimate combination, for example,

$$F + g = 0, \qquad (F - g)^2 + 2F = 0$$

In particular, let us combine the two equations so that C is eliminated, getting a result of the form

$$f\left(x, y, \frac{dy}{dx}\right) = 0 \qquad (1\text{-}15)$$

Since this equation does not contain C, it will be a legitimate combination of (1-13) and (1-14) whatever value of C we started with. Hence (1-15) is satisfied by the coordinates and slope at a point on any curve of the family. In other words, all curves of the family are integral curves of (1-15).

The process of finding the differential equation of the family then consists of the two steps of differentiating the equation of the family with respect to x (or y, if desired) and eliminating the parameter.

To get the differential equation of (1-11) we have, on differentiating,

$$2x + 2y \frac{dy}{dx} = 0$$

Since C is absent from this result the differential equation of the family is

$$x + y \frac{dy}{dx} = 0$$

In the family (1-12) we have

$$2x + 2y \frac{dy}{dx} + C = 0$$

Multiplying this by x and subtracting (1-12) we have the differential equation

$$x^2 - y^2 + 2xy \frac{dy}{dx} = 0$$

EXERCISES

Find the differential equation of:
1. All straight lines (a) parallel to and (b) perpendicular to the line $2x + 3y = 5$.
2. All straight lines through the point (1,1).
3. The parabolas $x = Cy^2$. 4. The hyperbolas $x + y = Cxy$.
5. The parabolas $(x + y + 3)^2 = C(x - y)$.
6. All circles through the points (1,0) and (−1,0).
7. The family $y = C \cos x + \sin x$. 8. The family $F(x - y, C) = 0$.
9. All straight lines tangent to the unit circle $x^2 + y^2 = 1$. Verify that the circle is also an integral curve of the equation.

10. All circles with unit radius and center on the line $3y = 4x$. Find two rectilinear integral curves of the resulting equation.

11. The family of confocal conics

$$\frac{x^2}{a^2+C} + \frac{y^2}{b^2+C} = 1$$

where C is the parameter. Show that the two lineal elements at each point are perpendicular to one another.

12. The family of straight lines normal to the parabola $y = x^2$. Verify that the evolute of the parabola

$$27x^2 = 2(2y-1)^3$$

is an integral curve.

13. Show that the differential equation of the family

$$y = \frac{a(x)C + b(x)}{c(x)C + d(x)}$$

has the form (Riccati's equation)

$$y' = P(x)y^2 + Q(x)y + R(x)$$

1-11. Clairaut's Equation. The straight line is the simplest of all curves, and it will be of interest to find what kind of differential equation we are led to in the case of a one-parameter family of straight lines. Let us put aside all families of parallel lines, namely, $y = mx + C$ and $x = C$, these having particularly simple differential equations.

If in the equation of the line

$$y = mx + b$$

we make m and b functions of a parameter c, so

$$y = m(c)x + b(c)$$

we have a one-parameter family of lines. Conversely, any nonparallel one-parameter family can be put in this form, for the slope and y intercept will depend upon, that is, be a function of, the parameter.

It will simplify matters if we change the parameter by putting

$$m(c) = C$$

We can do this since, by hypothesis, the slope is not fixed but varies with c. Then $b(c)$ is some function $f(C)$ of C, and the equation of the family is

$$y = Cx + f(C) \tag{1-16}$$

Differentiating, to find the differential equation of the family, we have

$$\frac{dy}{dx} = C$$

whence, eliminating C,

$$y = x\frac{dy}{dx} + f\left(\frac{dy}{dx}\right) \quad (1\text{-}17)$$

This equation is the most general differential equation of a nonparallel one-parameter family of straight lines. It is known as *Clairaut's equation*.

The solution of Clairaut's equation is a very simple matter. The general solution (1-16) is obtained from the differential equation (1-17) by replacing the slope dy/dx by an aribitrary constant. For example, the solutions of

(a) $\qquad y = x\dfrac{dy}{dx} - \dfrac{1}{4}\left(\dfrac{dy}{dx}\right)^2$

(b) $\qquad y = x\dfrac{dy}{dx} - \dfrac{dx}{dy}$

(c) $\qquad \left(y - x\dfrac{dy}{dx}\right)^2 + \dfrac{dy}{dx} = \left(\dfrac{dy}{dx}\right)^2$

are

(a') $\qquad y = Cx - \dfrac{1}{4}C^2$

(b') $\qquad y = Cx - \dfrac{1}{C}$

(c') $\qquad (y - Cx)^2 + C = C^2$

The third equation is not in the form (1-17) as it stands, but it is easily put in this form by writing it

$$y = x\frac{dy}{dx} \pm \sqrt{\left(\frac{dy}{dx}\right)^2 - \frac{dy}{dx}}$$

1-12. Singular Solutions. In Fig. 5 are plotted the integral curves $y = Cx - \frac{1}{4}C^2$ of the differential equation (a). It will be noted from the figure that these lines have an envelope.

It is easily seen that *the envelope of a family of integral curves is itself an integral curve*. At each of its points the envelope is tangent to one of the integral curves; the slope of the integral curve and of the envelope are the same. Hence, the coordinates of the point and the slope at the point satisfy the differential equation.

The envelope, if one exists, will not usually be a member of the one-parameter family of integral curves whose equation is the general solution, so that it cannot be obtained by giving the constant a particular value. It is called a *singular solution*.

The student will find the theory of envelopes on page 231 or in texts on the calculus. The method of finding envelopes is briefly as follows. Let $F(x,y,C) = 0$ be the family of curves. Differentiate this with respect to C, holding x and y constant. The envelope is obtained by eliminating C between the equations

$$F(x,y,C) = 0 \qquad \frac{\partial}{\partial C} F(x,y,C) = 0$$

This method may introduce other loci than the envelope; for example, loci of points where curves of the family have nodes or cusps. For this reason any curve found should be tested to see whether the differential equation is satisfied.

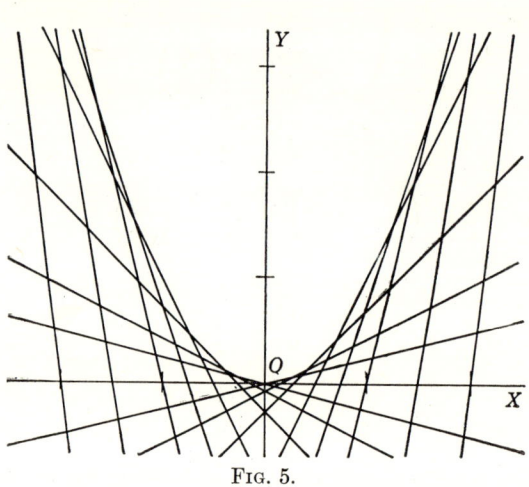

Fig. 5.

In the equation treated above we eliminate C between $y = Cx - \frac{1}{4}C^2$ and $0 = x - \frac{1}{2}C$. The result is $y = x^2$. The student should verify that this satisfies the differential equation.

EXERCISES

1. Solve:
$$y = xy' + \cos y'$$

2. Find the singular solution of Eq. (b), Sec. 1-11.
3. Find the singular solution of Eq. (c), Sec. 1-11.
4. Explain the special solutions found in Exercises 9, 10, and 12 of the preceding set of exercises.
5. Solve:
$$e^y \, dx + e^{xy'} \, dy = 0$$

Find the singular solution, and check.

6. A field is bounded by two fences meeting at right angles. A man walks across the field in such a way that at each point the tangent to his path forms with the fences a triangle whose area is 100 square rods. Find all possible paths.

7. Find all possible paths, given that the fences cut off on the tangent a length of 40 rods.

1-13. n-Parameter Families.

A family of curves

$$F(x, y, C_1, C_2, \ldots, C_n) = 0 \tag{1-18}$$

which depends upon n parameters, satisfies a differential equation of the nth order. On differentiating n times with respect to x we have expressions of the form

$$F_1\left(x, y, \frac{dy}{dx}, C_1, \ldots, C_n\right) = 0$$

$$F_2\left(x, y, \frac{dy}{dx}, \frac{d^2y}{dx^2}, C_1, \ldots, C_n\right) = 0$$

$$\cdots \cdots \cdots \cdots \cdots \cdots \cdots$$

$$F_n\left(x, y, \frac{dy}{dx}, \ldots, \frac{d^ny}{dx^n}, C_1, \ldots, C_n\right) = 0$$

If we eliminate $C_1, \ldots, C_n$ between these n equations and the original equation of the family, we have a differential equation of the nth order,

$$f\left(x, y, \frac{dy}{dx}, \frac{d^2y}{dx^2}, \ldots, \frac{d^ny}{dx^n}\right) = 0 \tag{1-19}$$

The reasoning to show that the curves of the family are integral curves of this equation is similar to that employed previously (Sec. 1-10) in the case of a one-parameter family and will not be repeated here.

Of course, (1-18) will satisfy differential equations of higher order than n; for example, equations obtained by differentiating (1-19), or equations got by differentiating (1-18) more than n times and using these higher derivatives in the elimination of the constants. When we speak of the differential equation satisfied by a family, we shall mean the differential equation of lowest order which the family satisfies. If the constants of (1-18) are not all essential but can be replaced by a smaller number, the family satisfies an equation of lower order than n.

Example. Find the differential equation of all unit circles. The family is

$$(x - a)^2 + (y - b)^2 = 1$$

where a and b, the coordinates of its center, are parameters. Differentiating twice, we have

$$x - a + (y - b)y' = 0$$
$$1 + (y - b)y'' + y'^2 = 0$$

Then

$$y - b = -\frac{1 + y'^2}{y''}$$

$$x - a = \frac{y'(1 + y'^2)}{y''}$$

On substituting in the original equation and simplifying, we have the differential equation of the family

$$\left(\frac{d^2y}{dx^2}\right)^2 = \left[1 + \left(\frac{dy}{dx}\right)^2\right]^3$$

The correctness of this result may be checked by writing it in the form

$$\frac{\left[1 + \left(\frac{dy}{dx}\right)^2\right]^{\frac{3}{2}}}{|y''|} = 1$$

which states that the radius of curvature of an integral curve at any point is 1.

EXERCISES

Find the differential equation of:
1. All straight lines in the plane.　　2. All circles through the origin.
3. All circles with centers on the x axis.
4. All circles in the plane.
5. All conics with the coordinate axes as axes.
6. All conics in the plane with the origin as center.

Find the differential equation of each of the following families:
7. $y = Ae^{-x} + Be^{2x}$.　　8. $y = A \cos x + B \sin x$.
9. $y = C \cos (2x + K)$.　　10. $y = e^x(A \cos x + B \sin x)$.
11. $y = A + B \sin x$.　　12. $y = A + \sin (B - x)$.
13. $y = x^2(A + Be^x + e^{2x})$.　　14. $y = A^{Bx}$.
15. $y = \dfrac{x + C}{x + K}$.

16. Write down five differential equations of the second order satisfied by $y = Ce^{2x}$.

17. The general equation of a parabola with vertical axis is $y = Ax^2 + Bx + C$. Find the differential equation of this family.

Find the differential equation of all parabolas with vertical axes which
18. Pass through the origin.
19. Pass through the points $(1,0)$ and $(-1,0)$.
20. Pass through the points $(1,1)$ and $(-1,-1)$.
21. Pass through the origin and touch the line $y = x$ there.

1-14. Geometrical Applications. At a point P on a curve let the tangent and the normal be drawn. Let T and N be the points of intersection of the tangent and the normal, respectively, with the x axis; and let M be the foot of the perpendicular from P to the x axis (Fig. 6). The several lines have been given the following names: $\underline{PT \text{ is the } \textit{tangent}; \ TM,}$ the $\textit{subtangent}; \ \underline{PN, \text{ the } \textit{normal}; \text{ and } \ MN,}$ the $\textit{subnormal}$. The ordinate MP is equal to y. There is also drawn in the figure the differential triangle at P, formed by a tangent to the curve and lines parallel to the axes and having the sides dx, dy, and ds, where

$$ds = \sqrt{dx^2 + dy^2}$$

INTRODUCTION TO DIFFERENTIAL EQUATIONS

The angles marked θ are easily shown by geometrical considerations to be equal.

By a consideration of similar triangles we are able to express the lengths in the figure in terms of y and dy/dx. Thus, to find the subnormal, we have

$$\frac{MN}{y} = \frac{dy}{dx} \qquad MN = y\frac{dy}{dx}$$

We are able to set up the differential equation of a family of curves whose tangent, subtangent, etc., satisfy specified conditions.

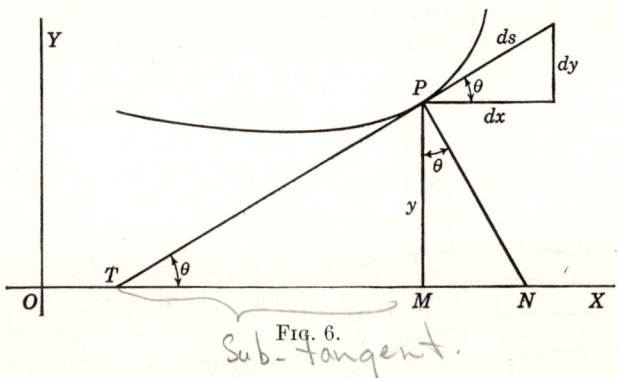

Fig. 6.

Example. Find the curves such that the subnormal at each point has a constant value a.

The differential equation of the family is

$$y\frac{dy}{dx} = a$$

Separating the variables and integrating, we have

$$y^2 = 2ax + C$$

a family of parabolas.

EXERCISES

Determine the curves for which:
1. The subtangent has a constant value a.
2. The tangent is equal to the normal.
3. The normal has a constant value a.
4. The mid-point of the normal lies on the y axis.
5. The tangent at each point is equal to the distance from the point to the origin.

6. Show that if a linear combination of the tangent, normal, subtangent, subnormal, and ordinate is zero,

$$a \cdot PT + b \cdot TM + c \cdot PN + d \cdot MN + e \cdot MP = 0$$

the solution, if one exists, is a family of straight lines.

1-15. Trajectories. A *trajectory* of a family of curves is a curve which cuts the members of the family according to a given law. An *orthogonal trajectory* is a curve which cuts the members of a family at right angles. Let

$$F(x,y,C) = 0$$

be a family of curves whose orthogonal trajectories we wish to find, and let

$$f\left(x, y, \frac{dy}{dx}\right) = 0 \qquad (1\text{-}20)$$

be the differential equation of the family. To avoid confusion let us represent the coordinates and slope of a trajectory by x_1, y_1, and dy_1/dx_1.

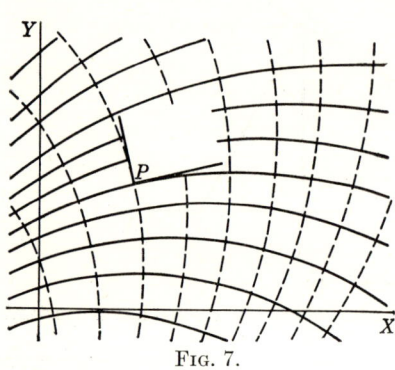

Fig. 7.

Consider a point $P(x,y)$, where a trajectory meets a curve of the family (Fig. 7). The coordinates and slope of the curve of the family satisfy (1-20). What relation exists between the coordinates and slope of the orthogonal trajectory? The coordinates are the same, $x = x_1$, $y = y_1$, and the slopes of the two curves are negative reciprocals,

$$\frac{dy}{dx} = -\frac{dx_1}{dy_1}$$

Hence, substituting in (1-20), we have the relation that must hold between the coordinates and slope of the trajectory,

$$f\left(x_1, y_1, -\frac{dx_1}{dy_1}\right) = 0$$

Dropping the subscripts, we can state the following procedure: *The differential equation of the orthogonal trajectories is got from the differential equation of the given family by replacing dy/dx by $-dx/dy$.* The solutions form a one-parameter family of orthogonal curves.

Example. Find the orthogonal trajectories of the family of straight lines through the origin.

The given family is $y = Cx$, and the differential equation of the family is

$$y = x\frac{dy}{dx}$$

The differential equation of the orthogonal trajectories is then

$$y = -x\frac{dx}{dy}$$

Separating the variables and integrating,

$$x\,dx + y\,dy = 0 \qquad x^2 + y^2 = C$$

The orthogonal trajectories are a family of circles with centers at the origin, as was probably foreseen by the reader.

Families of curves and surfaces which intersect at right angles are of widespread use in applied mathematics. The lines along which heat flows in a body are orthogonal to the isothermal surfaces. The lines of electric, magnetic, or gravitational force are orthogonal to the equipotential surfaces. Where the lines of flow or of force lie in a plane we are concerned with isothermal or equipotential *curves*, and the methods of this section apply.

Another application is to the lines of *steepest descent* on a hill. Let the equation of the surface be $F(x,y,z) = 0$, where the xy plane is horizontal. Then the lines sought are orthogonal to the contour lines cut on the surface by the planes $z = $ const. Projecting these lines on the xy plane, we see that we require the family orthogonal to $F(x,y,C) = 0$.

The differential equation of other kinds of trajectories may be found whenever we can express x, y, and dy/dx in terms of x_1, y_1, and dy_1/dx_1, for we have then but to substitute these values in (1-20). Let us find the differential equation of *oblique trajectories*, that is, trajectories that cut the curves of a given family at a constant angle α other than a right angle. Let θ and θ_1 be the angles which the tangents make with the x axis; whence $\theta_1 = \theta + \alpha$. Then

$$\frac{dy}{dx} = \tan\theta = \tan(\theta_1 - \alpha) = \frac{\tan\theta_1 - \tan\alpha}{1 + \tan\theta_1 \tan\alpha}$$
$$= \frac{dy_1/dx_1 - m}{1 + m\,dy_1/dx_1}$$

where $m = \tan\alpha$. We drop subscripts and substitute this for dy/dx in the differential equation of the given family:

$$f\left(x, y, \frac{dy/dx - m}{1 + m\,dy/dx}\right) = 0$$

EXERCISES

Find the orthogonal trajectories of the following families. Sketch the figures in the first seven exercises.

1. $3x + 2y = C$.
2. $y = Cx^2$.
3. $x^2 - y^2 = C$.
4. $y = Cx^4$.
5. $y^2 = Cx^3$.
6. $b^2x^2 + a^2y^2 = C$.
7. $Cx^2 + y^2 = 1$.
8. $y = (x - C)^2$.
9. $y = Ce^{-x^2}$.
10. $\cos y = Ce^{-x}$.

11. Show that the families

$$(x + C)(x^2 + y^2) + x = 0$$
$$(y + K)(x^2 + y^2) - y = 0$$

are orthogonal.

12. Show that the families

$$e^{x^2-y^2} \cos 2xy = A$$
$$e^{x^2-y^2} \sin 2xy = B$$

are orthogonal.

13. Show that the family of confocal parabolas

$$y^2 = 2Cx + C^2$$

is self-orthogonal. Sketch the curves.

14. Show that the family of confocal conics

$$\frac{x^2}{a^2 + C} + \frac{y^2}{b^2 + C} = 1$$

is self-orthogonal. Sketch the curves.

15. Show that a family whose differential equation has the form

$$\left(\frac{dy}{dx}\right)^2 + f(x,y)\frac{dy}{dx} = 1$$

is self-orthogonal.

16. Perpendicular planes bounding a solid are held at zero degrees and the temperature u inside the solid quadrant is $u = kxy$, where x and y are the distances from the two planes. Find the lines of heat flow.

17. The temperature in the metal of a hollow pipe is constant at a given distance from the axis. Find the lines of heat flow.

18. The temperature u in a slab is $u = ae^{-y} \sin x$. Find the lines of flow.

19. Find the curves of steepest descent on the saddle-shaped surface $z = xy$.

20. A perfectly smooth football in the form of an ellipsoid of revolution 16 in. long and 8 in. wide lies with its long axis horizontal. Find the paths along which raindrops will run down.

21. Find the family of curves which cut the members of the family of hyperbolas

$$y^2 + 2xy - x^2 = C$$

at an angle of 45°.

1-16. Equations of the Second Order Solved by First-order Methods. If either the dependent variable or the independent variable is absent from a differential equation of the second order, its solution can be reduced to the successive integration of two equations of the first order. The equation then has one of the two forms

$$f\left(x, \frac{dy}{dx}, \frac{d^2y}{dx^2}\right) = 0$$

or

$$g\left(y, \frac{dy}{dx}, \frac{d^2y}{dx^2}\right) = 0$$

INTRODUCTION TO DIFFERENTIAL EQUATIONS

We shall introduce the notation
$$v = \frac{dy}{dx}$$
Then
$$\frac{d^2y}{dx^2} = \frac{dv}{dx} = \frac{dv}{dy}\frac{dy}{dx} = v\frac{dv}{dy}$$
and we can put for the second derivative either of the expressions

(a)
$$\frac{d^2y}{dx^2} = \frac{dv}{dx}$$
or

(b)
$$\frac{d^2y}{dx^2} = v\frac{dv}{dy}$$

Using (a) in the equation which does not contain y, we have
$$f\left(x, v, \frac{dv}{dx}\right) = 0$$
This is an equation of the first order in x and v. Its solution,
$$F(x,v,C) = 0$$
becomes an equation of the first order in x and y when v is replaced by dy/dx. The solution of this equation introduces a second arbitrary constant.

Using (b) in the equation from which x is absent we have an equation of the first order in y and v,
$$g\left(y, v, v\frac{dv}{dy}\right) = 0$$
The solution of this,
$$G(y,v,C) = 0$$
is an equation of the first order when we put dy/dx for v.

These two kinds of equations are clearly very special types. However, a considerable number of the equations of the second order that arise in the simpler applications fall into one or the other of the two classes so that the method of solution is of practical importance.

Example. The following differential equation occurs very frequently in the applications and it is desirable to have the solution for reference. A shorter method of solving will be found in Chap. 3.
$$\frac{d^2y}{dx^2} + a^2y = 0$$
To solve we use (b),
$$v\frac{dv}{dy} + a^2y = 0$$

Separating variables and integrating, using C^2 for the constant, which is necessarily nonnegative,

$$\frac{v\,dv}{a^2} = -y\,dy \qquad \frac{v^2}{a^2} = C^2 - y^2$$

Putting $v = dy/dx$ and separating variables,

$$\left(\frac{dy}{dx}\right)^2 = a^2(C^2 - y^2) \qquad \frac{dy}{\sqrt{C^2 - y^2}} = a\,dx$$

whence, integrating,

$$\arcsin\frac{y}{C} = ax + K$$

or finally,

$$y = C\sin(ax + K)$$

Alternative general solutions, got by changes in the forms of the constants, are

$$y = C'\cos(ax + K') \qquad y = A\sin ax + B\cos ax$$

The Period. We observe that if ax is increased by 2π, or a multiple thereof, the value of y is unchanged. The change in x, namely, $2\pi/a$, is known as the period of the solution.

EXERCISES

Solve:

1. $\dfrac{d^2y}{dx^2} = y\dfrac{dy}{dx}$.

2. $x\dfrac{d^2y}{dx^2} = \dfrac{dy}{dx}$.

3. $y\dfrac{d^2y}{dx^2} - 2\left(\dfrac{dy}{dx}\right)^2 = 0$.

4. $\dfrac{d^2y}{dx^2} - a^2y = 0$.

5. Find the most general relation between x and y if d^2y/dx^2 and d^2x/dy^2 are reciprocals.

1-17. Motion of a Particle. In this section we shall study the motion of a particle along a known path under the action of known forces. The more general problem in which the path is not known depends for its solution upon simultaneous differential equations, but if the path is known we are led to a single differential equation with one dependent variable. Examples in which the path is known are the motion of a bead along a wire, where the bead is constrained to follow the particular path into which the wire is formed; the motion of the bob of a simple pendulum, where the cord constrains the bob to move in a circle; the motion of a body falling from rest under gravity, where the path is known to be vertical. Other examples will occur to the reader.

At the time t let s be the distance of the particle P from an origin O in the path (Fig. 8). The distance s is measured along the path and is called positive on one side of O and negative on the other. Let F be the force acting on the particle and let F_s be the component of F in the direction of the tangent to the path. F_s will be called positive if it acts in the direction in which s increases and negative if it acts in the opposite direction.

The differential equation of the motion is derived from the following consequence of Newton's second law of motion:

Mass of P × acceleration along the path = constant × F_s

Now the acceleration along the path is d^2s/dt^2, hence this equation can be written

$$m \frac{d^2s}{dt^2} = \lambda F_s \qquad (1\text{-}21)$$

where m is the mass of P and λ is a constant.

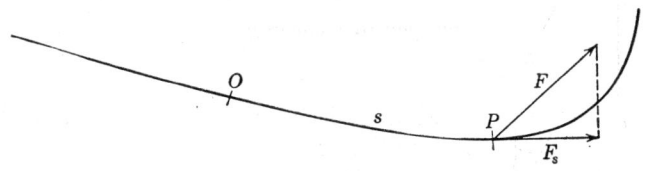

Fig. 8.

The value of λ depends upon the units used:

a. If t is measured in seconds, m in pounds, s in feet, and F_s in pounds weight, then $\lambda = g = 32$.

b. If t is measured in seconds, m in pounds, s in feet, and F_s in poundals, then $\lambda = 1$.

c. If t is measured in seconds, m in grams, s in centimeters, and F_s in dynes, then $\lambda = 1$.

The first step in finding the motion, that is, finding s in terms of t, so that the position of the particle at any time can be determined, is to find an expression for F_s. This force may vary with the time t with the position in the path s and may also depend upon the velocity ds/dt, with which the particle is moving. Having got F_s as a function of t, s, and ds/dt, or some of these quantities, and substituting into (1-21), we have a differential equation whose solution gives us the motion.

The general solution of the equation will contain two arbitrary constants. In order to evaluate these constants two additional conditions must be given by the problem. We may know the position of the particle and its velocity at a given time, or its position at two given times, or some other two conditions.

The following problems will illustrate the method of procedure.

The Simple Pendulum. The bob of a pendulum is pulled aside from its position of rest and released. Find its motion.

The path of the bob will be a circle (Fig. 9). Take the origin O at the lowest point, and let h be the distance along the path from O to the point of release.

We shall use the units (a) and count time from the moment of release. We have the two conditions that when $t = 0$, then $v = 0$ and $s = h$, where v is the velocity of the bob. The forces on the bob are two: the tension of the string, which has no component along the path, and a force vertically downward of m lb, where m is the mass of the bob. The angles marked θ in the figure are easily proved equal. We have then

$$F_s = -m \sin \theta$$

The equation of motion is

$$m \frac{d^2s}{dt^2} = -gm \sin \theta$$

From the relation

$$s = l\theta$$

where θ is measured in radians, we have the equation in the form

$$l \frac{d^2\theta}{dt^2} = -g \sin \theta \qquad (1\text{-}22)$$

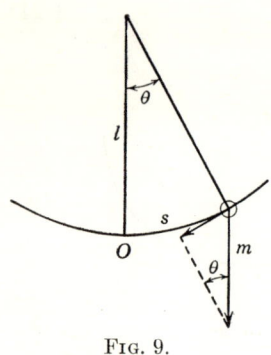

Fig. 9.

Approximate Solution. The problem is greatly simplified if, at this point, we introduce an approximation. We have

$$\sin \theta = \theta - \frac{\theta^3}{3!} + \cdots$$

Hence, if θ is always small, that is, if the pendulum has a small swing, the equation will be but slightly altered if we replace $\sin \theta$ by θ. We have then

$$l \frac{d^2\theta}{dt^2} = -g\theta$$

This equation has the form treated in the preceding section, with $a^2 = g/l$. Its solution is

$$\theta = A \sin at + B \cos at$$

The angular velocity is

$$\frac{d\theta}{dt} = a(A \cos at - B \sin at)$$

Let α be the angle at the time $(t = 0)$ of release, at which moment $d\theta/dt = 0$. Putting these values into the two preceding equations, we find

$$B = \alpha \qquad A = 0$$

Inserting these values into the general solution, we have the required result,
$$\theta = \alpha \cos \sqrt{\frac{g}{l}}\, t$$

The period of the pendulum is $2\pi \sqrt{l/g}$.

The Accurate Velocity. A first integration of the exact equation (1-22) offers no difficulty. Putting $d\theta/dt = u$, we have
$$lu\frac{du}{d\theta} = -g \sin \theta$$
$$lu^2 = 2g \cos \theta + C$$

For $\theta = \alpha$, $u = 0$ we have
$$C = -2g \cos \alpha$$
and
$$l\left(\frac{d\theta}{dt}\right)^2 = 2g(\cos \theta - \cos \alpha)$$

This equation gives the angular velocity $d\theta/dt$, or the actual velocity $l\, d\theta/dt$, at any point in the path, however large the swing. We observe that, as in the approximate solution, the bob is at rest whenever $\theta = \pm \alpha$, so that the bob does not pass these positions.

On separating variables, we have
$$\frac{d\theta}{\sqrt{\cos \theta - \cos \alpha}} = \pm \sqrt{\frac{2g}{l}}\, dt$$

The first member cannot be integrated in terms of elementary functions. It may be done by elliptic integrals, by series expansion, or by one of the methods of approximation to be described later.

Falling Raindrop. Find the motion of a raindrop falling from rest, given that the resistance of the air is proportional to the velocity.

Here the path is vertical. Measure s downward from the starting point. Measuring time from the start we have $s = 0$ and $v = 0$, when $t = 0$. Two forces act: a force of m lb, where m is the mass, acting downward; and a force of kv, where k is a constant, acting upward. The equation of motion is
$$m\frac{d^2s}{dt^2} = g(m - kv)$$
or, putting $h = kg/m$,
$$\frac{d^2s}{dt^2} = g - hv$$

Both s and t are absent from this equation, so either of the substitutions given in Sec. 1-16 may be used, t and s replacing x and y. Using the first, we have

$$\frac{dv}{dt} = g - hv$$

Separating the variables and integrating,

$$\frac{dv}{g - hv} = dt$$

$$-\frac{1}{h} \log (g - hv) = t + C$$

or

$$g - hv = Ke^{-ht}$$

When $t = 0$, $v = 0$, hence $K = g$, and we have

$$v = \frac{g}{h} (1 - e^{-ht})$$

As t increases without limit, e^{-ht} approaches zero since h is positive, and v approaches the limiting velocity g/h.

Replacing v by its value and separating variables, we have

$$ds = \frac{g}{h} (1 - e^{-ht}) \, dt$$

whence

$$s = \frac{g}{h} \left(t + \frac{1}{h} e^{-ht} \right) + C_2$$

When $t = 0$, $s = 0$, and we find $C_2 = -g/h^2$. The solution then is

$$s = \frac{g}{h} \left[t - \frac{1}{h} (1 - e^{-ht}) \right]$$

The work here could be considerably shortened by the methods to be learned in Chap. 3.

EXERCISES

1. Solve the problem of the falling raindrop if there is no friction.

2. A fish stops swimming and glides horizontally through the water, the only force being due to friction. Find the motion, given that the resisting force is proportional to the velocity. How far will it move? When will it stop?

3. Solve the like problem if the resistance is proportional to the square of the velocity.

4. Solve the problem if the resistance is proportional to the square root of the velocity.

5. A body outside the earth is attracted toward the earth's center with a force inversely proportional to the square of its distance. Knowing the force at the earth's surface (4,000 miles from the center), the factor of proportionality can be found.

Find the velocity with which a meteor will reach the surface falling from a distance of 10,000 miles from the surface.

6. With what velocity must a projectile be fired upward from the earth in order not to return?

7. Within the earth a body is attracted toward the center with a force which is proportional to the distance from the center. If a ball were dropped into a hole through the center, and if there were no friction, with what velocity would it reach the center?

When should you put your hand into the hole to take the ball out again?

8. Find the relation between the velocity and the distance in the raindrop problem of the text by replacing d^2s/dt^2 by $v\,dv/ds$.

9. Solve the problem of the falling raindrop if the resistance is proportional to the square of the velocity.

10. The rate at which the mass of a quantity of radium decreases is proportional to the mass. Find the mass as a function of the time.

11. Discuss the growth of a population in which the birth rate and death rate per thousand per year remain constant.

12. Discuss the growth if the birth rate per thousand, which is now higher than the death rate, decreases uniformly with the time.

13. An insecticide applied to a colony of insects produces an instantaneous death rate of 2,000 per 1,000 per hour. What fraction of the colony remains at the end of 1 hr?

14. If the food supply of a colony of insects is just sufficient to support a certain total and the increase per 1,000 per day is proportional to the number by which the population falls short of this total, discuss the growth of the population.

15. Assuming that a drop of liquid evaporates at a rate proportional to its surface area, find its radius as a function of the time.

16. What rate of interest payable annually is equivalent to 6 per cent continuously compounded?

17. A man walking in the positive direction along the x axis drags a weight in the upper half plane by a rope of length b. Show that the differential equation of the path (the tractrix) is

$$y\,dx + \sqrt{b^2 - y^2}\,dy = 0$$

18. The carbon in living matter contains a fairly definite minute proportion of C^{14}. This radiocarbon arises from cosmic-ray bombardment in the upper atmosphere and enters living systems by exchange processes, reaching an equilibrium concentration. After death, exchange stops, and the radiocarbon decreases at the rate of one part in 8,000 per year. Show that the amount of C^{14} in a gram of carbon t years later is given by the formula

$$y = y_0 e^{-t/8,000}$$

(a) A cypress beam from the tomb of Sneferu in Egypt contains 55 per cent of that expected in living matter. Show that its age is about 4,800 years.

(b) Charcoal from the Lascaux Cave in France (the cave with the remarkable prehistoric paintings) assayed 14.5 per cent. Find its age.

(c) Charcoal from a tree killed by the eruption of the volcano which formed Crater Lake in Oregon assayed 44.5 per cent. How long ago was this?

19. A tank contains 200 gal of water. Brine containing 2 lb of salt per gal runs in at the rate of 2 gal per min and the mixture runs out at the same rate, the contents being kept uniform by stirring. When will the tank contain 300 lb of salt?

20. Water will issue from an orifice with the velocity it would have in falling freely from the water level to the orifice. The stream contracts somewhat on coming forth so that its cross section has an area of approximately six-tenths that of the orifice. We call this smaller area the *effective* area of the orifice.

By equating the amount of water that flows through the orifice in a short interval of time to the decrease in the volume in a vessel, derive the differential equation for the height x of the water level above the orifice,

$$A(x) \frac{dx}{dt} = -a\sqrt{2gx}$$

where a is the effective area of the orifice and $A(x)$ is the area of the cross section of the vessel at height x.

21. Find the time required to empty a cube full of water of side 2 ft through an orifice in the bottom of 1 sq in. effective cross section. Find when the water is half emptied.

22. If in the preceding problem water is flowing in at the rate of 100 cu in. per sec, find the eventual water level.

23. Find the time to empty a right circular cone with vertex downward through an orifice in the vertex.

24. A hemispherical bowl 4 ft in diameter is full of water. Show that it will empty through an orifice of 1 sq in. effective cross section in the bottom in a little less than 5 min.

25. Prove that if a body slides from rest down a smooth curve of any shape the velocity at any point is

$$v = \sqrt{2gh}$$

where h is the vertical distance below the starting point.

26. The rate at which a body changes temperature is roughly proportional to the difference between its temperature and that of the surrounding medium (Newton's law of cooling). If a body in air at 60° cools from 200 to 120° in 30 min, show that the temperature at any time is

$$T = 60 + 140e^{-0.028243t}$$

When will its temperature be 90°? When 65°?

27. Two chemicals are brought together and combine, a molecules of one and b of the other uniting to form a molecule of the compound. The rate at which molecules are formed is proportional to the product of the numbers of molecules of the original chemicals remaining. Set up the differential equation from which the number of molecules of the compound at any time may be found, and solve.

28. Show that if a vessel of water is rotated about a vertical axis the surface assumes the form of a paraboloid of revolution.

Suggestion: The resultant force acting on a particle at the surface, owing to its weight and the centrifugal force, must be normal to the surface.

29. A coast-guard boat is hunting a rum runner in a fog. The fog rises disclosing the rum runner 4 miles distant and immediately descends. The speed of the boat is three times that of the rum runner; and it is known that the latter will immediately depart at full speed on a straight course of unknown direction. What course should the boat take in order to overtake the rum runner?

CHAPTER 2

SPECIAL METHODS FOR THE EQUATION OF THE FIRST ORDER

The schemes that have been devised for the solution of special differential equations of the first order are legion, and an industrious worker can add to their number at will. These devices have brought a considerable class of equations of simple form into tractability, and some of them are quite useful. In the present chapter we shall treat a few of the classic methods. They are applicable to numerous equations that arise in practice, and certain ones throw a good deal of light on the general problem. The reader should bear in mind, however, that the methods considered are limited in their scope.

In order not to distract the reader's mind from the main issue of acquiring the technique of the method, we shall take for granted matters of continuity, differentiability, and the like. We shall not pause to investigate the vanishing of a divisor and the consequent possible loss of a particular solution. Most of the functions appearing in practice are continuous and have continuous derivatives of all orders at most points. Unusual matters of rigor will be reserved for the existence proofs of a later chapter.

2-1. Linear Equations. A differential equation of the first order is called *linear* if it has the form

$$A(x)\frac{dy}{dx} + B(x)y + C(x) = 0$$

On division by the first coefficient, it can be put in the form

$$\frac{dy}{dx} + P(x)y = Q(x) \qquad (2\text{-}1)$$

This equation is important because of the frequency with which it is met in the simpler problems.

If the second member is zero, $Q(x) \equiv 0$, the equation can be solved by separating the variables.

$$\frac{dy}{dx} + Py = 0 \tag{2-2}$$

$$\frac{dy}{y} = -P\,dx$$

$$\log y = -\int P\,dx + C$$

$$y = Ke^{-\int P\,dx}$$

If $y = z(x)$ is a particular solution obtained by assigning a value to K, not zero, the general solution is $y = Cz(x)$; hence, if we have a particular solution, other than $y = 0$, we can write down the general solution at once.

In the general case (2-1) we shall make use of a multiplier suggested by the solution just found. We observe that

$$\frac{d}{dx} e^{\int P\,dx} y = e^{\int P\,dx} \left(\frac{dy}{dx} + Py\right)$$

Hence, if we multiply (2-1) by $e^{\int P\,dx}$,

$$e^{\int P\,dx}\left(\frac{dy}{dx} + Py\right) = e^{\int P\,dx} Q$$

the first member is now recognized as the derivative of $e^{\int P\,dx} y$ and we can get the general solution by an integration

$$e^{\int P\,dx} y = \int e^{\int P\,dx} Q\,dx + C$$

or

$$y = Ce^{-\int P\,dx} + e^{-\int P\,dx} \int e^{\int P\,dx} Q\,dx \tag{2-3}$$

The student should not memorize this complicated formula, but he should keep the process in mind. It may be stated as follows:

To solve the linear equation (2-1) *multiply by* $e^{\int P\,dx}$ *and integrate.*

Example 2-1. Solve

$$\frac{dy}{dx} + y = x$$

Here $P = 1$, and we multiply by e^x,

$$e^x\left(\frac{dy}{dx} + y\right) = xe^x$$

whence, integrating,

$$e^x y = (x - 1)e^x + C$$
$$y = Ce^{-x} + x - 1$$

Example 2-2. Solve

$$xy' + (1 - x)y = e^{2x}$$

We first write this in the form (2-1),

$$\frac{dy}{dx} + \left(\frac{1}{x} - 1\right) y = \frac{e^{2x}}{x}$$

Here
$$\int P\,dx = \int \left(\frac{1}{x} - 1\right) dx = \log x - x$$
$$e^{\log x - x} = xe^{-x}$$
On multiplying by this, we have
$$e^{-x}\left[x\frac{dy}{dx} + (1-x)y\right] = e^x$$
whence
$$e^{-x}xy = e^x + C$$

Certain simple properties of the linear equation should be mentioned. If we put
$$y = z + w$$
Eq. (2-1) takes the form
$$\left(\frac{dz}{dx} + Pz\right) + \left(\frac{dw}{dx} + Pw - Q\right) = 0$$
If now $y = w(x)$ is any solution of (2-1), the second term drops out, and we have
$$\frac{dz}{dx} + Pz = 0$$
and $y = z(x)$ is a solution of (2-2). Hence:

If $y = w(x)$ is a particular solution of (2-1) and if $y = z(x)$ is a particular nonzero solution of (2-2), then
$$y = Cz(x) + w(x) \qquad (2\text{-}4)$$
is the general solution of (2-1).

These facts may also be read immediately out of formula (2-3).

We see from this theorem that if we notice or can find in any way a particular solution of (2-1), the problem is then reduced to the solution of the less complicated Eq. (2-2). If further we know a nonzero solution of (2-2), the general solution of (2-1) can be written down at once.

Example 2-3. To solve
$$y' + y = 10$$
we note that $y = 10$ is a solution. The equation
$$y' + y = 0$$
has the solution $y = e^{-x}$. The general solution then is
$$y = Ce^{-x} + 10$$

Again if $y = u(x)$ and $y = v(x)$ are different particular solutions of (2-1),
$$u' + Pu = Q \qquad v' + Pv = Q$$

we have, subtracting,
$$u' - v' + P(u - v) = 0$$
In other words,
$$y = u(x) - v(x)$$
is a particular nonzero solution of (2-2). It thus suffices to know two different solutions of (2-1) in order to write down its general solution.

If $y = u(x)$ and $y = v(x)$ are different particular solutions of (2-1), then the general solution is
$$y = K(u - v) + v \qquad (2\text{-}5)$$

This can be put in various forms. Since we have also
$$y = K_1(u - v) + u$$
a division gives
$$\frac{y - u}{y - v} = C \qquad (2\text{-}6)$$

This could also be derived from the first form by putting $K = 1/(1 - C)$.

Example 2-4. The linear equation
$$(x^2 - x)y' + (1 - 2x)y + x^2 = 0$$
is found to be satisfied by $y = x$ and by $y = x^2$. The general solution is then
$$y = C(x^2 - x) + x$$

2-2. Bernoulli's Equation. Various equations can be made linear by changes of variable. Thus

$$2xy \frac{dy}{dx} + y^2 = x$$

becomes, on putting $Y = y^2$, the linear equation
$$x \frac{dY}{dx} + Y = x$$

Similarly
$$\frac{dy}{dx} + P(x) = Q(x)e^{my}$$
which may be written
$$e^{-my} \frac{dy}{dx} + Pe^{-my} = Q$$
becomes linear after the change of variable $Y = e^{-my}$,
$$-\frac{1}{m} \frac{dY}{dx} + PY = Q$$

The proper change of variable in each case was suggested by the presence of a function of y and its derivative.

THE EQUATION OF THE FIRST ORDER

Of a similar type is the equation of Bernoulli:

$$\frac{dy}{dx} + P(x)y = Q(x)y^n \qquad (2\text{-}7)$$

We rule out the cases $n = 0$ and $n = 1$, for which the equation is already linear. Writing this in the form

$$y^{-n}\frac{dy}{dx} + Py^{-n+1} = Q$$

we observe that the change of variable

$$Y = y^{-n+1} \qquad (2\text{-}8)$$

results in a linear equation

$$\frac{1}{1-n}\frac{dY}{dx} + PY = Q$$

Example. Solve

$$y' + y = xy^2$$

We have

$$y^{-2}y' + y^{-1} = x$$

and, setting $Y = y^{-1}$, we have

$$-Y' + Y = x$$

The solution is

$$Y = \frac{1}{y} = Ce^x + x + 1$$

EXERCISES

Find a particular solution by inspection; find a solution when the term not involving y is replaced by zero; and write down the general solution:

1. $y' - y = 2$.
2. $y' + y = 2e^x$.
3. $xy' - y = 1$.
4. $y' = y + 1$.
5. $y' + y = x + 1$.

6. Show that the differential equation of a family of the form $y = Cf(x) + g(x)$ is linear.

Solve:

7. $y' + y = x^2 + 2$.
8. $y' - 2y/x = x^4$.
9. $y' - 3y = e^{3x} + e^{-3x}$.
10. $y' - 4y = 2e^x\sqrt{y}$.
11. $y' = 2y + e^{2x} + 1$.
12. $(x^2 - y^2 - 1)y' = 2xy$.
13. $xy' + 2y = (3x + 2)e^{3x}$.
14. $2xy' - y = x^3 - x$.
15. $2xy' + y = (2ax^2 + 1)/x$.
16. $y' - y + y^2(x^2 + x + 1) = 0$.
17. $y' + y \tan x = \sin 2x$.

Investigate the maximum and minimum points of the integral curves directly from the differential equation in:
18. Exercise 2.   19. Exercise 5.   20. Exercise 7.

21. Find the orthogonal trajectories of the family
$$y = x + Ce^{-x}$$

In an electrical circuit the current i flowing at any time t depends upon the applied electromotive force E as follows
$$L\frac{di}{dt} + Ri = E$$

where L and R, the coefficient of self-induction and the resistance, are positive constants. In the following exercises find the current in terms of the time noting especially the current after a long time, the initial current being i_0.

22. The emf is cut off entirely, $E = 0$.

23. The emf is constant, $E = E_0$. For t large, show that Ohm's law holds very nearly $(i = E/R)$.

24. The emf falls off according to the law $E = E_0 e^{-kt}$.

25. The emf is the periodic function $E = E_0 \sin \omega t$.

26. Show that Ohm's law is satisfied for a variable emf when the current is a maximum or minimum. Show that when the current is a maximum the emf is decreasing.

27. When will the body of Exercise 26, page 32, reach 90°, if the room temperature is falling at the rate of 1° each 10 min?

28. If the room temperature is rising at the rate of 1° in 10 min, when will the body reach room temperature?

29. Solve:
$$xy' - 2y = 4x^3 \sqrt{y}$$

Make suitable changes of variable and solve:

30. $2xy \dfrac{dy}{dx} + (1 + x)y^2 = e^x$.   **31.** $\cos y \dfrac{dy}{dx} + ax \sin y = bx^2$.

32. $\dfrac{dy}{dx} + \dfrac{y + 3}{x} = x^2(y + 3)^3$.

2-3. Riccati's Equation. A first step in complexity beyond the linear equation is the equation in which y' is a quadratic function of y,

$$y' = Py^2 + Qy + R \qquad (2\text{-}9)$$

P, Q, and R being functions of x alone. This is known as Riccati's equation.

In general, this equation cannot be solved by quadratures (that is, a finite number of integrations). If, however, one solution, $y = u(x)$, is known, we can proceed to find the general solution. We have
$$u' = Pu^2 + Qu + R$$
Subtracting from (2-9),
$$y' - u' = P(y^2 - u^2) + Q(y - u) \qquad (2\text{-}10)$$
and putting $y - u = z$, we have
$$z' = P(2uz + z^2) + Qz$$

This is Bernoulli's equation with $n = 2$. It is solved by putting $Y = 1/z$.

THE EQUATION OF THE FIRST ORDER

We have then the following result.
If $y = u(x)$ is a solution of (2-9) the change of variable

$$y = u + \frac{1}{Y} \tag{2-11}$$

reduces (2-9) to a linear equation in Y.

If we know other solutions of (2-9) the labor of finding the general solution is correspondingly reduced. Let $y = v(x)$ be another solution. Writing (2-10) in the form

$$\frac{y' - u'}{y - u} = P(y + u) + Q$$

and subtracting the like expression involving the second solution

$$\frac{y' - v'}{y - v} = P(y + v) + Q$$

we have

$$\frac{y' - u'}{y - u} - \frac{y' - v'}{y - v} = P(u - v)$$

This can now be integrated,

$$\log (y - u) - \log (y - v) = \int P(u - v) \, dx + K_1$$

and from this we can state the following result.

If $y = u(x)$ and $y = v(x)$ are different solutions of (2-9), then the general solution is

$$\frac{y - u}{y - v} = K e^{\int P(u - v) dx} \tag{2-12}$$

Finally, if we know a third solution, $y = w(x)$, we can write down the general solution at once. We have

$$\frac{w - u}{w - v} = K' e^{\int P(u - v) dx}$$

whence a division gives the formula of the following theorem:

If $y = u(x)$, $y = v(x)$, $y = w(x)$ are different solutions of (2-9), then the general solution is

$$\frac{(y - u)(w - v)}{(y - v)(w - u)} = C \tag{2-13}$$

Example. Solve

$$y' = xy^2 + (1 - 2x)y + x - 1$$

We find that $y = 1$ is a solution. Substituting

$$y = 1 + \frac{1}{Y}$$

and simplifying, we have
$$Y' + Y = -x$$
This has the solution
$$Y = Ce^{-x} + 1 - x$$
whence
$$y = 1 + \frac{1}{Ce^{-x} + 1 - x}$$

EXERCISES

Find one or more solutions by trial and solve.
1. $y' = x^3(y - x)^2 + y/x$.
2. $y' = x^3(y^2 - x^2) + y/x$.
3. $y' = (y + x)(y + x - 2) - 1$.
4. $y' = (y - a)(y - b)$.
5. $y' = \dfrac{y^2 + (x^2 - 1)y - x^2}{x(x^2 - 1)}$, given that there are various solutions of the form $y = x^n$.

6. Show that the change of variable $y = -Y'/(PY)$ reduces Riccati's equation to the linear equation of the second order
$$Y'' - \left(Q + \frac{P'}{P}\right) Y' + PRY = 0$$

7. Use the preceding to transform and solve
$$y' = e^{-x}y^2 + y + e^x$$

2-4. Homogeneous Equations. A function $F(x,y)$ is called *homogeneous of degree n* if
$$F(tx,ty) \equiv t^n F(x,y) \tag{2-14}$$
that is, if x and y be replaced by tx and ty, t^n factors out of the resulting function, and the remaining factor is the original function. For example,
$$x^2 + 2xy \qquad \frac{1}{x+y}$$
$$\frac{\sqrt{x^2 + y^2}}{2x} + \log \frac{y}{x}$$
are homogeneous of degrees 2, -1, and 0, respectively, since
$$(tx)^2 + 2txty = t^2(x^2 + 2xy)$$
$$\frac{1}{tx + ty} = t^{-1} \frac{1}{x + y}$$
$$\frac{\sqrt{(tx)^2 + (ty)^2}}{2tx} + \log \frac{ty}{tx} = \frac{\sqrt{x^2 + y^2}}{2x} + \log \frac{y}{x}$$

The following properties of homogeneous functions follow immediately from the definition.

THE EQUATION OF THE FIRST ORDER   41

If $F(x,y)$ and $G(x,y)$ are homogeneous of degrees n and m, respectively, then the product $F(x,y)G(x,y)$ is homogeneous of degree $n+m$ and the quotient $F(x,y)/G(x,y)$ is homogeneous of degree $n-m$.

In particular, if $F(x,y)$ and $G(x,y)$ are homogeneous of the same degree, their quotient is homogeneous of degree zero.

Finally, a homogeneous function of degree zero is a function of y/x. For, if $n = 0$ in (2-14) and if we put $t = 1/x$, we get $F(x,y) = F(1, y/x)$.

A differential equation

$$M(x,y)\,dx + N(x,y)\,dy = 0 \qquad (2\text{-}15)$$

is called *homogeneous* if $M(x,y)$ and $N(x,y)$ are homogeneous functions of the same degree.

By a simple change of variable a homogeneous equation may be reduced to one in which the variables are separable. Writing the equation in the form

$$\frac{dy}{dx} = -\frac{M(x,y)}{N(x,y)}$$

the second member is now homogeneous of degree zero and so is a function of y/x,

$$\frac{dy}{dx} = f\left(\frac{y}{x}\right) \qquad (2\text{-}16)$$

This form of the equation suggests putting $y/x = v$, or $y = vx$. We have

$$v + x\frac{dv}{dx} = f(v)$$

in which the variables are readily separated.

We may state the method of solution as follows:

If we set $y = vx$ in a homogeneous equation in x and y, there results an equation in v and x in which the variables are separable.

After integration, we must, of course, replace v by its value y/x.

Example. Solve:

$$(x^2 + y^2)\,dx - 2xy\,dy = 0$$

The equation being homogeneous, we set

$$y = vx$$
$$dy = v\,dx + x\,dv$$

The resulting equation is

$$(1 - v^2)\,dx - 2vx\,dv = 0$$

whence

$$\frac{dx}{x} - \frac{2v\,dv}{1 - v^2} = 0$$
$$\log x + \log(1 - v^2) = K$$
$$x(1 - v^2) = C$$

or, finally,
$$x^2 - y^2 = Cx$$

The integral curves of a homogeneous differential equation have a simple geometric property. Let the xy plane be stretched from (or contracted toward) the origin in such a way that each point is moved to k times its original distance from the origin, its direction from the origin being unchanged. When this is done, *each integral curve is carried into an integral curve.*

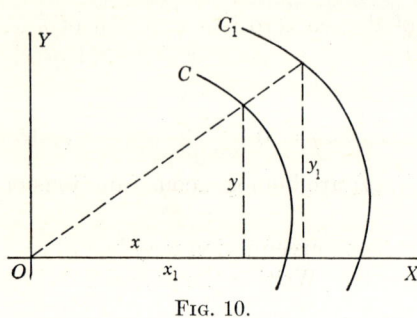

Fig. 10.

To show this, let C be an integral curve. Let (x,y) be a point of C at which the slope is y'. Let x_1, y_1, y_1' be the coordinates and slope of the corresponding point on C_1, the curve into which C is carried by the stretching. We have

$$x_1 = kx \qquad y_1 = ky \qquad y_1' = \frac{k\,dy}{k\,dx} = y'$$

Since C is an integral curve, we have

$$y' = f\left(\frac{y}{x}\right)$$

On C_1 then, replacing x and y by x_1/k and y_1/k,

$$y_1' = f\left(\frac{y_1}{x_1}\right)$$

The coordinates and slope on C_1 satisfy the differential equation, as was to be shown.

Conversely, let

$$\frac{dy}{dx} = f(x,y)$$

be the differential equation of a family of curves having the property of invariance when any stretching from the origin is made. Then the slope $f(x_1,y_1)$ at the point $x_1 = kx$, $y_1 = ky$ is always equal to the slope at (x,y). Hence

$$f(kx,ky) = f(x,y)$$

and the differential equation of the family is homogeneous.

Examples are the families in Figs. 3 and 4, page 14.

EXERCISES

Solve:

1. $(2x^2 - y^2) dx + 3xy \, dy = 0$. 2. $\dfrac{dy}{dx} = \dfrac{l_1 x - l_2 y}{l_2 x + m_2 y}$. 3. $2xy' = y - ax$.

4. $xy' = y(1 + \log y/x)$. 5. $x^2 y' + xy + 2y^2 = 0$.
6. $x^2 y' = n(x^2 + y^2) \arctan y/x + xy$.
7. $(5x^2 y + 30xy^2 + 4x^3) dy + (5xy^2 + 20y^3 + 6x^2 y) dx = 0$.

8. Find the orthogonal trajectories of the family

$$x^4 + 6x^2 y^2 + y^4 = C$$

9. Find the curves cutting the family $y = Cx$ at a constant angle.
10. Find the orthogonal trajectories of the family of circles tangent to the y axis at the origin.
11. Find the curves cutting the members of the preceding family at the constant angle α.
12. Prove that if the differential equation of a family of curves is homogeneous, the differential equation of the curves cutting the members at a constant angle is also homogeneous.
13. Prove that if polar coordinates be introduced in a homogeneous equation, the variables are separable.

2-5. The Equations $y' = f\left(\dfrac{l_1 x + m_1 y + q_1}{l_2 x + m_2 y + q_2}\right)$ **and** $y' = F(ax + by)$.
Among the innumerable equations which can be made homogeneous by a change of variables is the equation

$$\dfrac{dy}{dx} = f\left(\dfrac{l_1 x + m_1 y + q_1}{l_2 x + m_2 y + q_2}\right) \tag{2-17}$$

We put

$$\begin{cases} x = X + h \\ y = Y + k \end{cases} \tag{2-18}$$

then

$$l_1 x + m_1 y + q_1 = l_1 X + m_1 Y + l_1 h + m_1 k + q_1$$
$$l_2 x + m_2 y + q_2 = l_2 X + m_2 Y + l_2 h + m_2 k + q_2$$

If we choose h and k so that

$$l_1 h + m_1 k + q_1 = 0$$
$$l_2 h + m_2 k + q_2 = 0 \tag{2-19}$$

then Eq. (2-17) takes the form

$$\dfrac{dY}{dX} = f\left(\dfrac{l_1 X + m_1 Y}{l_2 X + m_2 Y}\right) \tag{2-20}$$

which is homogeneous.

Equations (2-19) can be solved for h and k, provided

$$D = l_1 m_2 - l_2 m_1 \neq 0 \tag{2-21}$$

By translating the origin to the point (h,k) we have removed the constant terms in the linear expressions.

If $D = 0$, the first-degree terms of numerator and denominator are proportional and the equation falls into the form

$$\frac{dy}{dx} = F(ax + by) \tag{2-22}$$

By putting
$$Y = ax + by$$
we have
$$\frac{dY}{dx} = a + bF(Y)$$

in which the variables are separable.

Example 2-5. Solve
$$\frac{dy}{dx} = 2\left(\frac{y+2}{x+y+1}\right)^2$$

Equations (2-19) are
$$k + 2 = 0 \qquad h + k + 1 = 0$$
whence $h = 1$, $k = -2$, and we put
$$x = X + 1 \qquad y = Y - 2$$
We have
$$\frac{dY}{dX} = \frac{2Y^2}{(X+Y)^2}$$

which is homogeneous. Setting $Y = VX$, we have
$$V + X\frac{dV}{dX} = \frac{2V^2}{(1+V)^2}$$

which reduces to
$$\left(\frac{1}{V} + \frac{2}{1+V^2}\right)dV + \frac{dX}{X} = 0$$

Integrating and returning to the original variables,
$$\log V + 2 \arctan V + \log X = C$$
$$\log Y + 2 \arctan \frac{Y}{X} = C$$
$$\log (y+2) + 2 \arctan \frac{y+2}{x-1} = C$$

Example 2-6. Solve
$$y' + \sin^2(x+y) = 0$$

Putting $Y = x + y$, we have
$$Y' = 1 + y' = 1 - \sin^2 Y = \cos^2 Y$$

whence
$$\sec^2 Y \, dY = dx$$
Integrating,
$$\tan Y = x + C$$
$$\tan(x + y) = x + C$$

2-6. The Linear Fractional Equation. We shall now consider the equation

$$\frac{dy}{dx} = \frac{l_1 x + m_1 y}{l_2 x + m_2 y} \quad (2\text{-}23)$$

This is the simplest case of (2-17) with $D \neq 0$, after the translation to remove the constant terms.

This equation is instructive, for one thing, because of the light it throws on the behavior of the solutions of the general equation of the first order,

$$M(x,y) \, dx + N(x,y) \, dy = 0 \quad (2\text{-}24)$$

near certain exceptional, or *singular*, points at which M and N are both zero. The slope, $y' = -M/N$, becomes indeterminate.

Consider the direction field near a singular point. The curve of points at which the slope has a constant value p, namely,

$$M(x,y) + pN(x,y) = 0$$

passes through the singular point. In its neighborhood the lineal elements usually have all possible directions.

Suppose the singular point to be the origin. Then M has ordinarily a series expansion of the form

$$M(x,y) = l_1 x + m_1 y + ax^2 + bxy + cy^2 + \cdots$$

and N has a similar expansion. If we neglect terms of the second and higher degrees, which are small near the origin, we are led, in general, to an equation of the form (2-23). The integral curves of (2-23) and (2-24) behave in much the same way.

Equation (2-23), being homogeneous, can be solved by the substitution $y = vx$. The solutions, however, fall into a great diversity of types, depending upon the particular values of the constants in the equation, and it is probably more instructive to make a geometric study of the situation.

Among special families whose equations are of this form are the similar conics with centers at the origin,

$$ax^2 + 2bxy + cy^2 = K$$

This is a family of hyperbolas (Fig. 11) if $b^2 - ac > 0$ and a family of ellipses (Fig. 12) if $b^2 - ac < 0$. On differentiating and solving, we find

$$ax + by + bxy' + cyy' = 0$$
$$\frac{dy}{dx} = \frac{-ax - by}{bx + cy} \qquad (2\text{-}25)$$

This has the form (2-23) with $l_2 + m_1 = 0$. Conversely, if $l_2 + m_1 = 0$, the integral curves of (2-23) are the family of similar conics
$$-l_1 x^2 + 2l_2 xy + m_2 y^2 = K$$
Here
$$D = l_1 m_2 - l_2 m_1 = l_2{}^2 + l_1 m_2$$

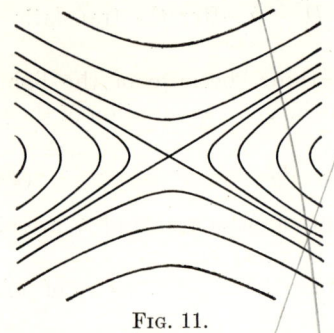

Fig. 11.

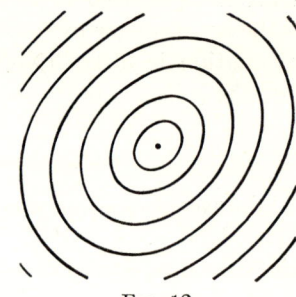

Fig. 12.

and we have a family of hyperbolas or a family of ellipses according as D is positive or negative.

The sign of D has a geometrical significance for the general case. Differentiating (2-23) and simplifying, we have

$$y'' = \frac{l_1 m_2 - l_2 m_1}{(l_2 x + m_2 y)^2}(y - xy') \qquad (2\text{-}26)$$

So y'' and $y - xy'$, if not zero, have the same or opposite signs according as D is positive or negative.

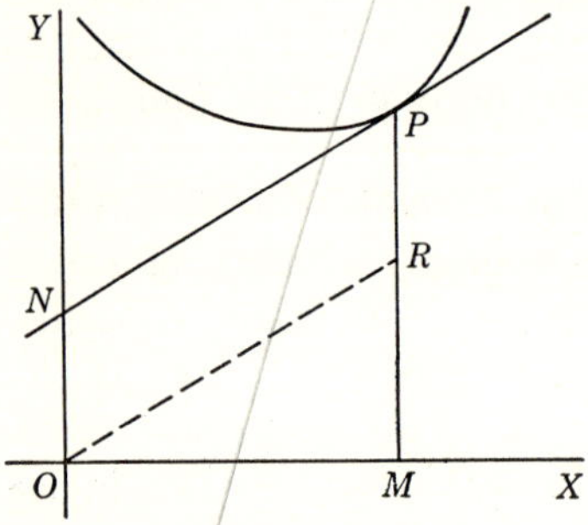

Now the integral curve near $P(x,y)$ lies above or below the tangent according as y'' is positive or negative; and the origin lies below or above the tangent at P, according as $y - xy'$ is positive or negative. For calling segments measured upward positive, we have from Fig. 13, in which OR is drawn parallel to the tangent NP,

$$ON = MP - MR = y - xy'$$

Fig. 13

Hence we see from (2-26) that, provided the tangent doesn't pass through the origin, *the integral curve near P and the origin lie on opposite sides of the tangent if $D > 0$ and on the same side of the tangent if $D < 0$.*

We know that if $y - xy' = 0$ at any point P, so that the tangent at P passes through the origin O, then the line OP is an integral curve. For the particular differential equation

$$y - xy' = 0$$

all integral curves are straight lines through the origin. We put this simple case aside.

We now consider straight-line solutions in the general case. From (2-23),

$$y - xy' = \frac{m_2 y^2 + (l_2 - m_1)xy - l_1 x^2}{l_2 x + m_2 y} \tag{2-27}$$

The vanishing of this expression depends upon the discriminant of the numerator

$$\Delta = (l_2 - m_1)^2 + 4l_1 m_2 = (l_2 + m_1)^2 + 4D \tag{2-28}$$

If $\Delta > 0$, the numerator breaks up into real distinct factors,

$$y - xy' = \frac{(ax + by)(cx + dy)}{l_2 x + m_2 y} \tag{2-29}$$

and there are two rectilinear solutions,

$$M_1: ax + by = 0 \qquad M_2: cx + dy = 0 \tag{2-30}$$

These two lines separate the plane into four regions in which the remaining integral curves lie.

If $\Delta = 0$, the factors in the numerator are equal and there is a single rectilinear solution. If $\Delta < 0$, the factors are imaginary and there is no real rectilinear solution.

Case 1. $D > 0$. We see from (2-28) that $\Delta > 0$, and the lines M_1 and M_2 exist. The integral curve and the origin lie on opposite sides of the tangent. The curves, while not usually hyperbolas, have the general shape of those in Fig. 11.

Before proceeding to the next case we shall derive a further geometric property of the solutions. We show first that each family of oblique trajectories has an equation of the form (2-23). Letting t be the tangent of the angle of intersection, the equation of the trajectories is (Sec. 1-15)

$$\frac{y' - t}{1 + ty'} = \frac{l_1 x + m_1 y}{l_2 x + m_2 y}$$

or

$$\frac{dy}{dx} = \frac{(tl_2 + l_1)x + (tm_2 + m_1)y}{(l_2 - tl_1)x + (m_2 - tm_1)y} \tag{2-31}$$

The determinant here,

$$D_t = (tl_2 + l_1)(m_2 - tm_1) - (l_2 - tl_1)(tm_2 + m_1) = D(t^2 + 1)$$

has the same sign as D.

Now among these oblique trajectories there is one and only one **family** of conics. We get this family when

$$(l_2 - tl_1) + (tm_2 + m_1) = 0$$

or

$$t = \frac{l_2 + m_1}{l_1 - m_2}$$

The conics are hyperbolas or ellipses according as D is positive or negative.

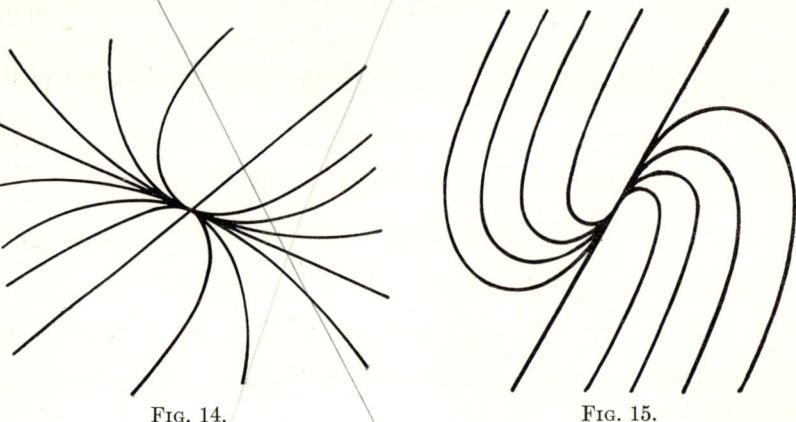

FIG. 14.   FIG. 15.

Case 2. $D < 0$. Each integral curve cuts a family of similar concentric ellipses (Fig. 12) at a constant angle. Unless this angle is zero, the curve when traced from a given point P in one direction approaches the origin, and when traced in the other direction recedes indefinitely. Further, the curve cannot cross one of the straight-line solutions, if any exist.

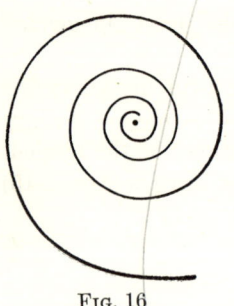

FIG. 16.

If $\Delta > 0$, the straight lines M_1 and M_2 separate the plane into four parts. Tracing an integral curve from a point P in one direction we approach the origin without crossing either of the lines. The curve terminates at the origin at which point it is tangent to one of the lines, M_1, say. A point moving in the other direction from P along the curve recedes so that the line joining it to the origin approaches coincidence with M_2 (see Fig. 14).

If $\Delta = 0$, there is a single line M. An integral curve comes into the origin, tangent to M, and the radius vector to a point receding along the curve approaches coincidence with M (see Fig. 15).

If $\Delta < 0$, there is no rectilinear solution. There is a positive lower bound to the angle between the curve and the radius vector so that the curve passes completely around the origin. Unless the curves are ellipses ($t = 0$), they are spirals (see Fig. 16).

EXERCISES

Solve, and sketch the integral curves:

1. $\dfrac{dy}{dx} = -\dfrac{2x}{3y}$. 2. $\dfrac{dy}{dx} = \dfrac{x}{y}$. 3. $\dfrac{dy}{dx} = \dfrac{3y}{x}$.

4. $\dfrac{dy}{dx} = \dfrac{2y - x}{x}$. 5. $\dfrac{dy}{dx} = \dfrac{x + y}{x - y}$.

6. Show that the family
$$(ax + by)^p (cx + dy)^q = K$$
satisfies an equation of the form (2-23). Discuss the rectilinear solutions.

Solve:

7. $y' = 2\sqrt{2x + y + 1}$. 8. $xy' = x + y$.

9. $y' = (x + y + 1)^2$. 10. $y' = \dfrac{x - 2y + 6}{2x + y + 2}$.

11. $y' = \dfrac{3x - y + 1}{3y - x + 5}$. 12. $y' = \left(\dfrac{x - y + 3}{x - y + 1}\right)^2$.

2-7. Exact Equations.

A differential equation

$$M\,dx + N\,dy = 0 \qquad (2\text{-}32)$$

is called *exact* if there exists a function $u(x,y)$ such that

$$du \equiv M\,dx + N\,dy \qquad (2\text{-}33)$$

that is, if the first member of the equation is the exact differential of a function of x and y. The differential equation then takes the form

$$du = 0 \qquad (2\text{-}34)$$

and has the immediate solution

$$u(x,y) = C \qquad (2\text{-}35)$$

Example.
$$2xy\,dx + x^2\,dy = 0$$
is exact, since it may be written
$$d(x^2 y) = 0$$
and its solution is
$$x^2 y = C$$

On the other hand the same equation written in the form
$$2y\,dx + x\,dy = 0$$
is not exact.

We shall now derive a test for exactness. We have

$$du = \dfrac{\partial u}{\partial x}dx + \dfrac{\partial u}{\partial y}dy \qquad (2\text{-}36)$$

This is identical with the first member of (2-32), since dx and dy vary independently, if and only if

$$\frac{\partial u}{\partial x} \equiv M \qquad \frac{\partial u}{\partial y} \equiv N \tag{2-37}$$

Since

we have

$$\frac{\partial^2 u}{\partial y\, \partial x} \equiv \frac{\partial^2 u}{\partial x\, \partial y}$$

$$\frac{\partial M}{\partial y} \equiv \frac{\partial N}{\partial x} \tag{2-38}$$

This is a necessary condition for exactness.

We shall show that, conversely, when (2-38) holds, the equation is exact. We prove this by finding a function u which satisfies both of Eqs. (2-37). The general solution of the first of these is, integrating,

$$u = \int_a^x M(x,y)\, \partial x + K(y)$$

where a is any convenient constant and K is an arbitrary function of y. Here ∂x is used to call attention to the fact that in the integration y is held constant.

We substitute this value of u into the second equation of (2-37),

$$\frac{\partial}{\partial y}\left[\int_a^x M\, \partial x + K\right] = N$$

Assuming that (2-38) holds, we have

$$\frac{\partial}{\partial y}\int_a^x M\, \partial x = \int_a^x \frac{\partial M}{\partial y}\, \partial x = \int_a^x \frac{\partial N}{\partial x}\, \partial x = N(x,y) - N(a,y)$$

and the equation becomes

$$N(x,y) - N(a,y) + K'(y) = N(x,y)$$

whence

$$K = \int N(a,y)\, dy$$

We have found u to satisfy both equations of (2-37),

$$u = \int_a^x M(x,y)\, \partial x + \int N(a,y)\, dy$$

We summarize the results in the following theorem.

THEOREM. *Equation* (2-32) *is exact if and only if*

$$\frac{\partial M}{\partial y} \equiv \frac{\partial N}{\partial x} \tag{2-38}$$

Mdx + Ndy = 0

THE EQUATION OF THE FIRST ORDER

The solution of the equation is

$$\int_a^x M(x,y)\, \partial x + \int N(a,y)\, dy = C \qquad (2\text{-}39)$$

where a is any convenient constant.

Example. Solve

$$(3x^2y + 8xy^2)\, dx + (x^3 + 8x^2y + 12y^2)\, dy = 0$$

Since

$$\frac{\partial}{\partial y}(3x^2y + 8xy^2) = 3x^2 + 16xy = \frac{\partial}{\partial x}(x^3 + 8x^2y + 12y^2)$$

the equation is exact. If we take $a = 0$ several terms drop out and (2-39) gives

$$\int_0^x (3x^2y + 8xy^2)\, \partial x + \int 12y^2\, dy = C$$

or

$$x^3y + 4x^2y^2 + 4y^3 = C$$

Line Integrals. The matters of the preceding paragraphs are connected with the so-called line integrals. The integral

$$I = \int_L M\, dx + N\, dy$$

where L is a curve extending from (a,b) to (c,d), is independent of the path connecting these two points if and only if (2-38) holds. That the integral depends only on the end points is obvious since

$$I = \int_{(a,b)}^{(c,d)} du = u(c,d) - u(a,b)$$

In solving the exact equation (2-32) the student has various procedures available. He may find the function u by employing any convenient line integral,

$$u(x,y) = \int_L M(X,Y)\, dX + N(X,Y)\, dY \qquad (2\text{-}40)$$

where L is some curve extending from a fixed point (a,b) to the point (x,y). Formula (2-39) results from integrating along the broken-line path formed by the vertical line $X = a$ and the horizontal line $Y = y$.

In the previous example, let L be the line segment from the origin to (x,y),

$$X = xt \qquad Y = yt \qquad 0 \leqslant t \leqslant 1$$

We have the solution

$$\int_0^1 [(3x^2yt^3 + 8xy^2t^3)x\, dt + (x^3t^3 + 8x^2yt^3 + 12y^2t^2)y\, dt]$$
$$= (\tfrac{3}{4}x^2y + 2xy^2)x + (\tfrac{1}{4}x^3 + 2x^2y + 4y^3)y$$
$$= x^3y + 4x^2y^2 + 4y^3 = C$$

In simple equations it is often possible to group terms into combinations which are exact and to solve by inspection. Thus in the preceding example we write

$$(3x^2y\,dx + x^3\,dy) + (8xy^2\,dx + 8x^2y\,dy) + (12y^2\,dy) = 0$$

Here each part in parentheses is an exact differential and we have immediately

$$x^3y + 4x^2y^2 + 4y^3 = C$$

EXERCISES

Prove exact and integrate:
1. $(x + y^2)\,dy + (y - x^2)\,dx = 0$.
2. $(lx + my + a)\,dx + (mx + ny + b)\,dy = 0$.
3. $\dfrac{xy+1}{y}\,dx + \dfrac{2y-x}{y^2}\,dy = 0$.
4. $(x - 2xy - 5)\,dx - (x^2 + 3y - 2)\,dy = 0$.
5. $(2x^2 - 6xy - 5y^2)\,dx - (3x^2 + 10xy - y^2)\,dy = 0$.
6. $(e^{y^2} - \csc y\,\csc^2 x)\,dx + (2xye^{y^2} - \csc y\,\cot y\,\cot x)\,dy = 0$.
7. $3\cos 3t\,\cos 2\theta\,dt - 2\sin 3t\,\sin 2\theta\,d\theta = 0$.
8. $(2xy^4 + \sin y)\,dx + (4x^2y^3 + x\cos y)\,dy = 0$.
9. Solve the problem in the text by integrating along the parabola $X = xt$, $Y = yt^2$.
10. Solve as an exact equation

$$\frac{4y^2 - 2x^2}{4xy^2 - x^3}\,dx + \frac{8y^2 - x^2}{4y^3 - x^2y}\,dy = 0.$$

11. Solve the preceding as a homogeneous equation.

2-8. Integrating Factors. We noted that the equation

$$2y\,dx + x\,dy = 0$$

is not exact, but that if we multiply by x we have an exact equation:

$$2xy\,dx + x^2\,dy = d(x^2y) = 0$$

We shall have again an exact equation if we multiply by $1/xy$,

$$\frac{2\,dx}{x} + \frac{dy}{y} = d(2\log x + \log y) = 0$$

or by x^3y,

$$2x^3y^2\,dx + x^4y\,dy = d\left(\frac{x^4y^2}{2}\right) = 0$$

or by $1/\sqrt{y}$,

$$2\sqrt{y}\,dx + \frac{x}{\sqrt{y}}\,dy = d(2x\sqrt{y}) = 0$$

THE EQUATION OF THE FIRST ORDER

The multipliers x, $1/xy$, x^3y, $1/\sqrt{y}$ are called *integrating factors* of the given equation.

DEFINITION. A nonzero function $\mu(x,y)$ is called an integrating factor of the equation

$$M\,dx + N\,dy = 0 \tag{2-41}$$

if the equation

$$\mu(M\,dx + N\,dy) = 0 \tag{2-42}$$

is exact.

It is frequently possible to note integrating factors by inspection. Certain combinations of differentials suggest trial factors. Thus in

$$(y + x^3y^2)\,dx + x\,dy = 0$$

the combination $y\,dx + x\,dy = d(xy)$ is suggestive, and we have

$$\frac{y\,dx + x\,dy}{x^2y^2} + x\,dx = 0$$

$$-\frac{1}{xy} + \frac{x^2}{2} = C$$

Again, $x\,dy - y\,dx$ reminds one of

$$d\left(\frac{y}{x}\right) = \frac{x\,dy - y\,dx}{x^2}$$

or

$$d\left(\frac{x}{y}\right) = \frac{y\,dx - x\,dy}{y^2}$$

etc. Thus

$$(4x^2 + y)\,dx - x\,dy = 0$$

may be written

$$4\,dx + \frac{y\,dx - x\,dy}{x^2} = 0$$

whence

$$4x - \frac{y}{x} = C$$

Skill in the matter depends upon familiarity with the formulas of the differential calculus.

We shall now show that the multiplicity of integrating factors exhibited by our first example is not exceptional. We shall assume—what will be proved later under quite general conditions—that the differential equation has a general solution.

Let the general solution be solved for the arbitrary constant:

$$u(x,y) = C \tag{2-43}$$

Differentiating,

$$\frac{\partial u}{\partial x}\,dx + \frac{\partial u}{\partial y}\,dy = 0 \tag{2-44}$$

Equation (2-43) is a solution of (2-41) if and only if the slopes are the same,

$$\frac{dy}{dx} = -\frac{\partial u}{\partial x}\bigg/\frac{\partial u}{\partial y} = -\frac{M}{N}$$

We can state from this the general result:

THEOREM. *The necessary and sufficient condition that $u(x,y) = C$, where u is nonconstant, be the general solution of (2-41) is that*

$$M\frac{\partial u}{\partial y} - N\frac{\partial u}{\partial x} \equiv 0 \qquad (2\text{-}45)$$

We can now demonstrate the existence of integrating factors. We write (2-45) in the form

$$\frac{\partial u/\partial x}{M} = \frac{\partial u/\partial y}{N} = \mu, \text{ say}$$

Here μ is some function of x and y. We have

$$\frac{\partial u}{\partial x} = \mu M \qquad \frac{\partial u}{\partial y} = \mu N$$

$$\mu(M\,dx + N\,dy) = \frac{\partial u}{\partial x}dx + \frac{\partial u}{\partial y}dy = du$$

and the equation is exact. Equation (2-41) thus has an integrating factor.

Let $F(u)$ be any function of u. We have

$$\mu F(u)(M\,dx + dy) = F(u)\,du = d[\textstyle\int F(u)\,du]$$

Hence $\mu F(u)$ is an integrating factor. Since $F(u)$ is an arbitrary function of u, we have proved the following result.

THEOREM. *The differential equation (2-41) has an infinite number of integrating factors.*

It can be proved that $\mu F(u)$ is the most general integrating factor.

We now derive the differential equation of the integrating factors. From the condition that (2-42) be exact, we have

$$\frac{\partial}{\partial y}(\mu M) \equiv \frac{\partial}{\partial x}(\mu N)$$

or

$$\mu\frac{\partial M}{\partial y} + M\frac{\partial \mu}{\partial y} = \mu\frac{\partial N}{\partial x} + N\frac{\partial \mu}{\partial x}$$

This may be written in the following forms:

$$\frac{1}{\mu}\left(N\frac{\partial \mu}{\partial x} - M\frac{\partial \mu}{\partial y}\right) = \frac{\partial M}{\partial y} - \frac{\partial N}{\partial x}$$

$$\boxed{N\frac{\partial}{\partial x}\log \mu - M\frac{\partial}{\partial y}\log \mu = \frac{\partial M}{\partial y} - \frac{\partial N}{\partial x}} \qquad (2\text{-}46)$$

THE EQUATION OF THE FIRST ORDER 55

The integrating factor is thus a solution of a partial differential equation of the first order. It would appear that we have replaced our original problem by a less simple one. It may be remarked, however, that we do not need a general solution of (2-46)—any nonzero solution will suffice. Also, as will appear in the next section, Eq. (2-46) may be used to discover what types of differential equations will have particular kinds of integrating factors.

Let ν be a second integrating factor. Then

$$N \frac{\partial}{\partial x} \log \nu - M \frac{\partial}{\partial y} \log \nu = \frac{\partial M}{\partial y} - \frac{\partial N}{\partial x}$$

Subtracting (2-46) from this, we have

$$N \frac{\partial}{\partial x} \log \frac{\nu}{\mu} - M \frac{\partial}{\partial y} \log \frac{\nu}{\mu} = 0$$

Referring to the first theorem of this section, we see that if ν/μ is not constant, the general solution of (2-41) is

$$\log \frac{\nu}{\mu} = K$$

or

$$\frac{\nu}{\mu} = C$$

THEOREM. *If μ and ν are integrating factors of (2-41) whose ratio is not constant, the general solution of the equation is*

$$\nu = C\mu$$

2-9. On the Making of Rules. If, reversing the process, we start with the integrating factor, we shall be able to construct innumerable rules for the solution of particular types of differential equations. For, if μ is given, Eq. (2-46) provides a criterion that a given equation have μ as integrating factor. Many of the rules found in textbooks may be derived in this way.

For example, let us find the most general differential equation having an integrating factor which is a function of x alone. We have

$$\frac{\partial}{\partial y} \log \mu = 0 \qquad \frac{\partial}{\partial x} \log \mu = \frac{d}{dx} \log \mu$$

and (2-46) becomes

$$\boxed{\frac{d}{dx} \log \mu = \frac{1}{N} \left(\frac{\partial M}{\partial y} - \frac{\partial N}{\partial x} \right)}$$

Here the first member is a function of x alone, and this equation can be

satisfied if and only if the second member is a function of x alone, say $f(x)$. We have then
$$d \log \mu = f(x) \, dx$$
whence
$$\log \mu = \int f(x) \, dx + C$$
$$\mu = K e^{\int f(x) dx}$$

Knowing an integrating factor, the equation can be made exact and hence can be solved by quadratures.

We have then the following rule:

If, in the equation
$$M \, dx + N \, dy = 0$$
we have
$$\frac{1}{N}\left(\frac{\partial M}{\partial y} - \frac{\partial N}{\partial x}\right) = f(x)$$
a function of x alone, then
$$\mu = e^{\int f(x) dx}$$
is an integrating factor.

As an example consider the linear equation
$$y' + P(x)y = Q(x)$$
or
$$(Py - Q) \, dx + dy = 0$$
Here
$$\frac{1}{N}\left(\frac{\partial M}{\partial y} - \frac{\partial N}{\partial x}\right) = P$$
which is a function of x alone. So we have the integrating factor found earlier, $\mu = e^{\int P dx}$.

More generally, let μ be a function of some function of x and y, $w(x,y)$, let us say. Then substituting in (2-46),
$$N \frac{d}{dw} \log \mu \frac{\partial w}{\partial x} - M \frac{d}{dw} \log \mu \frac{\partial w}{\partial y} = \frac{\partial M}{\partial y} - \frac{\partial N}{\partial x}$$
or
$$\frac{d}{dw} \log \mu = \frac{\dfrac{\partial M}{\partial y} - \dfrac{\partial N}{\partial x}}{N \dfrac{\partial w}{\partial x} - M \dfrac{\partial w}{\partial y}}$$

This is possible if and only if the second member is a function of w, say $f(w)$. We have then readily
$$\mu = K e^{\int f(w) dw}$$

As an example, letting
$$w = xy$$

we have, *if*

$$\frac{\frac{\partial M}{\partial y} - \frac{\partial N}{\partial x}}{Ny - Mx} = f(xy)$$

a function of the product xy, then

$$\mu = e^{\int f(xy)d(xy)}$$

is an integrating factor.

Problem. If (2-41) is homogeneous show that $1/(Mx + Ny)$ is an integrating factor, provided $Mx + Ny$ is not zero.

We wish to show that

$$\frac{M\,dx + N\,dy}{Mx + Ny} = \frac{dx}{x} + \frac{N(x\,dy - y\,dx)}{x(Mx + Ny)}$$

is an exact differential. It is a question of the last term, which may be written

$$\frac{(x\,dy - y\,dx)/x^2}{M/N + y/x} = \frac{dv}{f(v)}$$

where $v = y/x$. This is an exact differential.

We have ruled out essentially the equation $y\,dx - x\,dy = 0$.

EXERCISES

1. Show that if an equation is both homogeneous and exact, its solution is

$$Mx + Ny = C$$

provided the first member is not a constant.

Show exact and write down the solutions:

$$(ax + by)\,dx + (bx + cy)\,dy = 0$$
$$(3x^2 + y^2)\,dx + (2xy + 3y^2)\,dy = 0$$

2. Find an integrating factor, given that

$$\frac{1}{M}\left(\frac{\partial N}{\partial x} - \frac{\partial M}{\partial y}\right)$$

is a function of y alone.

3. Find the condition satisfied by M and N in order that there be an integrating factor which is a function of $x + y$.

4. Solve the like problem for an integrating factor which is a function of $x^2 + y^2$.

5. From the list of homogeneous equations on an earlier page solve

$$(ax - y)\,dx + 2x\,dy = 0$$

by finding an integrating factor.

6. Solve likewise

$$(xy + 2y^2)\,dx + x^2\,dy = 0$$

7. Solve

$$(1 - xy)\,dx + (xy - x^2)\,dy = 0$$

58   DIFFERENTIAL EQUATIONS

8. Show that if $AD - BC \neq 0$, the equation

$$x^p(Ay\, dx + Bx\, dy) + y^q(Cy\, dx + Dx\, dy) = 0$$

has an integrating factor of the form $\mu = x^m y^n$.

9. Solve $x(2y\, dx + 3x\, dy) + y^2(y\, dx + x\, dy) = 0$.
10. Solve $x^2 y^3 (y\, dx + x\, dy) + y\, dx - x\, dy = 0$.
11. Show that if μ is an integrating factor of (2-41) and if $u = C$ is the general solution, then the most general integrating factor can be written $\mu F(u)$.

REVIEW EXERCISES

Find the general solution in each case:

1. $\dfrac{dy}{dx} - \dfrac{y}{x} = 1 + \sqrt{x}$. **2.** $\dfrac{dy}{dx} - \dfrac{y}{x} = \dfrac{dx}{x\, dy}$. **3.** $\dfrac{dy}{dx} - \dfrac{y}{x} = \dfrac{x}{y}$.

4. $x\, dx + y\, dy = (x^2 + y^2)\, dx$.   **5.** $x\, dx - y\, dy = (x^2 + y^2)\, dx$.
6. $y\, dx - x\, dy = (x^2 + y^2)\, dx$.
7. $\sqrt{x^4 + y^4}\,(x\, dy + y\, dx) + x^3\, dx + y^3\, dy = 0$.
8. $y(y + \cos xy)\, dx + x(2y + \cos xy)\, dy = 0$.
9. $y' = (x + y + 1)^{-2}$.
10. $2(x + y^3 - y)\, dx + (6xy^2 - 3y^2 - 2x)\, dy = 0$.
11. $y^2 = x^2 \left(\dfrac{dy}{dx}\right)^2 + 1$.
12. $(10x + 8y + 2)\, dx + (7x + 5y + 2)\, dy = 0$.
13. $[\cos x + \cos(x + y)]\, dx + [\cos y + \cos(x + y)]\, dy = 0$.
14. $2xy' + y = (2ax^2 + 1)/x$.   **15.** $y' = y + 2x(y - e^x)^2$.
16. $y' - 2y = e^x$.
17. $(2x - 2y^3 + y)\, dx + (x - 6xy^2)\, dy = 0$.
18. $x\, dy = (x + y + 1)\, dx$.
19. $(x^2 + 6xy + y^2)\, dx + (2x^2 + xy)\, dy = 0$.
20. $(y^3 - y)\, dx + (3xy^2 - x)\, dy = 0$.   **21.** $\log(1 + y)\, dx + \dfrac{dy}{1 + y} = 0$.
22. $x\, dy = (x - y + 1)\, dx$.   **23.** $y^2 = (1 - x^2)y'^2 + 2xyy'$.
24. $(x^2 y - 1)\, dx/x^2 + (xy^2 - 1)\, dy/y^2 = 0$.
25. $dy = (y \tan x + 2e^x)\, dx$.
26. $(x - y - 5)\, dx - (x + 3y - 2)\, dy = 0$.
27. $(x - 2y + 10)\, dx + (2x - y + 8)\, dy = 0$.
28. $2xy' - y = x^3 - x$.
29. $(2x^2 - 6xy - 5y^2)\, dx - (3x^2 + 10xy - y^2)\, dy = 0$.
30. $(2x - y + 6)\, dx + (4x - 2y - 3)\, dy = 0$.
31. $x + yy' = \sqrt{x^2 + y^2}$.   **32.** $(1 + x^2)\, dy = (1 + y^2)\, dx$.
33. $y' - \dfrac{ny}{x} = kx^3$.   **34.** $x(2x + 3y)y' = 3y(x + 2y)$.
35. $y' + y^2 = 1 + x^2$.

36. Four flies sit at the corners of a card table, facing inward. They start simultaneously walking at the same rate, each directing its motion steadily toward the fly on its right. Find the path of each. Can you find without calculus the distance traveled?

37. It has been observed that a bug will move in a plane containing two lights in such a way that the intensities of the lights on its two sides are the same. Show that the equation of the path may be written

$$k_1 \cos \theta_1 - k_2 \cos \theta_2 = C$$

where k_1 and k_2 are the strengths of the lights, and θ_1 and θ_2 are interior angles at the lights in the triangle whose vertices are the bug and the lights. Show that the bug, if approaching, will enter the first or second light according as $C < k_1 - k_2$ or $C > k_1 - k_2$.

CHAPTER 3

LINEAR EQUATIONS OF THE SECOND ORDER

3-1. Linear Equations. An equation of the form

$$A_0(x)\frac{d^n y}{dx^n} + A_1(x)\frac{d^{n-1}y}{dx^{n-1}} + \cdots + A_{n-1}(x)\frac{dy}{dx} + A_n(x)y + B(x) = 0 \qquad (A_0(x) \not\equiv 0) \qquad (3\text{-}1)$$

is called a *linear differential equation of order n*. The linear differential equation of the second order has the form (setting $n = 2$)

$$A_0(x)\frac{d^2y}{dx^2} + A_1(x)\frac{dy}{dx} + A_2(x)y + B(x) = 0 \qquad (3\text{-}2)$$

On division by $A_0(x)$, we can put this in the form which we shall most frequently use:

$$\frac{d^2y}{dx^2} + P(x)\frac{dy}{dx} + Q(x)y = R(x) \qquad (3\text{-}3)$$

It would be difficult to exaggerate the importance of the linear equation of the second order. It has had wide application to theoretical problems in the sciences. It has been studied by the ablest mathematicians of the past. Its solutions have furnished many important new functions, with the most diverse and interesting properties. Out of its study the most beautiful and profound theories have arisen. The present elementary chapter is, therefore, an introduction to much that is important in mathematics.

The existence of solutions will be taken for granted. Our subsequent existence theorems will establish the following theorem, which will be assumed here.

THEOREM. *Let x_0 lie in an interval ab in which $P(x)$, $Q(x)$, $R(x)$ are continuous and let y_0, y_0' be given, then there is one and only one solution $y = y(x)$ in ab such that $y(x_0) = y_0$, $y'(x_0) = y_0'$.*

If $y = f(x)$ is a solution of the equation, we shall say, for the sake of brevity, that $f(x)$ is a solution.

We shall call (3-3) the *complete equation* and we shall call the equation obtained by replacing $R(x)$ by zero the *reduced equation*:

LINEAR EQUATIONS OF THE SECOND ORDER

$$\frac{d^2y}{dx^2} + P\frac{dy}{dx} + Qy = 0 \tag{3-4}$$

Let $y(x)$ be a solution of (3-3) and let us write it as the sum of two functions,
$$y(x) = u(x) + v(x)$$
one of which may be chosen arbitrarily. On substituting in (3-3) we have

$$\left(\frac{d^2u}{dx^2} + P\frac{du}{dx} + Qu\right) + \left(\frac{d^2v}{dx^2} + P\frac{dv}{dx} + Qv - R\right) = 0 \tag{3-5}$$

If we choose for v some solution of the complete equation, then the second parenthesis of (3-5) vanishes and the first must vanish also. Hence u, or $y - v$, is a solution of the reduced equation. We can then state the following theorem.

THEOREM. *The difference of any two solutions of the complete equation is a solution of the reduced equation.*

Holding the particular solution v fixed, any solution u of the reduced equation will result in the vanishing of both parentheses of (3-5) and y is a solution of (3-3). Hence:

THEOREM. <u>*The general solution of the complete equation is got by adding to any particular solution of the complete equation the general solution of the reduced equation.*</u>

3-2. The Reduced Equation. We now consider the solutions of the reduced equation.

THEOREM. *If u_1 and u_2 are solutions of* $R(x) = 0$

$$\frac{d^2y}{dx^2} + P\frac{dy}{dx} + Qy = 0 \tag{3-4}$$

then
$$c_1 u_1 + c_2 u_2$$
where c_1 and c_2 are constants, is a solution.

This appears immediately on substituting in the equation,

$$\frac{d^2}{dx^2}(c_1 u_1 + c_2 u_2) + P\frac{d}{dx}(c_1 u_1 + c_2 u_2) + Q(c_1 u_1 + c_2 u_2)$$
$$= c_1\left(\frac{d^2 u_1}{dx^2} + P\frac{du_1}{dx} + Qu_1\right) + c_2\left(\frac{d^2 u_2}{dx^2} + P\frac{du_2}{dx} + Qu_2\right) = 0$$

both parentheses vanishing by virtue of the fact that u_1 and u_2 are solutions.

It thus appears that if we can find two particular solutions, u_1 and u_2, then $y = c_1 u_1 + c_2 u_2$, which contains two arbitrary constants, should be the general solution. Of course, u_2 should not be a multiple of u_1, since then there is but a single essential constant. This brings us to the matters treated in the following section.

Example 3-1. Solve
$$y'' - y' = 0$$
By inspection $y = e^x$ and $y = 1$ are solutions, so the general solution is
$$y = Ae^x + B$$
Example 3-2. Solve
$$y'' + a^2y = bx + c$$
Here $y = (bx + c)/a^2$ is an obvious solution. The reduced equation, $y'' + a^2y = 0$, has solutions $y = \sin ax$ and $y = \cos ax$. The general solution then is
$$y = A \sin ax + B \cos ax + \frac{bx + c}{a^2}$$

EXERCISES

Find particular solutions of the complete and reduced equations by inspection and write down the general solution:
1. $y'' - 3y' = 6$.
2. $y'' - 4y = \sin x$.
3. $y'' + 4y = e^x + 1$.
4. $xy'' - y' = x^4$.
5. $(x - 1)y'' - xy' + y = 0$.

Find the differential equations of the following families:
6. $y = Ax + Bx^2$.
7. $y = Ax + B/x$.
8. $y = Ae^x + Be^{2x}$.
9. $y = Ae^x + Be^{-x} + e^{2x}$.

3-3. Linear Dependence. Wronskians. A set of functions $u_1(x)$, $u_2(x)$, ..., $u_n(x)$ are called *linearly dependent* in an interval ab of the x axis if there exist constants c_1, c_2, ..., c_n, not all zero, such that in the interval
$$c_1u_1 + c_2u_2 + \cdots + c_nu_n \equiv 0$$
If no such identity holds, the functions are called *linearly independent*.

We shall consider here the case of two functions. If $f(x)$ and $g(x)$ are linearly dependent in ab, we have
$$c_1f(x) + c_2g(x) \equiv 0$$
where $c_2 \neq 0$, say. This can be written, setting $k = -c_1/c_2$,
$$g(x) \equiv kf(x) \tag{3-6}$$
Hence, if two functions are linearly dependent, one is merely a constant times the other in the interval considered (for example, x^2 and $2x^2$).

On differentiating,
$$g'(x) \equiv kf'(x)$$

LINEAR EQUATIONS OF THE SECOND ORDER 63

and eliminating k, we have

$$f(x)g'(x) - g(x)f'(x) \equiv 0$$

in the interval. The first member of this identity is known as the *Wronskian* of $f(x)$ and $g(x)$ and will be represented by $W(f,g)$. We have proved the first part of the following theorem.

THEOREM. *If two functions, $f(x)$ and $g(x)$, are linearly dependent in an interval, then the Wronskian of the two functions,*

$$W(f,g) = f(x)g'(x) - g(x)f'(x) \tag{3-7}$$

vanishes identically in the interval.

Conversely, if the Wronskian vanishes identically in an interval and if one of the functions does not vanish in the interval, then the functions are linearly dependent in the interval.

To prove the second part of the theorem, suppose $f(x) \neq 0$ in the interval. Dividing the Wronskian by f^2, we have

$$\frac{fg' - gf'}{f^2} \equiv 0$$

whence, integrating,

$$\frac{g}{f} \equiv C \qquad g \equiv Cf$$

which was to be proved.

If $f(x)$ is zero at certain points of the interval, a linear relation holds in each subinterval between roots of $f(x)$. It may happen that different values of C appear for different intervals.

3-4. The Wronskian of Two Solutions. We shall now consider solutions of the reduced equation in an interval ab of the x axis at all points of which the coefficients $P(x)$ and $Q(x)$ are continuous. Let $u_1(x), u_2(x)$ be two solutions of (3-4),

$$\begin{aligned} u_1'' + Pu_1' + Qu_1 &= 0 \\ u_2'' + Pu_2' + Qu_2 &= 0 \end{aligned} \tag{3-8}$$

The Wronskian of the two solutions,

$$W = u_1 u_2' - u_2 u_1' \tag{3-9}$$

has the derivative

$$W' = u_1 u_2'' - u_2 u_1''$$

Multiplying the first of Eqs. (3-8) by $-u_2$ and the second by u_1 and adding, we have

$$\frac{dW}{dx} + PW = 0$$

This has the solution

$$W = Ke^{-\int P dx}$$

a result which is due to Abel. Letting x_0 be any point of ab and W_0 the corresponding value of W, we find

$$W = W_0 e^{-\int_{x_0}^{x} P\,dx} \qquad (3\text{-}10)$$

The factor $e^{-\int_{x_0}^{x} P\,dx}$ is positive in the interval ab, hence W has throughout ab the same sign. W can vanish only if $W_0 = 0$, in which case $W \equiv 0$. The theorem follows.

THEOREM. *Throughout the interval ab the Wronskian of two solutions is everywhere positive, or everywhere negative, or identically zero.*

If $W_0 \neq 0$, then $W \neq 0$ throughout ab and the two solutions are not linearly dependent in any subinterval of ab.

THEOREM. *If the Wronskian of two solutions of the reduced equation vanishes at any point of ab, then the solutions are linearly dependent in ab.*

We have $W \equiv 0$ in ab. Unless both solutions are identically zero, in which case they are linearly dependent, there is an interval $a'b'$ in ab in which one of them does not vanish. Then they are linearly dependent,

$$c_1 u_1 + c_2 u_2 \equiv 0 \qquad (3\text{-}11)$$

in $a'b'$. The two members of (3-11) are solutions of the reduced equation which have the same values and the same derivatives, namely, zero, at a point of $a'b'$. According to the existence theorem they are identical throughout the whole of ab.

COROLLARY 1. *If two solutions of the reduced equation have a common root in ab they are linearly dependent in ab.*

For, at the root $u_1 = u_2 = 0$, and hence $W = 0$.

COROLLARY 2. *If two solutions of the reduced equation have maxima or minima at the same point x_0 of ab, then the solutions are linearly dependent in ab.*

For, at x_0, $u_1' = u_2' = 0$, and $W = 0$.

We prove finally the fundamental theorem concerning the general solution.

THEOREM. *Any solution $y(x)$ of the reduced equation in the interval ab is expressible as a linear combination,*

$$y = c_1 u_1 + c_2 u_2 \qquad (3\text{-}12)$$

of any two solutions $u_1(x)$, $u_2(x)$, which are linearly independent in ab.

At a point x_0 of ab let us try to determine c_1 and c_2 so that the two members of (3-12) shall have the same values and the same derivatives.

$$c_1 u_1(x_0) + c_2 u_2(x_0) = y(x_0)$$
$$c_1 u_1'(x_0) + c_2 u_2'(x_0) = y'(x_0)$$

LINEAR EQUATIONS OF THE SECOND ORDER

These equations can be solved for c_1 and c_2, since

$$u_1(x_0)u_2'(x_0) - u_2(x_0)u_1'(x_0) = W_0 \neq 0$$

With these values of c_1 and c_2 the two members of (3-12) are solutions of the reduced equation which have the same values and the same derivatives at x_0. Hence they are identical throughout ab.

It follows from this theorem that we can write down the general solution of the reduced equation as soon as we can find in some way two linearly independent solutions.

Example. Show that the functions $2x^2$, $(x+1)^2$, $x+5$, and $3x$ are linearly dependent.

Using the multipliers 5, -10, 2, and 6, respectively, we have the following relation of dependence,

$$10x^2 - 10(x+1)^2 + 2(x+5) + 18x = 0$$

EXERCISES

Show that the following sets of functions are linearly dependent:
1. e^x, e^{-x}, $\cosh x$.　　　　2. $\sin^2 x$, $\cos^2 x$, 100.

3. Show that four polynomials in x of degree 2 or less are necessarily linearly dependent.

Find the Wronskian of each pair of functions:
4. x^m, x^n.　　　　5. e^{mx}, e^{nx}.
6. 0, $\arctan x$.　　　　7. $e^{mx} \sin ax$, $e^{mx} \cos ax$.

8. Show that the Wronskian of the two functions x^3 and $|x|^3$ vanishes identically, but that the functions are not linearly dependent in any interval enclosing the origin.

9. Show that if the change of variable $y = e^{-(1/2)\int P dx} Y$ be made in Eq. (3-4), there results a linear equation in which the second term is missing,

$$Y'' + (Q - \tfrac{1}{2}P' - \tfrac{1}{4}P^2)Y = 0$$

10. Make the preceding change of variable in

$$y'' + 2xy' + (1 + x^2)y = 0$$

and solve.

3-5. The Reduced Equation with Constant Coefficients. We proceed to solve the reduced equation when $P(x) = p$, $Q(x) = q$, p and q being constants:

$$\frac{d^2y}{dx^2} + p\frac{dy}{dx} + qy = 0 \tag{3-13}$$

The coefficients p and q are continuous everywhere, so the interval ab of the previous section may be taken to be the whole x axis.

The solution is arrived at by an ingenious device. We try a solution of the form

$$y = e^{mx} \tag{3-14}$$

On substituting in (3-13), we have

$$e^{mx}(m^2 + pm + q) = 0$$

This equation will hold, and (3-14) will be a solution if m be chosen so that

$$m^2 + pm + q = 0 \qquad (3\text{-}15)$$

Equation (3-15) is called the *auxiliary equation*.

Real Unequal Roots. If (3-15) has real and distinct roots, m_1 and m_2, we have two solutions

$$y = e^{m_1 x} \qquad y = e^{m_2 x}$$

These are readily found to be linearly independent, and the general solution is

$$\boxed{y = Ae^{m_1 x} + Be^{m_2 x}} \qquad (3\text{-}16)$$

where A and B are arbitrary constants.

Imaginary Roots. If the roots of (3-15) are not real, we still have solutions if we know how to interpret imaginary exponents. We digress a moment to recall the formula

$$e^{i\theta} = \cos\theta + i\sin\theta \qquad (3\text{-}17)$$

where $i = \sqrt{-1}$. The functions appearing here are defined for all real and imaginary values of θ by their ordinary series expansions,

$$\sin\theta = \theta - \frac{\theta^3}{3!} + \frac{\theta^5}{5!} - \cdots$$

$$\cos\theta = 1 - \frac{\theta^2}{2!} + \frac{\theta^4}{4!} - \cdots$$

$$e^{i\theta} = 1 + i\theta + \frac{(i\theta)^2}{2!} + \frac{(i\theta)^3}{3!} + \frac{(i\theta)^4}{4!} + \cdots$$

$$= 1 + i\theta - \frac{\theta^2}{2!} - \frac{i\theta^3}{3!} + \frac{\theta^4}{4!} + \frac{i\theta^5}{5!} - \cdots$$

If the first of these equations be multiplied by i and added to the second, the last series results, which establishes (3-17). That the usual properties of exponents hold could be established.

If (3-15) has a complex root $m = \alpha + i\beta$, where $\beta \neq 0$, we have a complex solution

$$y = e^{(\alpha+i\beta)x} = e^{\alpha x}(\cos\beta x + i\sin\beta x) \qquad (3\text{-}18)$$

If p and q are real then the conjugate imaginary $m_2 = \alpha - i\beta$ is also a root, and we have the solution

$$y = e^{(\alpha-i\beta)x} = e^{\alpha x}(\cos\beta x - i\sin\beta x)$$

LINEAR EQUATIONS OF THE SECOND ORDER 67

Let us get simpler solutions from these two, first by adding and dividing by 2, second by subtracting and dividing by $2i$. There result

$$y = e^{\alpha x} \cos \beta x \qquad y = e^{\alpha x} \sin \beta x \qquad (3\text{-}19)$$

which are linearly independent. The general solution then is

$$y = e^{\alpha x}(A \sin \beta x + B \cos \beta x) \qquad (3\text{-}20)$$

We could look at this in another way. A complex function $u + iv$ satisfies a *real* reduced equation, as we find on substituting, if and only if u and v are both solutions. A complex solution thus yields two real solutions. The solution (3-18) gives the two solutions (3-19) immediately.

Double Roots. A useful change of variable, which keeps the coefficients of the equation constant, is the following,

$$y = e^{mx} Y \qquad (3\text{-}21)$$

where m is any constant. We have

$$y' = e^{mx}(Y' + mY)$$
$$y'' = e^{mx}(Y'' + 2mY' + m^2 Y)$$

Equation (3-13) then becomes, on canceling e^{mx} and collecting terms,

$$Y'' + (2m + p)Y' + (m^2 + pm + q)Y = 0 \qquad (3\text{-}22)$$

If now α is a double root of the auxiliary equation, both the quadratic in (3-15) and its derivative vanish when we set $m = \alpha$,

$$\alpha^2 + p\alpha + q = 0 \qquad 2\alpha + p = 0$$

Equation (3-22) becomes, with this value of m,

$$Y'' = 0$$

which has the solution

$$Y = A + Bx$$

The general solution of (3-13) then is

$$y = e^{\alpha x}(A + Bx) \qquad (3\text{-}23)$$

We state the preceding results in the following theorem:

THEOREM. *Given the differential equation*

$$\frac{d^2 y}{dx^2} + p \frac{dy}{dx} + qy = 0 \qquad (3\text{-}13)$$

in which p and q are constants. Consider the auxiliary equation

$$m^2 + pm + q = 0 \qquad (3\text{-}15)$$

a. *If the auxiliary equation has real and distinct roots m_1, m_2, the general solution of the differential equation is*

$$y = Ae^{m_1 x} + Be^{m_2 x} \qquad (3\text{-}16)$$

b. If the auxiliary equation has imaginary roots, $\alpha \pm \beta \sqrt{-1}$, the general solution of the differential equation is

$$y = e^{\alpha x}(A \sin \beta x + B \cos \beta x) \qquad (3\text{-}20)$$

c. If the auxiliary equation has a double root, $m = \alpha$, the general solution of the differential equation is

$$y = (A + Bx)e^{\alpha x} \qquad (3\text{-}23)$$

Example 3-3. Solve
$$9y'' - 12y' + 4y = 0$$
The auxiliary equation
$$9m^2 - 12m + 4 = 0$$
has the double root $m = \frac{2}{3}$. The solution then is
$$y = e^{2x/3}(A + Bx)$$

Example 3-4. Solve
$$y'' - 4y' + 29y = k$$
The complete equation has the particular solution $y = k/29$. The auxiliary equation
$$m^2 - 4m + 29 = 0$$
has the roots $2 \pm 5i$. The solution is
$$y = e^{2x}(A \sin 5x + B \cos 5x) + k/29$$

Example 3-5. Solve
$$y'' + 5y' + 4y = 0$$
Find the particular solution for which $y = 3$, $y' = 0$ when $x = 0$.
The auxiliary equation
$$m^2 + 5m + 4 = 0$$
has roots -1 and -4. The general solution is
$$y = Ae^{-x} + Be^{-4x}$$
Putting $x = 0$, $y = 3$, $y' = 0$ in the general solution and in its derivative
$$y' = -Ae^{-x} - 4Be^{-4x}$$
we have
$$A + B = 3 \qquad -A - 4B = 0$$
whence $A = 4$, $B = -1$. The desired particular solution is
$$y = 4e^{-x} - e^{-4x}$$

EXERCISES

Solve:
1. $y'' + 8y' + 15y = 0$.
2. $y'' + 4y' + 5y = 0$.
3. $y'' - 6y' + 9y = 0$.
4. $y'' = 0$.
5. $y'' - 3y' + 2y = 0$.
6. $2y'' + y' - 3y = 0$.
7. $2y'' + 3y' = 0$.
8. $2y'' + 3y = 0$.
9. $4y'' + y' + y = 0$.
10. $2y'' + 3y' + 4y + 5 = 0$.

11. Solve
$$y'' - 6y' + 5y = 0$$
and find the particular solution such that $y = 3$, $y' = 7$ when $x = 0$.

12. Solve
$$y'' + 4y' + 4y + 4 = 0$$
and find the particular solution such that $y = 0$, $y' = 0$ when $x = 0$.

13. Solve the pendulum equation
$$l \frac{d^2\theta}{dt^2} = -g\theta$$
of Sec. 1-17 by present methods, evaluating the constants.

14. Treat the case of complex roots $\alpha \pm i\beta$ by solving Eq. (3-22) with $m = \alpha$, noting that $\alpha = -p/2$, $\beta^2 = q - p^2/4$.

3-6. Roots of Solutions of the Reduced Equation. We return to the general reduced equation,

$$\frac{d^2y}{dx^2} + P \frac{dy}{dx} + Qy = 0 \tag{3-4}$$

We shall consider solutions in an interval $a < x < b$, in which P and Q are continuous. We consider the roots of a solution in ab, that is, the values of x for which the solution is zero.

THEOREM. *At a root of a solution, not identically zero, of (3-4) in the interval ab, the derivative of the solution is different from zero.*

This is an immediate consequence of the existence theorem. At a root of a solution $u(x)$ suppose we have not only $u(x_0) = 0$ but also $u'(x_0) = 0$. These conditions determine the solution uniquely in ab. It must therefore coincide with the known solution $y \equiv 0$ in ab. Hence, $u(x) \equiv 0$, and the theorem is established.

It follows that the graph of a solution which is not identically zero is not tangent to the x axis at any point of ab.

Linearly dependent solutions, neither of which is identically zero, have the same roots. Concerning linearly independent solutions we have the following interesting theorem.

THEOREM. *Let $u_1(x)$, $u_2(x)$ be solutions of (3-4), linearly independent in ab. Between consecutive roots of $u_1(x)$ in ab there is one and only one root of $u_2(x)$.*

Suppose, on the contrary, that $u_2(x)$ has no root between consecutive roots x_1 and x_2 of $u_1(x)$. The quotient $u_1(x)/u_2(x)$ is continuous in the interval $x_1 \leqslant x \leqslant x_2$ and vanishes at the end points. Then the derivative

$$\frac{d}{dx} \frac{u_1}{u_2} = \frac{u_2 u_1' - u_1 u_2'}{u_2^2} = -\frac{W(u_1, u_2)}{u_2^2}$$

must vanish at an intermediate point. This is impossible since $W \neq 0$, and the existence of a root of $u_2(x)$ is established.

70   DIFFERENTIAL EQUATIONS

Likewise there is a root of $u_1(x)$ between consecutive roots of $u_2(x)$. If there were two roots of $u_2(x)$ in the interval, $u_1(x)$ would have a root between them and x_1 and x_2 would not be consecutive roots of $u_1(x)$.

Let us apply this result to the equation with constant coefficients. Here ab may be taken to be the whole x axis.

If the roots of the auxiliary equation are complex, we have the solution

$$y = e^{\alpha x} \sin \beta x$$

This has the equally spaced roots

$$x = 0, \pm \frac{\pi}{\beta}, \pm \frac{2\pi}{\beta}, \pm \frac{3\pi}{\beta}, \cdots$$

Every solution, other than a constant times this one, has a single root in each of these intervals.

If the roots of the auxiliary equation are real, we have a solution of the form

$$y = e^{mx}$$

This solution has no roots. It follows that no solution, excepting the identically zero solution, has more than one root.

EXERCISES

1. Prove that the quotient of two linearly independent solutions is an increasing function (or decreasing function) of x at all points of ab at which the quotient exists.

2. Prove that two linearly independent solutions cannot have common points of inflection for x_0 of ab unless P and Q vanish simultaneously there.

3. If at x_0 of ab two solutions satisfy

$$u_1(x_0) = u_2(x_0) > 0 \qquad u_1'(x_0) > u_2'(x_0)$$

show that u_2 will have a root to the right of x_0 before u_1 does.

4. Discuss the behavior of the roots of a solution u determined by

$$u(x_0) = A > 0 \qquad u'(x_0) = B > 0$$

(a) as A is held fixed and B is steadily decreased;
(b) as B is held fixed and A is steadily decreased.

3-7. The Complete Equation. Method of Undetermined Coefficients. A great variety of methods has been given for finding particular solutions. We continue with the equation with constant coefficients,

$$\frac{d^2y}{dx^2} + p\frac{dy}{dx} + qy = R(x) \qquad (3\text{-}24)$$

and we shall describe a method which is effective for certain forms of the function $R(x)$. The *method of undetermined coefficients* is applicable to certain combinations of polynomials, exponentials, sines, and cosines. When it can be used, it is probably the handiest method to employ.

It is clear that if $R(x)$ is a polynomial of degree n,

$$R(x) = R_n(x) = r_0 x^n + r_1 x^{n-1} + \cdots$$

and if $q \neq 0$, we can satisfy the equation with a polynomial of the form

$$y = P_n(x) = ax^n + bx^{n-1} + \cdots$$

For, on substituting into (3-24), the coefficient of x^n gives $qa = r_0$, the coefficient of x^{n-1} determines qb, etc., and the constants are found in order.

If $q = 0$, the presence in the first member of a term in x^n will require a trial polynomial of degree $n + 1$ if $p \neq 0$ or of degree $n + 2$ if $p = 0$.

Suppose next that the second member has the form $R(x) = e^{mx} R_n(x)$. To see what the situation is, let us make the change of variable $y = e^{mx} Y$. We have—referring to (3-22)—

$$Y'' + (2m + p)Y' + (m^2 + pm + q)Y = R_n(x)$$

and the problem has been reduced to the preceding case, in which we use a trial polynomial. We have the following result: If m is not a root of the auxiliary equation, we use a trial solution of the form $y = e^{mx} P_n(x)$; if m is a simple or double root of the auxiliary equation, we increase the degree of the trial polynomial by 1 or 2, respectively.

Functions of the forms $e^{\alpha x} \sin \beta x \, R_n(x)$ and $e^{\alpha x} \cos \beta x \, S_n(x)$ fall under the preceding case, since sines and cosines are linear combinations of exponentials. We use the trial solution

$$y = e^{\alpha x}[\sin \beta x \, P_n(x) + \cos \beta x \, Q_n(x)]$$

increasing the degrees of the trial polynomials by 1 if $\alpha \pm i\beta$ satisfies the auxiliary equation.

Example. Solve

(i) $$y'' + 3y' + 2y = x^3 + x$$

The second member is a polynomial of the third degree. We take

(ii) $$y = ax^3 + bx^2 + cx + d$$

and substitute into (i),

$$(6ax + 2b) + 3(3ax^2 + 2bx + c) + 2(ax^3 + bx^2 + cx + d) = x^3 + x$$

Equating coefficients of like power of x,

$$2a = 1$$
$$9a + 2b = 0$$
$$6a + 6b + 2c = 1$$
$$2b + 3c + 2d = 0$$

whence
$$a = \tfrac{1}{2} \quad b = -\tfrac{9}{4} \quad c = \tfrac{23}{4} \quad d = -\tfrac{51}{8}$$

Putting these values into (ii), we have the particular solution

(iii) $\qquad y = \tfrac{1}{8}(4x^3 - 18x^2 + 46x - 51)$

3-8. The Method of the Variation of Parameters. We now give a method of complete generality for the determination of a particular solution of the complete equation

$$\frac{d^2y}{dx^2} + P(x)\frac{dy}{dx} + Q(x)y = R(x) \qquad (3\text{-}3)$$

It is applicable whenever we can solve the reduced equation.

Let $u_1(x)$, $u_2(x)$ be linearly independent solutions of the reduced equation,

$$\begin{aligned} u_1'' + Pu_1' + Qu_1 &= 0 \\ u_2'' + Pu_2' + Qu_2 &= 0 \end{aligned} \qquad (3\text{-}25)$$

We shall try a solution of the form

$$y = t_1 u_1 + t_2 u_2 \qquad (3\text{-}26)$$

where t_1 and t_2, called *parameters*, are functions of x to be determined. We have, differentiating,

$$y' = t_1 u_1' + t_2 u_2' + u_1 t_1' + u_2 t_2'$$

We shall set

$$u_1 t_1' + u_2 t_2' \equiv 0$$

so that we have

$$y' = t_1 u_1' + t_2 u_2'$$

Differentiating again,

$$y'' = t_1 u_1'' + t_2 u_2'' + u_1' t_1' + u_2' t_2'$$

Substituting these values of y' and y'' into (3-3), we have

$$t_1 u_1'' + t_2 u_2'' + u_1' t_1' + u_2' t_2' + P(t_1 u_1' + t_2 u_2') + Q(t_1 u_1 + t_2 u_2) = R$$

Making use of the relations (3-25), this reduces to

$$u_1' t_1' + u_2' t_2' = R$$

We can now state the following procedure.

Solve the equations

$$\begin{aligned} u_1 t_1' + u_2 t_2' &= 0 \\ u_1' t_1' + u_2' t_2' &= R \end{aligned} \qquad (3\text{-}27)$$

for t_1' and t_2', and get t_1 and t_2 by integrating. Then (3-26) *is a particular solution.*

LINEAR EQUATIONS OF THE SECOND ORDER

Example. Solve
$$y'' + y = \csc x$$
Since $\sin x$ and $\cos x$ are solutions of the reduced equation, we take
$$y = t_1 \cos x + t_2 \sin x$$
and we are led to the equations
$$t_1' \cos x + t_2' \sin x = 0$$
$$-t_1' \sin x + t_2' \cos x = \csc x$$
Solving these, and integrating, we have
$$t_1' = -1 \qquad t_2' = \cot x$$
$$t_1 = -x \qquad t_2 = \log \sin x$$
The resulting particular solution is
$$y = -x \cos x + \sin x \log \sin x$$

The result of the preceding process may be put in a compact formula if we wish. Solving (3-27) and integrating, we have
$$t_1 = -\int \frac{R u_2}{W} dx \qquad t_2 = \int \frac{R u_1}{W} dx$$
where W is the Wronskian $u_1 u_2' - u_2 u_1'$. We insert these values into (3-26). Changing the variable of integration to t, particularizing the formula by writing a lower limit of integration, and writing everything under the sign of integration, we have
$$y = \int_a^x \frac{u_1(t) u_2(x) - u_2(t) u_1(x)}{u_1(t) u_2'(t) - u_2(t) u_1'(t)} R(t) \, dt \tag{3-28}$$
In the integration, x is held constant. It is not essential, of course, that a lower limit be inserted.

3-9. The Use of a Known Solution of the Reduced Equation. The complete equation can be solved, as we shall now show, if we know a single nonzero solution of the reduced equation.

In the complete equation,
$$\frac{d^2 y}{dx^2} + P \frac{dy}{dx} + Qy = R \tag{3-3}$$
let us put
$$y = tu \tag{3-29}$$
where t and u are functions of x, one of which may be chosen arbitrarily. We have
$$y' = tu' + ut'$$
$$y'' = tu'' + 2t'y' + ut''$$

DIFFERENTIAL EQUATIONS

Substituting into (3-3), we have

$$ut'' + (2u' + Pu)t' + (u'' + Pu' + Qu)t = R \qquad (3\text{-}30)$$

For any choice of u this is a linear equation of the second order for the determination of t. If u is a solution of the reduced equation, the last parenthesis in (3-30) vanishes and we have

$$u\frac{d^2t}{dx^2} + \left(2\frac{du}{dx} + Pu\right)\frac{dt}{dx} = R \qquad (3\text{-}31)$$

If we put

$$\frac{dt}{dx} = v \qquad (3\text{-}32)$$

(3-31) becomes

$$u\frac{dv}{dx} + \left(2\frac{du}{dx} + Pu\right)v = R \qquad (3\text{-}33)$$

This is a linear equation of the first order in v. It can be solved by the methods of the preceding chapter. Having got v, we have t by integrating,

$$t = \int v\,dx$$

and thence we have, from (3-29), the value of y.

Example. Solve

$$xy'' - 2(x+1)y' + (x+2)y = x^3 e^{2x}$$

observing that e^x is a solution of the reduced equation.
We put

$$y = te^x$$

and get, on substituting into the equation,

$$e^x(xt'' - 2t') = x^3 e^{2x}$$
$$v' - \frac{2v}{x} = x^2 e^x$$

This has the integrating factor $1/x^2$,

$$\frac{v'}{x^2} - \frac{2v}{x^3} = e^x$$

whence, integrating,

$$\frac{v}{x^2} = e^x + C$$
$$t' = v = x^2 e^x + Cx^2$$

Integrating again,

$$t = (x^2 - 2x + 2)e^x + Ax^3 + B$$

and finally

$$y = e^x(Ax^3 + B) + (x^2 - 2x + 2)e^{2x}$$

LINEAR EQUATIONS OF THE SECOND ORDER

EXERCISES

Solve:
1. $y'' + y = x^3 + x$.
2. $y'' + y' = x^2 + 2x$.
3. $y'' + 2y = e^{2x} + x$.
4. $y'' + 5y' + 4y = x^2 + \sin 2x$.
5. $y'' + y = \sin x$.
6. $y'' - 4y' + 4y = x + e^x$.
7. $y'' + 4y = x^2 + \cos x$.
8. $y'' - 2y' + y = 2xe^x$.
9. $y'' + y = e^{2x} \cos 3x$.

Solve by the method of the variation of parameters:
10. $y'' - y = 2e^x$.
11. $y'' + 4y = \sec 2x$.
12. $y'' + y + \tan x = 0$.
13. $y'' + y = \cot^2 x$.
14. $y'' + 2y' + y = e^{-x}/x^2$.

15. Find the linear equation of the second order which has the solutions $y = 1$, $y = x$, $y = x^2$. What is the general solution?

16. Solve completely the problem of the falling raindrop (Sec. 1-17) by the methods of the present chapter.

17. Show that
$$y'' + n^2 y = R(x)$$
has the particular solution
$$y = \frac{1}{n} \int_a^x \sin n(x - t) R(t) \, dt$$

18. If m is a double root of the auxiliary equation of (3-24), show that we have the particular solution
$$y = e^{mx} \int_a^x (x - t) e^{-mt} R(t) \, dt$$

19. Solve Exercise 8 by the preceding formula.
20. Solve Exercise 9 by solving
$$y'' + y = e^{(2+3i)x}$$
and taking the real part.

21. Noting that e^x is a solution, solve
$$xy'' - (2x + 1)y' + (x + 1)y = 0$$

22. Noting that x is a particular solution, solve
$$(1 - x^2)y'' - xy' + y = 0$$

23. Solve the preceding by making the change of variable $x = \cos \theta$.

24. Show that
$$y'' + (x + a)y' + (bx + c)y = 0$$
has a solution of the form $y = e^{mx}$ if $b^2 - ab + c = 0$, and solve.

25. Given that there is a solution of the form e^{mx}, solve
$$(2 - \cot x)y'' - 5y' + 2(1 + 2 \cot x)y = 0$$

26. Show that $1/x$ is a particular solution, and solve
$$(x^2 - x)y'' + (x - 2)y' - y = 0$$

76   DIFFERENTIAL EQUATIONS

27. Show that $1/(x-1)$ satisfies the equation, and solve

$$x(1-x)y'' + [b - (b+2)x]y' - by = 0$$

28. Observing certain simple solutions of the reduced equation, solve

$$(1-x)y'' + xy' - y = (1-x)^2$$

29. Given that the reduced equation has a solution of the form x^n, solve

$$x^2 y'' - (x+4)xy' + 2(x+3)y = 2x^4 e^{2x}$$

30. Change the independent variable to t, where $t = e^x$, and solve

$$y'' + (2e^x - 1)y' + e^{2x} y = e^{4x}$$

31. Solve by putting $y = e^{x^2} Y$,

$$y'' - 4xy' + (4x^2 - 18)y = xe^{x^2}$$

3-10. The Euler Equation.
The linear equation

$$x^2 \frac{d^2 y}{dx^2} + px \frac{dy}{dx} + qy = R(x) \tag{3-34}$$

where p and q are constants, may be reduced to an equation with constant coefficients by the change of variable

$$x = e^t$$

since

$$\frac{dy}{dx} = \frac{dy}{dt}\frac{dt}{dx} = \frac{1}{x}\frac{dy}{dt}$$

$$\frac{d^2 y}{dx^2} = \frac{1}{x^2}\left(\frac{d^2 y}{dt^2} - \frac{dy}{dt}\right)$$

It is unnecessary, however, actually to make the change of variable. The reduced equation has solutions of the form $e^{mt} = x^m$; we can substitute $y = x^m$ into the reduced equation, cancel the powers of x, which are the same for all terms, and solve the resulting quadratic in m.

In the case of a double root we have a solution

$$y = te^{mt} = x^m \log x$$

In the case of imaginary roots solutions are furnished by the real and imaginary parts of

$$y = e^{(\alpha + i\beta)t} = x^\alpha [\cos(\beta \log x) + i \sin(\beta \log x)]$$

Example. Solve

$$x^2 y'' - 2y = 0$$

Putting $y = x^m$,

$$x^m [m(m-1) - 2] = 0$$

LINEAR EQUATIONS OF THE SECOND ORDER

The roots of the expression in brackets are $m = 2$, $m = -1$; and the general solution is

$$y = Ax^2 + \frac{B}{x}$$

EXERCISES

Solve
1. $6x^2y'' - 5xy' + 4y = 0.$
2. $x^2y'' + 3xy' + y = 0.$
3. $2x^2y'' + 3xy' - y = 0.$
4. $xy'' - 2y' = x^4.$
5. $y'' - 2y/x^2 = x + 1/x.$
6. $x^2y'' - xy' + y = x^2.$
7. $(x+1)^2y'' - 4(x+1)y' + 6y = 0.$
8. $x^2y'' - 4xy' + 6y = 2x + 5.$

9. Show that the most general equation of the second order whose reduced equation has independent solutions of the form x^m and x^n has the form (3-34).

3-11. Exact Equations. The notion of exactness and the concept of the integrating factor can be extended to the linear equation. Consider the equation

$$x^2 \frac{d^2y}{dx^2} + 3x \frac{dy}{dx} + y = 3x^2$$

It happens that the first member is the derivative of a linear expression, so that, as can be readily verified, the equation may be written

$$\frac{d}{dx}\left(x^2 \frac{dy}{dx} + xy\right) = 3x^2$$

This can be integrated at once,

$$x^2 \frac{dy}{dx} + xy = x^3 + C$$

The solution of this linear equation of the first order gives the general solution.

The linear equation

$$P_0(x) \frac{d^2y}{dx^2} + P_1(x) \frac{dy}{dx} + P_2(x)y = R(x) \qquad (3\text{-}35)$$

is said to be *exact* if the first member is the derivative of a linear expression.

THEOREM. *The necessary and sufficient condition that (3-35) be exact is that*

$$\frac{d^2P_0}{dx^2} - \frac{dP_1}{dx} + P_2 \equiv 0 \qquad (3\text{-}36)$$

Suppose that (3-35) is exact; then there exist functions $A(x)$, $B(x)$ such that

$$P_0 \frac{d^2y}{dx^2} + P_1 \frac{dy}{dx} + P_2 y \equiv \frac{d}{dx}\left(A \frac{dy}{dx} + By\right)$$

$$\equiv A \frac{d^2y}{dx^2} + \left(\frac{dA}{dx} + B\right)\frac{dy}{dx} + \frac{dB}{dx} y$$

Equating coefficients,

$$P_0 \equiv A \qquad P_1 \equiv \frac{dA}{dx} + B \qquad P_2 \equiv \frac{dB}{dx}$$

Putting these expressions into (3-36), we find that (3-36) is identically satisfied.

Conversely, suppose that (3-36) holds. We have

$$\frac{d}{dx}\left[P_0 \frac{dy}{dx} + \left(P_1 - \frac{dP_0}{dx}\right)y\right] = P_0 \frac{d^2y}{dx^2} + P_1 \frac{dy}{dx} + \left(\frac{dP_1}{dx} - \frac{d^2P_0}{dx^2}\right)y$$

$$= P_0 \frac{d^2y}{dx^2} + P_1 \frac{dy}{dx} + P_2 y \qquad (3\text{-}37)$$

the coefficient of y being altered by virtue of (3-36). The first member of (3-35) is thus the derivative of a linear expression, and the equation is consequently exact.

We have in the first member of (3-37) a formula for the immediate integration of the equation. We have

$$P_0 \frac{dy}{dx} + \left(P_1 - \frac{dP_0}{dx}\right)y = \int R \, dx + C \qquad (3\text{-}38)$$

3-12. Integrating Factors. We call a function $\mu(x)$ an *integrating factor* of (3-35) if after multiplication by $\mu(x)$ the equation is exact.

THEOREM. *The necessary and sufficient condition that $\mu(x)$ be an integrating factor of* (3-35) *is that μ satisfy the differential equation of the second order*

$$\frac{d^2(P_0 \mu)}{dx^2} - \frac{d(P_1 \mu)}{dx} + P_2 \mu = 0 \qquad (3\text{-}39)$$

For μ is an integrating factor if and only if

$$\mu P_0 \frac{d^2y}{dx^2} + \mu P_1 \frac{dy}{dx} + \mu P_2 y = \mu R$$

is exact. Applying the condition (3-36), we have (3-39) at once.

Since (3-39) has an infinite number of solutions, there is an infinite number of integrating factors.

Equation (3-39) for the integrating factor, which expanded is

$$P_0 \frac{d^2\mu}{dx^2} + \left(2 \frac{dP_0}{dx} - P_1\right)\frac{d\mu}{dx} + \left(\frac{d^2P_0}{dx^2} - \frac{dP_1}{dx} + P_2\right)\mu = 0 \qquad (3\text{-}40)$$

is called the *adjoint equation* of the reduced equation

$$P_0 \frac{d^2y}{dx^2} + P_1 \frac{dy}{dx} + P_2 y = 0 \tag{3-41}$$

If we compute the adjoint equation of (3-40), we get Eq. (3-41). The adjoint relation is a reciprocal one.

The use of the adjoint equation is of doubtful assistance in trying to solve the equation, for the adjoint is likely to be as difficult to solve as the original equation. The adjoint equation, however, is of theoretical interest, particularly in connection with the study of the properties of the solutions of the equation.

EXERCISES

Show exact and solve:
1. $(x^2 + 2)y'' + 4xy' + 2y = \sin x$.
2. $(x + 3)y'' + (2x + 8)y' + 2y = 2$.
3. Find an integrating factor and solve

$$x^2 y'' + (3x^2 + 4x)y' + 2(x^2 + 3x + 1)y = 0$$

4. Prove that the equation

$$\frac{d}{dx}\left[K(x) \frac{dy}{dx} \right] + G(x)y = 0$$

is self-adjoint.

5. Show that any equation may be multiplied by such a function that it becomes self-adjoint.

6. Representing the first members of (3-40) and (3-41) by $M(\mu)$ and $L(y)$, derive Green's theorem

$$\int_{x_1}^{x_2} [vL(u) - uM(v)]\, dx = [P_0(vu' - uv') + (P_1 - P_0')uv]_{x_1}^{x_2}$$

7. Making the equation self-adjoint, use Green's theorem to show that the Wronskian of two solutions u and v does not change sign in ab.

3-13. Vibrations. A vast number of elastic vibrations, electrical oscillations, musical performances in pipes, and so on, are connected with equations of the form

$$\frac{d^2s}{dt^2} + 2k \frac{ds}{dt} + n^2 s = f(t) \tag{3-42}$$

and the more important cases of this equation will now be studied. Here t is the time, and $k \geqslant 0$.

As an example, let a weight hang from a spring of natural length l (Fig. 17). We have a downward force of mg, together with an upward pull proportional to the extension of the spring:

$$mg - \lambda(y - l) = -\lambda s$$

where we put $s = y - l - mg/\lambda$. Here s is the distance below the point at which the weight would hang in equilibrium.

Assume further the existence of a force of friction proportional to the velocity, $-av$. Finally, suppose that a force $E(t)$, depending only on the time, is applied. The equation of motion now is, from Newton's second law,

$$ms'' = -\lambda s - av + E(t)$$

Putting $k = a/2m$, $n^2 = \lambda/m$, $f(t) = E(t)/m$, we have Eq. (3-42).

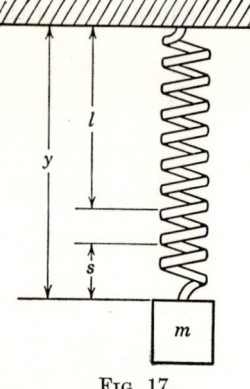

Fig. 17.

Students of electrical engineering are familiar with the equations of the simple circuit,

$$\frac{d^2q}{dt^2} + \frac{R}{L}\frac{dq}{dt} + \frac{q}{LC} = \frac{E(t)}{L}$$

$$\frac{d^2i}{dt^2} + \frac{R}{L}\frac{di}{dt} + \frac{i}{LC} = \frac{E'(t)}{L}$$

where R, L, C are the constant resistance, inductance, and capacitance, respectively; $E(t)$ is the impressed electromotive force, supposed known; and q and i are respectively the quantity of electricity and the current. These equations are of type (3-42).

The pendulum equation at the end of Chap. 1 has this form, with the use of an approximation which is allowable when the amplitudes are small. Friction and other forces can be introduced.

Simple Harmonic Motion. If there is no friction and no forcing function,

$$\frac{d^2s}{dt^2} + n^2 s = 0 \qquad (3\text{-}43)$$

we have the case treated earlier. The general solution is

$$s = A \sin (nt + B) \qquad (3\text{-}44)$$

an oscillatory motion of amplitude A and period $2\pi/n$. The constants A and B are determined by the initial conditions. The period of vibration is independent of these.

Damping. If a friction term is added,

$$\frac{d^2s}{dt^2} + 2k\frac{ds}{dt} + n^2 s = 0 \qquad (3\text{-}45)$$

the motion dies out. The roots of the auxiliary equation,

$$m^2 + 2km + n^2 = 0$$

are $-k \pm \sqrt{k^2 - n^2}$.

If the friction is small, so that $k < n$, the roots are imaginary, $-k \pm i\beta$, where $\beta = \sqrt{n^2 - k^2}$, and the solution is

$$s = Ae^{-kt} \sin (\beta t + B) \tag{3-46}$$

The factor e^{-kt} becomes small as t increases, while the factor $\sin (\beta t + B)$ is periodic. Thus the motion is oscillatory with a steadily decreasing swing. Since $\beta < n$, the period $2\pi/\beta$ is larger than if there is no friction. Here again the period is independent of the initial conditions.

If $k \geqslant n$, the auxiliary equation has negative real roots. If $k > n$, the solution is

$$s = Ae^{-(k+\sqrt{k^2-n^2})t} + Be^{-(k-\sqrt{k^2-n^2})t} \tag{3-47}$$

if $k = n$, it is

$$s = e^{-kt}(A + Bt) \tag{3-48}$$

In each case—like a pendulum swinging in molasses—the motion is not oscillatory.

Forced Vibrations. So far we have considered reduced equations. If there is a force function $f(t)$, we add to the previous general solution a particular solution of the complete equation. If there is damping, this previous solution is a transient, which dies out.

We shall treat the case of a periodic force which is a pure sine function. This is often the important case; for example, in the electrical problem in which an alternating electromotive force is applied.

$$\frac{d^2s}{dt^2} + 2k \frac{ds}{dt} + n^2 s = E \sin pt \tag{3-49}$$

Using a linear combination of $\sin pt$ and $\cos pt$, or, which is the same,

$$s = a \sin (pt - \alpha)$$

we have, substituting into (3-49),

$$a[(n^2 - p^2) \sin (pt - \alpha) + 2kp \cos (pt - \alpha)] = E \sin pt$$

The part in brackets produces $\sin pt$ if we choose α so that $\sin \alpha$ and $\cos \alpha$ are proportional to $2kp$ and $n^2 - p^2$; that is,

$$\sin \alpha = \frac{2kp}{\sqrt{(n^2 - p^2)^2 + 4k^2p^2}} \qquad \cos \alpha = \frac{n^2 - p^2}{\sqrt{(n^2 - p^2)^2 + 4k^2p^2}}$$

With these values, the first member of the preceding equation becomes

$$a \sqrt{(n^2 - p^2)^2 + 4k^2p^2} \sin pt$$

We determine a by equating the coefficients of $\sin pt$ in the two members. The general solution is then

$$s = Ae^{-kt} \sin (\beta t + B) + \frac{E}{\sqrt{(n^2 - p^2)^2 + 4k^2p^2}} \sin (pt - \alpha) \tag{3-50}$$

Here the first term is transient, and the motion takes on the period of the impressed force. Thus a tuning fork when subjected to a continued external vibration will presently give off the impressed note. In the ultimate periodic motion the maximum displacement lags behind the maximum of the impressed force by the time α/p.

Resonance. What can we say of the amplitude of the final vibration? Its value, $E/\sqrt{(n^2 - p^2)^2 + 4k^2p^2}$, depends upon all the constants in (3-49). If k is small and p is nearly equal to n, the amplitude can be very large. This is known as *resonance*. Classic examples are the violent motion of a bridge under the impact of marching feet descending in the natural period of the structure, and the result of applying a periodic force to the pendulum consisting of a child in a swing.

If $p = n$ exactly, we have $\sin \alpha = 1$, and the lag in the vibration is $t = \alpha/p = \pi/2p$. This is a quarter period.

Beats. The addition of two simple harmonic terms with comparable amplitudes and with periods which are nearly, but not exactly, equal produces the phenomenon known as *beats*. We add two sine functions, for example, in which the graphs both lie above the axis for a time, thus reinforcing each other, and later lie on opposite sides and cancel out. If the system produces sound, we have loud periods alternating with periods of quiet.

This situation arises also in the case of (3-50) if p is near to β and if the coefficients of the sines are of comparable size. With the passage of time the first term damps out and the phenomenon disappears.

The explanation of beats could, of course, be given an analytic formulation.

EXERCISES

1. Obtain the particular solution (3-50) starting with the trial solution

$$s = a \sin pt + b \cos pt$$

2. If $p = n$ in (3-49) and if there is no friction, show that the solution contains the disruptive term $-(Et \cos nt)/2n$.

3. With fixed E find p so that the amplitude of the forced vibration in (3-50) shall be a maximum.

4. A 1-lb weight hangs at rest by an elastic string of negligible weight and of natural length 5 ft, stretching it 1 ft. It is raised 1 ft and released. Find the motion, neglecting friction, and state the period. (Use $g = 32$.)

5. Solve the preceding exercise if there is friction of such a sort that $k = 4$ in Eq. (3-45).

6. Solve the like problem if $k = 6$.

7. The string of Exercise 4 is attached in a vertical position to the floor and the ceiling of a room 10 ft high. The 1-lb weight is fastened to the middle of the string and released. Find the motion, neglecting friction, and state the period.

8. Place the preceding string in a horizontal position with its ends attached to walls 10 ft apart. Attach a $\frac{1}{2}$-lb weight to the middle and release. Neglect friction and the variation in the tension of the string, and use sin θ and tan θ interchangeably, where θ is the angle between the string and the horizontal. Find the motion and state the period.

9. A cylindrical buoy 12 in. in diameter and weighing 200 lb floats on the water with axis vertical. Find its period of oscillation, given that the upward force is equal to the weight of the displaced water.

10. Show that a train would move under its own weight in a straight tunnel joining two points on the earth's surface in simple harmonic motion, assuming the density of the earth to be uniform, and neglecting friction and the earth's rotation and oblateness. Show that it could go from one point to the other in about 42 min.

CHAPTER 4

GENERAL LINEAR EQUATIONS

4-1. The Linear Equation. The linear differential equation of the nth order,

$$\frac{d^n y}{dx^n} + P_1(x)\frac{d^{n-1}y}{dx^{n-1}} + \cdots + P_n(x)y = R(x) \qquad (4\text{-}1)$$

has been solved completely for $n = 1$ (Sec. 2-1) and has been discussed at some length for $n = 2$ (Chap. 3). Most of the results in Chap. 3 can be generalized readily and will be presented rather briefly here. For an adequate study a knowledge of determinants is requisite.

We shall fix our attention on an interval $a < x < b$, which we shall call I, in which the coefficients $P_1(x), \ldots, P_n(x), R(x)$ are continuous. We state the existence theorem of Chap. 6.

THEOREM. *There exists one and only one solution $y = y(x)$ of (4-1), continuous together with its first n derivatives in I, such that the function and its first $n - 1$ derivatives take on prescribed values at a point x_0 of I.*

We observe that a solution can fail to be continuous only if one of the coefficients has a singularity.

For the sake of brevity we shall, as hitherto, speak of solutions $y(x)$, $u(x)$, x^2, etc., meaning thereby $y = y(x)$, $y = u(x)$, $y = x^2$, and so on.

4-2. The Reduced Equation. The linear equation resulting from replacing $R(x)$ by zero,

$$\frac{d^n y}{dx^n} + P_1(x)\frac{d^{n-1}y}{dx^{n-1}} + \cdots + P_n(x)y = 0 \qquad (4\text{-}2)$$

is called the reduced equation.

THEOREM. *If a solution $y(x)$ of (4-2) and its first $n - 1$ derivatives vanish at a point of ab, then $y(x) \equiv 0$ in ab.*

For, $y(x) \equiv 0$ is obviously a solution of (4-2) which vanishes together with its first $n - 1$ derivatives at the point, and there is but one such solution.

THEOREM. *If $y_1(x), \ldots, y_k(x)$ are solutions of the reduced equation (4-2), then $c_1 y_1(x) + \cdots + c_k y_k(x)$, where the c's are constants, is a solution.*

This is proved, as in Chap. 3 for $n = 2$, by direct substitution in the equation.

We also prove by direct substitution the following.

THEOREM. *If $y_0(x)$ and $y_1(x)$ are any two solutions of* (4-1), *then $y_0(x) - y_1(x)$ is a solution of* (4-2).

If $y_1(x)$ is any solution of (4-1) *and $u(x)$ is any solution of* (4-2), *then $u(x) + y_1(x)$ is a solution of* (4-1).

We have then as an immediate consequence of the preceding theorem the following result.

THEOREM. *The general solution of the complete equation,* (4-1), *is got by adding to a particular solution of* (4-1) *the general solution of the reduced equation* (4-2).

The general solution of the complete equation (4-1) thus raises two problems: first, the problem of finding the general solution of the reduced equation; second, the problem of finding a single solution of the complete equation.

EXERCISES

1. Show that a linear equation remains linear after any change of the independent variable, $x = \varphi(t)$.
2. Show that if $u(x) + iv(x)$ is a complex solution of (4-2), the coefficients in the equation being real, then $u(x)$ and $v(x)$ are solutions.
3. Verify that e^{ix^2} satisfies

$$xy'' - y' + 4x^3 y = 0$$

and write down the general solution.

4. Show that if $R(x) = U(x) + iV(x)$, a complex quantity, whereas the coefficients in the first member are real, and if $u(x) + iv(x)$ is a solution, then $u(x)$ is a solution when the second member is $U(x)$, and $v(x)$ is a solution when the second member is $V(x)$.
5. Show that if $y(x)$ is a solution of (4-1) and $z(x)$ is a solution with second member $S(x)$, then $y(x) + z(x)$ is a solution with second member $R(x) + S(x)$.

4-3. Wronskians. We now proceed to generalize the matters treated in Secs. 3-3 and 3-4. We call n functions $u_1(x), \ldots, u_n(x)$ *linearly dependent* in an interval $a < x < b$ if there exist constants $c_1, \ldots, c_n$ not all zero such that

$$c_1 u_1(x) + \cdots + c_n u_n(x) = 0 \qquad (4\text{-}3)$$

holds identically in the interval.

Closely allied with the subject of linear dependence is the determinant known as the *Wronskian*. The Wronskian of n functions having derivatives of the first $n - 1$ orders is

$$W(u_1, \ldots, u_n) = \begin{vmatrix} u_1 & u_2 & \ldots & u_n \\ u_1' & u_2' & \ldots & u_n' \\ u_1'' & u_2'' & \ldots & u_n'' \\ \vdots & \vdots & & \vdots \\ u_1^{(n-1)} & u_2^{(n-1)} & \ldots & u_n^{(n-1)} \end{vmatrix} \quad (4\text{-}4)$$

THEOREM. *If $u_1(x), \ldots, u_n(x)$ are linearly dependent in ab and possess derivatives of order $n - 1$, then the Wronskian of the functions vanishes identically in ab.*

Differentiating (4-3) $n - 1$ times, we have in ab:

$$c_1 u_1'(x) + \cdots\cdots\cdots + c_n u_n'(x) \equiv 0$$
$$\cdots\cdots\cdots\cdots\cdots\cdots\cdots\cdots\cdots\cdots$$
$$c_1 u_1^{(n-1)}(x) + \cdots + c_n u_n^{(n-1)}(x) \equiv 0$$

Eliminating $c_1, \ldots, c_n$ from these equations and Eq. (4-3), we have

$$W(u_1, \ldots, u_n) \equiv 0$$

We now consider solutions of the reduced equation.

THEOREM. *If the Wronskian of n solutions of the reduced equation (4-2) vanishes at one point of an interval $a < x < b$ in which $P_1(x), \ldots, P_n(x)$ are continuous, then the solutions are linearly dependent in ab and the Wronskian vanishes identically.*

Let $u_1(x), \ldots, u_n(x)$ be the solutions. Let the Wronskian be zero at x_0; and consider the equations

$$\begin{aligned} c_1 u_1(x_0) + \cdots\cdots\cdots + c_n u_n(x_0) &= 0 \\ c_1 u_1'(x_0) + \cdots\cdots\cdots + c_n u_n'(x_0) &= 0 \\ \cdots\cdots\cdots\cdots\cdots\cdots\cdots\cdots\cdots\cdots & \\ c_1 u_1^{(n-1)}(x_0) + \cdots + c_n u_n^{(n-1)}(x_0) &= 0 \end{aligned} \quad (4\text{-}5)$$

The determinant $W[u_1(x_0), \ldots, u_n(x_0)]$ of these linear equations in $c_1, \ldots, c_n$ is zero. The equations can then be satisfied by values of $c_1, \ldots, c_n$, not all zero.

The equations just written show that the solution

$$c_1 u_1(x) + \cdots + c_n u_n(x)$$

vanishes together with its first $n - 1$ derivatives at x_0. It is therefore identically zero in ab:

$$c_1 u_1(x) + \cdots + c_n u_n(x) \equiv 0$$

Since the solutions are linearly dependent, the Wronskian vanishes identically in ab.

THEOREM. *Every solution of (4-2) is expressible in ab as a linear combination of n linearly independent solutions.*

GENERAL LINEAR EQUATIONS   87

Let $u_1(x), \ldots, u_n(x)$ be linearly independent in ab. The Wronskian is nowhere equal to zero, and the determinant in (4-5) is not zero. Let $y(x)$ be any solution. If we put $y(x_0), y'(x_0), \ldots, y^{(n-1)}(x_0)$ in the second members of (4-5), the resulting equations determine $c_1, \ldots, c_n$ uniquely. Having thus found the constants, we have

$$y(x) \equiv c_1 u_1(x) + \cdots + c_n u_n(x) \qquad (4\text{-}6)$$

for the two members are solutions of (4-2) which are equal and have the same derivatives up to the $(n-1)$th at x_0.

We shall have in Eq. (4-6) the general solution of the reduced equation, when we have found in some way n particular solutions, provided these are linearly independent. That such solutions exist is easily seen. There exists a unique solution $u_1(x)$ for which $u_1(x_0), u_1'(x_0), \ldots, u_1^{(n-1)}(x_0)$ have prescribed values. Similarly we can prescribe the remaining columns in the Wronskian at x_0. We have but to prescribe values not making the Wronskian vanish at x_0 in order to have n linearly independent solutions.

We shall prove finally a sufficient condition for the linear dependence of a set of functions. That the identical vanishing of the Wronskian is not alone sufficient has been shown by an example for the case $n = 2$. We can, however, establish the following:

THEOREM. *Given a set of functions $u_1(x), \ldots, u_n(x)$ with continuous $(n-1)th$ derivatives in an interval $a < x < b$. If the Wronskian of the n functions vanishes identically in ab and if the n Wronskian formed from $n-1$ of the functions do not vanish simultaneously at any point of ab, then the functions are linearly dependent in ab.*

Suppose, first, that one of the Wronskians of $n - 1$ of the functions, say, $W_n(u_1, \ldots, u_{n-1})$, is not zero anywhere in ab. Consider the differential equation

$$W(u_1, \ldots, u_{n-1}, y) = \begin{vmatrix} u_1 & \cdots & u_{n-1} & y \\ u_1' & \cdots & u_{n-1}' & y' \\ \cdots & \cdots & \cdots & \cdots \\ u_1^{(n-1)} & \cdots & u_{n-1}^{(n-1)} & y^{(n-1)} \end{vmatrix} = 0 \qquad (4\text{-}7)$$

This is an equation of order $n - 1$. The coefficient of $y^{(n-1)}$, the highest derivative, is W_n. Since W_n is nowhere zero, we can divide by it, putting the equation in the form (4-2) with coefficients which are continuous in ab.

We observe that $u_1, \ldots, u_{n-1}$ are solutions of (4-7), for the substitution of any one for y makes two columns alike in the determinant. These solutions are linearly independent; hence any solution is a linear combination of these. Also u_n is a solution, owing to the identical vanishing of the Wronskian. Hence

$$u_n \equiv c_1 u_1 + \cdots + c_{n-1} u_{n-1} \qquad (4\text{-}8)$$

and the functions are linearly dependent.

If no Wronskian of order $n-1$ is different from zero throughout ab, this treatment requires modification. Let W_n be different from zero in an interval $a' < x < b'$ of ab, where at a', say, $W_n = 0$. In $a'b'$ relation (4-8) holds. In an interval $a'' < x < b''$ about the point a', one of the other Wronskians, say $W_1(u_2, \ldots, u_n)$, is different from zero. Hence in $a''b''$ we have a relation of the form

$$u_1 \equiv k_2 u_2 + \cdots + k_n u_n \qquad (4\text{-}9)$$

This relation is the same as (4-8). For, if not, in the interval $a'b''$ in which both hold we can replace u_n in (4-9) by its value from (4-8) and get a linear relation between $u_1, \ldots, u_{n-1}$, which is impossible.

The relation (4-8) can thus be extended to left and right from the interval $a'b'$, so long as no point is encountered at which all Wronskians of order $n-1$ vanish simultaneously. It then holds throughout the whole of ab.

4-4. Gramians. Necessary conditions for linear dependence may be set up in great variety. From (4-3) we can get linear equations in the constants in many ways—by differentiation, by integration between arbitrary limits, by differencing, by giving x fixed values, and so on—and the elimination of the c's from n of these equations gives a necessary condition. As a rule, the resulting condition is not sufficient.

A condition which is both necessary and sufficient is the vanishing of the determinant known as the *Gramian*. Let $u_1(x), \ldots, u_n(x)$ be continuous in the closed interval $a \leqslant x \leqslant b$. The Gramian of the functions is defined as follows:

$$G = \begin{vmatrix} \int_a^b u_1{}^2\, dx & \int_a^b u_1 u_2\, dx & \cdots & \int_a^b u_1 u_n\, dx \\ \int_a^b u_2 u_1\, dx & \int_a^b u_2{}^2\, dx & \cdots & \int_a^b u_2 u_n\, dx \\ \cdots & \cdots & \cdots & \cdots \\ \int_a^b u_n u_1\, dx & \int_a^b u_n u_2\, dx & \cdots & \int_a^b u_n{}^2\, dx \end{vmatrix}$$

THEOREM. *A necessary and sufficient condition that $u_1, \ldots, u_n$ be linearly dependent in ab is that $G = 0$.*

The proof of the necessity of the condition being easily made, we limit ourselves to establishing its sufficiency. Since G, a determinant with constant terms, vanishes, we can choose constants $c_1, \ldots, c_n$, not all zero, so that

$$c_1 \int_a^b u_1{}^2 \, dx + c_2 \int_a^b u_1 u_2 \, dx + \cdots + c_n \int_a^b u_1 u_n \, dx = 0$$

$$\cdots\cdots\cdots\cdots\cdots\cdots\cdots\cdots\cdots\cdots\cdots\cdots\cdots\cdots\cdots\cdots$$

$$c_1 \int_a^b u_n u_1 \, dx + c_2 \int_a^b u_n u_2 \, dx + \cdots + c_n \int_a^b u_n{}^2 \, dx = 0$$

Multiplying these equations by $c_1, \ldots, c_n$, respectively, and adding, we have

$$\int_a^b (c_1 u_1 + \cdots + c_n u_n)^2 \, dx = 0$$

The first member is positive unless the integrand vanishes identically in ab,

$$c_1 u_1 + \cdots + c_n u_n \equiv 0$$

which was to be proved.

EXERCISES

1. Evaluate the Wronskian of the functions

$$(x+1)^2 \qquad 3x^2 + 2x - 3 \qquad x^2 - 2$$

2. Evaluate the Gramian of the preceding functions for the interval 01.

3. Prove that the Wronskian of any $n+1$ solutions of (4–2) vanishes identically in ab.

4. Show that the Wronskian of a set of n solutions of (4–2) satisfies the equation

$$W = W_0 e^{-\int_{x_0}^x P_1 \, dx}$$

5. If the functions $u_1, \ldots, u_n$ are complex, show that a necessary and sufficient condition for linear dependence in ab is the vanishing of the determinant of order n having in the ith row and jth column the element

$$\int_a^b u_i \bar{u}_j \, dx$$

where the bar denotes the conjugate imaginary.

6. Show that if the coefficients $P_i(x)$ of (4–2) are periodic functions of period h and $y_1(x)$ is a solution, then $y_1(x+h)$ is a solution. Hence $y_1(x+h)$ is expressible as a linear combination of independent solutions $y_1(x), \ldots, y_n(x)$.

7. Show that, in the preceding, certain solutions can be found with the property $u_i(x+h) = s_i u_i(x)$, where s_i is a constant.

4-5. The Reduced Equation with Constant Coefficients. If the reduced equation has the form

$$p_0 \frac{d^n y}{dx^n} + p_1 \frac{d^{n-1} y}{dx^{n-1}} + \cdots + p_n y = 0 \tag{4-10}$$

where the p's are constants and $p_0 \neq 0$, the discovery of particular solutions offers no difficulty.

We try a solution of the form
$$y = e^{mx}$$
where m is a constant. Substituting into (4-10) and observing that
$$y^{(k)} = m^k e^{mx} \tag{4-11}$$
we have, after canceling e^{mx},
$$f(m) = p_0 m^n + p_1 m^{n-1} + \cdots + p_n = 0 \tag{4-12}$$

We have a solution, provided m is a root of (4-12), which is called the *auxiliary equation*. This equation has n roots $m_1, m_2, \ldots, m_n$. Each gives a solution $e^{m_k x}$.

Real Distinct Roots. If the roots of (4-12) are real and distinct, we have in this way n real and linearly independent solutions. The general solution then is
$$y = C_1 e^{m_1 x} + \cdots + C_n e^{m_n x} \tag{4-13}$$

Repeated Real Roots. If one of the roots m_k is a multiple root, say, an r-fold root, the preceding solutions are not all different, and (4-13) is not the general solution. If we put
$$y = e^{mx} Y$$
we have
$$y' = e^{mx}(mY + Y')$$
$$y'' = e^{mx}(m^2 Y + 2mY' + Y'')$$
$$y''' = e^{mx}(m^3 Y + 3m^2 Y' + 3m Y'' + Y''')$$
and so on. Substituting into (4-10) and canceling e^{mx}, we have
$$f(m)Y + f'(m)Y' + \cdots + \frac{1}{r!} f^{(r)}(m) Y^{(r)} + \cdots + \frac{1}{n!} f^{(n)}(m) Y^{(n)} = 0$$

Since m_k is an r-fold root, we have
$$f(m_k) = f'(m_k) = \cdots = f^{(r-1)}(m_k) = 0$$

The differential equation will be satisfied if $m = m_k$ and if $Y^{(r)}$ and higher derivatives vanish identically; that is, if Y is a polynomial of degree less than r. The r terms in (4-13) due to the repeated root m_k may then be replaced by
$$(B_1 + B_2 x + \cdots + B_r x^{r-1}) e^{m_k x}$$
where the B's are arbitrary constants.

Complex Roots. If some of the roots of (4-12) are complex, the preceding expressions still give us the general solution, but complex quantities are involved. If the coefficients in the equation are real, we can rewrite the particular solutions so that they are real.

GENERAL LINEAR EQUATIONS 91

If $\alpha + i\beta$ is a complex root so also is $\alpha - i\beta$, and we have in (4-13) the two terms

$$C_j e^{(\alpha+i\beta)x} + C_k e^{(\alpha-i\beta)x}$$

Using the relations

$$e^{(\alpha+i\beta)x} = e^{\alpha x}(\cos \beta x + i \sin \beta x)$$
$$e^{(\alpha-i\beta)x} = e^{\alpha x}(\cos \beta x - i \sin \beta x)$$

and putting

$$A = C_j + C_k \qquad B = i(C_j - C_k)$$

the two terms may be written

$$e^{\alpha x}(A \cos \beta x + B \sin \beta x) \quad \text{single complex roots}$$

If $\alpha + i\beta$ is a complex root of order r, so also is $\alpha - i\beta$. With the preceding reduction the $2r$ roots contribute the following to the general solution:

$$e^{\alpha x}[(A_1 + A_2 x + \cdots + A_r x^{r-1}) \cos \beta x$$
$$+ (B_1 + B_2 x + \cdots + B_r x^{r-1}) \sin \beta x] \quad \text{multiple complex roots}$$

which contains $2r$ arbitrary constants.

Alternative forms, which are sometimes useful, are

$$A \cos \beta x + B \sin \beta x = K \cos (\beta x + \gamma)$$
$$= H \sin (\beta x + \delta)$$

These involve mere changes in the definition of the constants.

EXERCISES

Solve:
1. $y^{(4)} - a^4 y = 0$.
2. $y''' - 3y'' + 2y' = 0$.
3. $y''' - 3y'' + 4y' - 2y = 0$.
4. $y^{(4)} + 4y''' + 6y'' + 4y' + y = 0$.
5. $y^{(4)} + 5y'' + 4y = 0$.
6. $y^{(4)} + 2y''' + 2y'' + 2y' + y = 0$.
7. $y^{(4)} + y''' - 3y'' - 5y' - 2y = 0$.
8. $y^{(4)} + a^4 y = 0$.
9. $y^{(8)} = y$.

10. Verify that the Wronskian of four independent solutions of Exercise 1 is identically constant.

11. Show that any derivative of a solution of the reduced equation with constant coefficients is also a solution. What can be said of the integral of a solution?

4-6. The Complete Equation with Constant Coefficients. Method of Undetermined Coefficients. Having solved the reduced equation, we require one solution of the complete equation. This can be accomplished in various ways. Without loss of generality we take $p_0 = 1$,

$$\frac{d^n y}{dx^n} + p_1 \frac{d^{n-1} y}{dx^{n-1}} + \cdots + p_n y = R(x) \qquad (4\text{-}14)$$

If $R(x)$ is made up of certain types of terms, we can use the method of undetermined coefficients described in the preceding chapter. If $R(x)$ is a polynomial of degree k and if $p_n \neq 0$, then we take the trial solution

$$y = P_k(x) = ax^k + bx^{k-1} + \cdots$$

substitute into the equation, and determine the coefficients $a, b, \ldots$ in order. If $p_n = 0$, we must clearly increase the degree of the polynomial. If the last r p's are zero but p_{n-r} is not, we use a polynomial of degree $k + r$.

If $R(x) = e^{mx}P(x)$, where $P(x)$ is a polynomial of degree k, we get the previous case by putting $y = e^{mx}Y$. We see then, without the necessity of changing the variable, that we shall usually have a solution of the form

$$y = e^{mx}(ax^k + bx^{k-1} + \cdots)$$

If m is a root of the auxiliary equation of multiplicity r, we increase by r the degree of the polynomial in the trial solution.

The preceding covers the case of terms of the type $e^{\alpha x}P(x) \sin \beta x$ and $e^{\alpha x}P(x) \cos \beta x$. We use as trial solutions $e^{\alpha x}[P_k(x) \sin \beta x + Q_k(x) \cos \beta x]$.

In all cases where the degree of the polynomial must be raised we can shorten the polynomial correspondingly at the lower end, since it is not necessary to include terms which are solutions of the reduced equation.

Example. Solve

$$y^{(4)} + y'' = 20x^3 + 60x^2 + e^x(2x + 8) + 2 \sin x$$

Here 1, x, $\sin x$, $\cos x$ are solutions of the reduced equation. We put

$$y = ax^5 + bx^4 + cx^3 + fx^2 + e^x(gx + h) + x(m \sin x + n \cos x)$$

into the equation, getting for the first member

$$20ax^3 + 12bx^2 + (120a + 6c)x + 24b + 2f + e^x(2gx + 6g + 2h) + 2n \sin x$$
$$- 2m \cos x$$

Equating coefficients of like terms in the two members, we have

$$20a = 20 \qquad 2g = 2$$
$$12b = 60 \qquad 6g + 2h = 8$$
$$120a + 6c = 0 \qquad 2n = 2$$
$$24b + 2f = 0 \qquad 2m = 0$$

Solving,

$$a = 1 \qquad b = 5 \qquad c = -20 \qquad f = -60 \qquad g = h = n = 1 \qquad m = 0$$

and we have the solution

$$y = x^5 + 5x^4 - 20x^3 - 60x^2 + e^x(x + 1) + x \cos x$$

4-7. Symbolic Methods. The reader is referred to a later chapter (Sec. 8-15) for an account of operators. Their use in finding particular solutions will be sketched very briefly here. Representing derivatives by powers of D,

$$Dg(x) = g'(x) \qquad D^2g(x) = g''(x) \qquad \cdots \qquad D^0g(x) = g(x)$$

we can form operator polynomials with constant coefficients. Thus (4-14) can be written in the form

$$f(D)y = (D^n + p_1 D^{n-1} + \cdots + p_n)y = R(x)$$

The successive application of two or more such operators may be found by first multiplying the operators together according to the usual laws of algebra and then applying the product operator. For example, $f(D)$ may be broken into factors,

$$f(D) = (D - m_1)(D - m_2) \cdots (D - m_n)$$

where $m_1, \ldots, m_n$ are the roots of the auxiliary equation. These factors may be applied successively in any order to yield the same result as a single application of $f(D)$.

We may introduce the notation

$$u = \frac{1}{D - m} g(x)$$

to mean any function u such that

$$(D - m)u = g(x)$$

This is an equation of the first order whose general solution is

$$u = e^{mx} \int e^{-mx} g(x) \, dx + C e^{mx}$$

We take a particular one of the solutions,

$$\boxed{\frac{1}{D - m} g(x) = e^{mx} \int e^{-mx} g(x) \, dx}$$

1. *A First Symbolic Method.* The particular solution may then be written in the symbolic form

$$y = \frac{1}{f(D)} R(x) = \frac{1}{D - m_1} \frac{1}{D - m_2} \cdots \frac{1}{D - m_n} R(x)$$

Here we apply the n operators in any order. The process involves n integrations. That the resulting function is a solution is readily seen; for on applying in succession to y the factors of $f(D)$, we undo the results of the successive integrations and arrive at $R(x)$.

Example 4-1.

(i) $$y'' - 3y' + 2y = xe^x$$

We have
$$(D^2 - 3D + 2)y = xe^x$$
$$y = \frac{1}{D-1}\frac{1}{D-2} xe^x$$
$$\frac{1}{D-2} xe^x = e^{2x}\int e^{-2x}xe^x\,dx = -(1+x)e^x$$
$$y = \frac{1}{D-1}[-(1+x)e^x] = -e^x\int e^{-x}(1+x)e^x\,dx$$
$$= -\tfrac{1}{2}(1+x)^2 e^x$$ Part. Sol. to D.E.

2. Symbolic Partial Fractions. The successive integrations of the preceding method are likely, even with relatively simple functions, to become complicated and tedious to carry out. A symbolic method in which the integrations are hitched abreast instead of tandem avoids this difficulty. The operator $1/f(D)$ may be broken into partial fractions. If the factors of $f(D)$ are distinct, we have the form

$$y = \frac{1}{f(D)} R(x) = \left[\frac{a_1}{D-m_1} + \cdots + \frac{a_n}{D-m_n}\right] R(x)$$

Here $f(D)$ applied after the operator in brackets gives 1, so that
$$f(D)y = R(x)$$
as required.

In Example 4-1
$$y = \left[\frac{1}{D-2} - \frac{1}{D-1}\right] xe^x$$
$$= e^{2x}\int e^{-2x}xe^x\,dx - e^x\int e^{-x}xe^x\,dx$$
$$= -(1+x)e^x - \frac{x^2}{2}e^x$$
$$= -\left(1 + x + \frac{x^2}{2}\right)e^x$$

If some of the factors are repeated, the form of the partial fraction is altered. If $D - m_k$ is an r-fold factor, we have in the partial fractions expansion the terms

$$\frac{a_k}{D-m_k} + \frac{a_{k+1}}{(D-m_k)^2} + \cdots + \frac{a_{k+r-1}}{(D-m_k)^r}$$

The application of these operators to $R(x)$ may be made in order, each involving an integration based on the preceding.

If in a real equation, two roots, say m_1 and m_2, are conjugate imaginaries, then a_1 and a_2 are conjugates. The evaluation of $[a_2/(D-m_2)]R(x)$ may be got from $[a_1/(D-m_1)]R(x)$ by taking the conjugate function, and the sum of the two functions will be real.

3. Symbolic Series Expansions. The symbolic operator may be thrown into various useful forms by skillful algebraic manipulation. A form

GENERAL LINEAR EQUATIONS   95

which is particularly useful when $R(x)$ is a polynomial is the expansion of $1/f(D)$ in a power series in D:

$$y = \frac{1}{f(D)} R(x) = (1 + b_1 D + b_2 D^2 + \cdots) R(x)$$

Example 4-2.

(ii) $\qquad y''' + y'' + y = x^4 + 2x + 1$

$$y = \frac{1}{1 + D^2 + D^3} (x^4 + 2x + 1)$$

On dividing 1 by the denominator we find

$$\begin{aligned} y &= (1 - D^2 - D^3 + D^4 + 2D^5 + \cdots)(x^4 + 2x + 1) \\ &= (x^4 + 2x + 1) - (12x^2) - (24x) + (24) \\ &= x^4 - 12x^2 - 22x + 25 \end{aligned}$$

Example 4-3. The following illustrates the possibilities of manipulation.

(iii) $\qquad y''' + y'' + y' + y = x^4 + 2x + 1$

$$\begin{aligned} y &= \frac{1}{1 + D + D^2 + D^3} (x^4 + 2x + 1) \\ &= \frac{1}{1 - D^4} (1 - D)(x^4 + 2x + 1) \\ &= \frac{1}{1 - D^4} [(x^4 + 2x + 1) - (4x^3 + 2)] \\ &= (1 + D^4 + D^8 + \cdots)(x^4 - 4x^3 + 2x - 1) \\ &= (x^4 - 4x^3 + 2x - 1) + (24) \\ &= x^4 - 4x^3 + 2x + 23 \end{aligned}$$

4. *Exponential Terms.* The special properties of the operator D when applied to an exponential function enable us to introduce simplifications when a term of $R(x)$ contains a factor of the form e^{kx}. We have

$$\begin{aligned} (D - m)e^{kx}V &= e^{kx}DV + ke^{kx}V - me^{kx}V \\ &= e^{kx}(D + k - m)V \end{aligned}$$

Applying this formula to a succession of factors we have for the polynomial $f(D)$,

$$f(D)e^{kx}V = e^{kx}f(D + k)V$$

that is, we may remove the factor e^{kx} to the left of the operator if we replace D by $D + k$ in the operator.

The same procedure holds if the operator is of the form $1/f(D)$. Suppose $R(x) = e^{kx}V(x)$; then

$$y = \frac{1}{f(D)} e^{kx} V = e^{kx} \frac{1}{f(D + k)} V$$

That any value of the last member also satisfies the second member is seen by operating with $f(D)$,

$$f(D)e^{kx}\frac{1}{f(D+k)}V = e^{kx}f(D+k)\frac{1}{f(D+k)}V = e^{kx}V$$

Example.
$$y'' - 3y' + 2y = xe^x$$

$$y = \frac{1}{D^2 - 3D + 2}xe^x = e^x\frac{1}{(D+1)^2 - 3(D+1) + 2}x$$

$$= e^x\frac{1}{D^2 - D}x = -e^x\left(\frac{1}{D}\frac{1}{1-D}\right)x$$

$$= -e^x\left(\frac{1}{D} + 1 + D + D^2 + \cdots\right)x$$

$$= -e^x\left(\frac{x^2}{2} + x + 1\right)$$

a solution already found by the method of 2.

EXERCISES

Find a particular solution of
$$y'' - y = x^2 e^{2x}$$
by the use of various symbolic methods:
  1. Method 1.   2. Method 2.   3. Method 4.
  4. Find a solution of Eq. (i) by the method of undetermined coefficients.

Find a solution by the use of symbolic series for:
  5. $y''' - 2y' + y = x^3 + (x+1)^2$.   6. $y''' + y'' = 6x^2 + 4$.

Find a particular solution by any convenient method:
  7. $y''' - 8y = x^2$.   8. $y^{(4)} - y = x^2 - 1$.
  9. $y''' + 2y'' + y' = x$.   10. $y''' = 1/x^3$.
  11. $y''' + a^2 y' = x$.   12. $y''' + by'' = x$.
  13. $y''' + y + 1 = 2 \sin x$.   14. $y^{(6)} - 4y^{(4)} + 4y'' = e^{2x}$.

  15. Show that a solution of $f(D)y = e^{ax}$ is $e^{ax}/f(a)$, provided $f(a) \neq 0$.
  16. Find a particular solution of
$$y''' + y = e^{2x}\cos 3x$$
(a) by the method of undetermined coefficients;
(b) by replacing the second member by $e^{(2+3i)x}$ and taking the real part of the resulting solution.
  17. Get particular solutions of
$$y''' + y = \cos x \quad \text{and} \quad y''' + y = \sin x$$
by the use of imaginary exponents.

4-8. Method of the Variation of Parameters. We can generalize the method of the preceding chapter and find a particular solution of the complete equation whenever we can find the general solution of the reduced equation. Unlike the methods of the preceding sections, we no longer require that the coefficients be constants.

Let $u_1(x), \ldots, u_n(x)$ be linearly independent solutions of (4-2), and let us try a solution of (4-1) of the form

$$y = t_1 u_1 + \cdots + t_n u_n \qquad (4\text{-}15)$$

We have replaced the constants in the solution of (4-2) by variables $t_1, \ldots, t_n$, which are to be found. We shall impose on the t's the following conditions:

$$\begin{aligned} u_1 t_1' + \cdots + u_n t_n' &= 0 \\ u_1' t_1' + \cdots + u_n' t_n' &= 0 \\ &\cdots \cdots \cdots \cdots \cdots \\ u_1^{(n-2)} t_1' + \cdots + u_n^{(n-2)} t_n' &= 0 \end{aligned} \qquad (4\text{-}16)$$

On differentiating (4-15) we can cancel half of each derivative up to the $(n-1)$th by virtue of Eqs. (4-16),

$$\begin{aligned} y' &= t_1 u_1' + \cdots + t_n u_n' \\ y'' &= t_1 u_1'' + \cdots + t_n u_n'' \\ &\cdots \cdots \cdots \cdots \cdots \\ y^{(n-1)} &= t_1 u_1^{(n-1)} + \cdots + t_n u_n^{(n-1)} \\ y^{(n)} &= t_1 u_1^{(n)} + \cdots + t_n u_n^{(n)} + u_1^{(n-1)} t_1' + \cdots + u_n^{(n-1)} t_n' \end{aligned}$$

We now substitute y, y', etc., into Eq. (4-1). The terms in $t_1, \ldots, t_n$ drop out since the u's are solutions of (4-2), and there remains

$$u_1^{(n-1)} t_1' + \cdots + u_n^{(n-1)} t_n' = R(x) \qquad (4\text{-}17)$$

Equations (4-16) and (4-17) are n equations linear in the unknowns $t_1', \ldots, t_n'$. Their determinant is the Wronskian, which is not zero; hence they can be solved,

$$t_1' = \varphi_1(x), \ldots, t_n' = \varphi_n(x)$$

Integrating and substituting into (4-15), we have the particular solution

$$y = u_1(x) \int \varphi_1(x)\, dx + \cdots + u_n(x) \int \varphi_n(x)\, dx \qquad (4\text{-}18)$$

The determination of the particular integral thus involves n integrations.

4-9. Reduction of the Order of the Equation. There remains the question of finding n particular solutions of the reduced equation. Except for series, numerical processes, and the like, there are no methods applicable to the general case. If, however, we can find in some way one or more linearly independent solutions, we can reduce the order of the equation to be solved.

Let $u_1(x)$ be a solution of

$$y^{(n)} + P_1(x)y^{(n-1)} + \cdots + P_n(x)y = 0 \qquad (4\text{-}2)$$

not identically zero, and put

$$y = u_1 t \qquad (4\text{-}19)$$

On substituting into (4-2), we have an equation of the form

$$u_1 t^{(n)} + Q_1 t^{(n-1)} + \cdots + Q_{n-1} t' = 0$$

the coefficient of t dropping out. This follows from the fact that $t = 1$ is a solution, as we see from (4-19). If t' be replaced by a new variable w, we have a linear equation of order $n - 1$,

$$u_1 w^{(n-1)} + Q_1 w^{(n-2)} + \cdots + Q_{n-1} w = 0$$

If this can be solved for w, an integration gives t, and multiplication by u_1 gives y.

More generally, if p solutions are known, we may use a Wronskian to set up a differential expression vanishing for some of the solutions of (4-2) and thence reduce the order of the equation.

THEOREM. *If $u_1(x), \ldots, u_p(x)$ are linearly independent solutions of* (4-2), *then the substitution*

$$w = \begin{vmatrix} u_1 & \cdots & u_p & y \\ u_1' & \cdots & u_p' & y' \\ \cdots & \cdots & \cdots & \cdots \\ u_1^{(p)} & \cdots & u_p^{(p)} & y^{(p)} \end{vmatrix} \varphi(x) \qquad (4\text{-}20)$$

where $\varphi(x)$ is some convenient function, reduces (4-2) *to a linear reduced equation of order $n - p$ in w.*

Here w can be written in the form

$$w = A(x)y^{(p)} + B(x)y^{(p-1)} + \cdots \qquad A(x) \neq 0 \qquad (4\text{-}21)$$

and its derivatives have the form

$$w' = Ay^{(p+1)} + \cdots, \ldots, w^{(n-p)} = Ay^{(n)} + \cdots$$

These equations can be used to eliminate $y^{(p)}, \ldots, y^{(n)}$, and (4-2) takes the form

$$Q_0 w^{(n-p)} + \cdots + Q_{n-p} w + V = 0$$

where V is linear in $y, y', \ldots, y^{(p-1)}$.

Now $y = u_1, \ldots, y = u_p$ satisfy this last equation and also make $w = 0$, so the differential equation $V = 0$ is satisfied. Since its order is less than p, this is possible only if V vanishes identically. It follows that (4-2) has been reduced to the desired form,

$$Q_0 w^{(n-p)} + \cdots + Q_{n-p} w = 0 \qquad (4\text{-}22)$$

If the last equation can be solved, we put the resulting value of w into (4-20), thus obtaining a complete equation in y of order p. We know the solution of the reduced equation, so the complete equation can be solved by the method of the preceding section.

It will be observed that the substitution (4-19) falls under this case. We have

$$w = t' = \frac{d}{dx}\frac{y}{u_1} = \frac{u_1 y' - y u_1'}{u_1^2}$$

which has the form (4-20) with $p = 1$ and $\varphi = 1/u_1^2$.

EXERCISES

1. Find a particular solution by the method of the variation of parameters,

$$y''' - y' = x^2$$

Solve the following Euler equations, which have solutions of the form x^m:

2. $x^3 y''' + x^2 y'' - 2xy' + 2y = 0$.
3. $x^4 y^{(4)} + 6x^3 y''' + 6x^2 y'' - 12xy' + 12y = 0$.
4. $x^3 y''' - 3x^2 y'' + 6xy' - 6y = 0$.

5. If in the preceding the second member is replaced by $\log x$, find a particular solution.
6. Solve $(2x + 3)^3 y''' + 4(2x + 3)^2 y'' + 16y = 0$.
7. Solve

$$x^3 y''' - 6x^2 y'' + x(x^2 + 18)y' - 2(x^2 + 12)y = 0$$

given that there is a solution of the form x^n.

8. Solve $x(y''' + y') - 4(y'' + y) = 0$.

LAPLACE TRANSFORMS

If we apply to the two members of a differential equation, or indeed to any two equal quantities, the same process, we get results which are equal. In this way we could find innumerable relations which a solution must satisfy. Such a process is the Laplace transformation, which involves multiplication by a certain function and integration between limits. It leads to a method of solving linear differential equations, which has come much into vogue in recent years, a method which yields particular solutions without the necessity of finding general solutions and evaluating constants, and which is, in general, shorter than our earlier methods. The treatment here is necessarily brief.

4-10. The Laplace Transform. Given a function $f(t)$ defined for positive values of t. Multiplying by e^{-pt} and integrating, we have, provided the integral exists,

$$F(p) = \int_0^\infty e^{-pt} f(t)\, dt \qquad (4\text{-}23)$$

This function $F(p)$ is known as the *Laplace transform* of $f(t)$. Various notations for the function have been used, such as, $L\{f\}$, $\bar{f}(p)$, which we shall employ at times, but we shall often use capital letters for the Laplace transforms of functions represented by lower-case letters.

In the following examples the integrals converge for $p > 0$, except in the third case which requires that $p > a$.

$$L\{1\} = \int_0^\infty e^{-pt}\, dt = -\frac{1}{p} e^{-pt} \Big]_0^\infty = \frac{1}{p}$$

$$L\{t\} = \int_0^\infty e^{-pt} t\, dt = -\frac{pt+1}{p^2} e^{-pt} \Big]_0^\infty = \frac{1}{p^2}$$

$$L\{e^{at}\} = \int_0^\infty e^{-pt} e^{at}\, dt = \frac{1}{a-p} e^{(a-p)t} \Big]_0^\infty = \frac{1}{p-a}$$

$$L\{\sin at\} = \int_0^\infty e^{-pt} \sin at\, dt$$

$$= -\frac{e^{-pt}(p \sin at + a \cos at)}{p^2 + a^2} \Big]_0^\infty = \frac{a}{p^2 + a^2}$$

These results will be found among the entries in the table on page 107.

We shall assume that $f(t)$ is continuous in $0 \leqslant t < \infty$, or at least is sectionally continuous. By this latter we mean that in any finite interval $0 \leqslant t \leqslant b$, $f(t)$ is continuous, except possibly for a finite number of points at which there are finite jumps at which the function approaches limits from either side. This covers the vast majority of functions which arise in practice and provides for those which appear in the instantaneous application of or removal of forces, voltages, and the like.

We say that $f(t)$ is of *exponential order* if for sufficiently large t, $t > c$, and for M and a suitably chosen we have

$$|f(t)| < M e^{at}$$

All the solutions of reduced linear differential equations with constant coefficients are of this type, as well as all their derivatives, and so also are most of the functions that arise in practice in the case of complete equations.

THEOREM. *If $f(t)$ is sectionally continuous in the interval $0 \leqslant t < \infty$ and if it is of exponential order, then $f(t)$ has a Laplace transform.*

For a function of this type we have, for t sufficiently large,

$$|e^{-pt} f(t)| < M e^{-(p-a)t}$$

GENERAL LINEAR EQUATIONS   101

and for $p > a$ the integral of the second member of this inequality converges. Hence (4-23) converges absolutely by the comparison test.

THEOREM. *If (4-23) converges absolutely for $p = p_0$, it converges absolutely for all $p > p_0$.*

This follows from the decrease of the absolute value of the integrand with increase of p,

$$|e^{-pt}f(t)| \leqslant |e^{-p_0 t}f(t)|$$

which assures absolute convergence by the comparison test.

Finally, we may point out that the Laplace transformation is a *linear operator;* that is,

$$L\{C_1 f_1(t) + C_2 f_2(t)\} = C_1 L\{f_1(t)\} + C_2 L\{f_2(t)\}$$

This follows immediately from the definition.

We might have used this property to get $L\{\sin at\}$ above from the transform which precedes it:

$$L\{\sin at\} = L\left\{\frac{e^{iat} - e^{-iat}}{2i}\right\} = \frac{1}{2i}(L\{e^{iat}\} - L\{e^{-iat}\})$$

$$= \frac{1}{2i}\left(\frac{1}{p - ai} - \frac{1}{p + ai}\right) = \frac{a}{p^2 + a^2}$$

4-11. Laplace Transforms of Differential Equations. The Laplace transform of a derivative may be simply expressed in terms of the transform of the function itself, a fact which is especially useful in connection with differential equations. If we integrate

$$L\{f'(t)\} = \int_0^\infty e^{-pt} f'(t)\, dt$$

by parts [putting $u = e^{-pt}$, $v = f(t)$ in the formula $\int u\, dv = uv - \int v\, du$], we have

$$\int_0^\infty e^{-pt} f'(t)\, dt = e^{-pt} f(t) \Big]_0^\infty + p \int_0^\infty e^{-pt} f(t)\, dt$$

Now if $e^{-pt}f(t) \to 0$ as $t \to \infty$, as will be the case if $f(t)$ is of exponential order and p is sufficiently large, we have

$$L\{f'(t)\} = pL\{f(t)\} - f(0) \qquad (4\text{-}24)$$

$f(0)$ meaning $\lim f(t)$ as t approaches 0 from above.

Consider now the differential equation with constant coefficients

$$a_0 \frac{d^n y}{dt^n} + a_1 \frac{d^{n-1} y}{dt^{n-1}} + \cdots + a_{n-1} \frac{dy}{dt} + a_n y = r(t) \qquad (4\text{-}25)$$

We wish to find the solution $y(t)$ with the initial conditions

$$y(0) = y_0,\ y'(0) = y_1,\ \ldots,\ y^{(n-1)}(0) = y_{n-1} \qquad (4\text{-}26)$$

We take the Laplace transform of both members of (4-25). By repeated applications of (4-24), we have

$$L\{y'(t)\} = pY(p) - y_0$$
$$L\{y''(t)\} = p^2Y(p) - py_0 - y_1$$

etc., where $Y(p)$ is the transform of $y(t)$. Designating by $R(p)$ the transform of $r(t)$, we have, after transposing,

$$(a_0p^n + a_1p^{n-1} + \cdots + a_n)Y(p)$$
$$= y_0(a_0p^{n-1} + a_1p^{n-2} + \cdots + a_{n-1})$$
$$+ y_1(a_0p^{n-2} + a_1p^{n-3} + \cdots + a_{n-2}) + \cdots$$
$$+ y_{n-2}(a_0p + a_1) + y_{n-1}(a_0) + R(p) \quad (4\text{-}27)$$

Formula (4-27) is known as the *subsidiary equation*. $R(p)$ must be found by integration or from a table; the rest can be written down.

The subsidiary equation is used in the following way: Solving for $Y(p)$, let us suppose that the result can be written as the sum of terms—partial fractions, for instance—which are the Laplace transforms of known functions, such as those appearing in the table. We can then write down $y(t)$ itself, and the solution has been found.

Example 4-4. Solve
$$y''' - y' = 2\cos t$$
given that $y = 3$, $y' = 2$, $y'' = 1$ when $t = 0$.

Since $L\{2\cos t\} = 2p/(p^2 + 1)$, the subsidiary equation is

$$(p^3 - p)Y = 3(p^2 - 1) + 2(p) + 1(1) + \frac{2p}{p^2 + 1}$$

whence
$$Y = \frac{(3p^2 + 2p - 2)(p^2 + 1) + 2p}{(p^3 - p)(p^2 + 1)}$$

Breaking this into partial fractions, we find

$$Y = \frac{2}{p} + \frac{2}{p-1} - \frac{1}{p+1} - \frac{1}{p^2 + 1}$$

A reference to the table of transforms produces the solution

$$y = 2 + 2e^t - e^{-t} - \sin t$$

Example 4-5. Solve the preceding equation with the conditions $y = y' = y'' = 0$ when $t = 0$.

The subsidiary equation now is

$$(p^3 - p)Y = \frac{2p}{p^2 + 1}$$

from which
$$Y = \frac{\frac{1}{2}}{p-1} - \frac{\frac{1}{2}}{p+1} - \frac{1}{p^2 + 1}$$

whence
$$y = \tfrac{1}{2}(e^t - e^{-t}) - \sin t = \sinh t - \sin t$$

We can write down a formula for the solution in a very common case. If Y is a rational function
$$Y = \frac{P(p)}{Q(p)}$$
in which the polynomial $Q(p)$ has *distinct* roots $p_1, \ldots, p_k$, the numerator being of lower degree, we can use a well-known formula for the partial fractions,
$$Y = \sum \frac{P(p_j)}{(p - p_j)Q'(p_j)}$$
We can now write down the inverse function, getting an expansion due to Heaviside,
$$y = \frac{P(p_1)}{Q'(p_1)} e^{p_1 t} + \cdots + \frac{P(p_k)}{Q'(p_k)} e^{p_k t}$$

Example 4-6. Solve
$$y'' - 3y' + 2y = 2e^{3t} \qquad y_0 = 2 \qquad y_1 = 3$$
The subsidiary equation is
$$(p^2 - 3p + 2)Y = 2(p - 3) + 3(1) + \frac{2}{p - 3}$$
whence
$$Y = \frac{2p^2 - 9p + 11}{p^3 - 6p^2 + 11p - 6}$$
Here
$$\frac{P(p)}{Q'(p)} = \frac{2p^2 - 9p + 11}{3p^2 - 12p + 11}$$
and this has at the roots 1, 2, 3 the values 2, -1, 1, respectively. Hence
$$y = 2e^t - e^{2t} + e^{3t}$$

The validity of all these methods depends on the fact that there is only one possible function yielding a given transform; in other words, that the inverse of the Laplace transform is unique. This is indeed the case, and the matter will be discussed in the next section.

4-12. Uniqueness of the Inverse Transformation. As we proceed from the transform back to the function, it is essential to know that there is only one function producing a given Laplace transform. We shall prove the following.

THEOREM. *Let $F(p)$ be the Laplace transform of a continuous function $f(t)$, then no other continuous function has the same Laplace transform.*

In the proof of this theorem we shall make use of the following preliminary result.

LEMMA. *If $u(x)$ is continuous in the interval $0 \leqslant x \leqslant 1$, and if*

$$\int_0^1 x^n u(x)\, dx = 0 \qquad n = 0, 1, 2, \ldots$$

then $u(x) \equiv 0$ in the interval.

The proof of the lemma may be based on the fact that a continuous function in a finite interval can be approximated to with arbitrary exactness by a polynomial. For $\epsilon > 0$, however small, a polynomial $P(x)$ may be found so that

$$|u(x) - P(x)| < \epsilon$$

in the interval. Since $P(x)$ is made up of terms in powers of x, we have $\int_0^1 u(x) P(x)\, dx = 0$, so

$$\int_0^1 u^2(x)\, dx = \int_0^1 u(x)[u(x) - P(x)]\, dx \leqslant \epsilon \int_0^1 |u(x)|\, dx$$

The final member of this inequality may be made arbitrarily small, so the first member is zero; whence $u(x) \equiv 0$, as was to be proved.

For the proof of the main theorem, suppose that the continuous functions $f(t)$ and $g(t)$ have the same Laplace transform for $p \geqslant p_0$. Then the difference

$$h(t) = f(t) - g(t)$$

has the transform zero:

$$\int_0^\infty e^{-pt} h(t)\, dt = 0 \qquad p \geqslant p_0$$

We wish to show that $h(t)$ is zero.

Put $p = p_0 + n$, where n is any positive integer, and integrate by parts:

$$\int_0^\infty e^{-nt} e^{-p_0 t} h(t)\, dt = e^{-nt} \int_0^t e^{-p_0 \tau} h(\tau)\, d\tau \Big]_0^\infty$$
$$+ n \int_0^\infty e^{-nt} \left[\int_0^t e^{-p_0 \tau} h(\tau)\, d\tau \right] dt = 0$$

The first term of the second member vanishes, so that the final term is zero.

Putting

$$v(t) = \int_0^t e^{-p_0 \tau} h(\tau)\, d\tau$$

we write the preceding in the form

$$\int_0^\infty e^{-nt} v(t)\, dt = 0$$

By the change of variable $x = e^{-t}$, this becomes

$$\int_0^1 x^{n-1} v(-\log x)\, dx = 0 \qquad n = 1, 2, 3, \ldots$$

GENERAL LINEAR EQUATIONS 105

By the lemma
$$v(-\log x) = v(t) \equiv 0$$
$$v'(t) = e^{-p_0 t}h(t) \equiv 0$$

and $h(t) \equiv 0$, as was to be proved.

We could extend our results in various ways. Suppose that $f(t)$ and $g(t)$, and hence $h(t)$, are required merely to be sectionally continuous. Any discontinuities are smoothed out by the integration, and $v(t)$ is continuous; so the lemma still applies. Then $h(t) \equiv 0$ except for isolated points.

Also we have used the equality of the transforms only for $p_0 + n$. By a slight revision of the proof we can show that if the transforms are the same for infinitely many values equally spaced along the p axis, the functions are identical. The transforms are then equal everywhere.

4-13. Further Properties. Certain simple relations will be useful both in constructing entries for a table of Laplace transforms and in applying these entries to varied situations. We derive transforms involving functions connected with $f(t)$ and its transform $F(p)$.

The differentiation of (4-23) any number of times with respect to p can be justified in the case of convergence; this yields

$$F^{(n)}(p) = (-1)^n \int_0^\infty e^{-pt} t^n f(t)\, dt$$

THEOREM. *The Laplace transform of $t^n f(t)$, $n = 1, 2, 3, \ldots,$ is $(-1)^n F^{(n)}(p)$.*

In an earlier section we found the transform of $f'(t)$. By replacing $f(t)$ in (4-24) by $\int_0^t f(\tau)\, d\tau$ we have, since this integral is zero when $t = 0$,

$$L\{f(t)\} = pL\left\{\int_0^t f(\tau)\, d\tau\right\}$$

THEOREM. *The Laplace transform of $\int_0^t f(\tau)\, d\tau$ is $F(p)/p$.*

Shifting Theorems. We shall derive results involving the functions $F(p + a)$ and $f(t + b)$. From (4-23),

$$F(p + a) = \int_0^\infty e^{-pt} e^{-at} f(t)\, dt$$

THEOREM. *The Laplace transform of $e^{-at}f(t)$ is $F(p + a)$.*

Let us consider next the step function which is equal to zero for $t < b$, where $b > 0$, and to $f(t)$ for $t \geq b$. Call it $f_b(t)$. Its transform, using τ as the variable of integration, is

$$F_b(p) = \int_0^\infty e^{-p\tau} f(\tau)\, d\tau$$

If we put $\tau = t + b$, we have

$$F_b(p) = e^{-pb} \int_0^\infty e^{-pt} f(t + b) \, dt$$

THEOREM. *The Laplace transform of $f(t + b)$, $b > 0$, is equal to e^{pb} multiplied by the transform of a function which is zero to the left of b and $f(t)$ to the right of b.*

As an application of this theorem we consider the function of impulsive character which has the large value $1/\epsilon$ over a short range $0 < t < \epsilon$, and which is zero elsewhere. The area under its graph is unity. Its Laplace transform is

$$L\left\{\frac{1}{\epsilon}\right\} - e^{-p\epsilon}L\left\{\frac{1}{\epsilon}\right\} = \frac{1 - e^{-p\epsilon}}{p\epsilon}$$

If $p\epsilon$ is very small, the transform is near the limiting value 1.

We shall state a final theorem. Its proof, involving double integration, is omitted.

THEOREM. *If $F(p)$ and $G(p)$ are the Laplace transforms of $f(t)$ and $g(t)$, then the product $F(p)G(p)$ is the transform of $\int_0^t f(\tau)g(t - \tau) \, d\tau$, and this is equal to $\int_0^t f(t - \tau)g(\tau) \, d\tau$.*

Either of these integrals is called the *convolution* of $f(t)$ and $g(t)$.

Example. Solve in integral form

$$y'' + 2y' + 2y = f(t) \qquad y_0 = y_1 = 0$$

The subsidiary equation is

$$(p^2 + 2p + 2)Y = F(p)$$

$$Y = \frac{F(p)}{p^2 + 2p + 2} = F(p)L\{e^{-t} \sin t\}$$

From the preceding theorem we can write the solution in either of the forms

$$y = \int_0^t f(t - \tau)e^{-\tau} \sin \tau \, d\tau$$

$$= e^{-t} \int_0^t f(\tau)e^{\tau} \sin (t - \tau) \, d\tau$$

For a more complete treatment of Laplace transforms the reader is referred to books on the subject; for example, "Operational Methods in Applied Mathematics," by Carslaw and Jaeger; "Modern Operational Mathematics in Engineering," by Churchill; or the "Laplace Transformation," by Thompson. Here will be found more extensive tables, a wide variety of applications, and a broad theory, which includes the use of functions of complex variables.

GENERAL LINEAR EQUATIONS   107

A Short Table of Laplace Transforms

$$F(p) = \int_0^\infty e^{-pt}f(t)\,dt$$

Formula	$F(p)$	$f(t)$
1	$\dfrac{1}{p}$	1
2	$\dfrac{1}{p-a}$	e^{at}
3	$\dfrac{1}{p^n}$	$\dfrac{t^{n-1}}{(n-1)!}\quad n=1,2,3,\ldots$*
4	$\dfrac{1}{(p-a)^n}$	$\dfrac{t^{n-1}e^{at}}{(n-1)!}\quad n=1,2,3,\ldots$*
5	$\dfrac{1}{p^2+a^2}$	$\dfrac{\sin at}{a}$
6	$\dfrac{p}{p^2+a^2}$	$\cos at$
7	$\dfrac{1}{p^2-a^2}$	$\dfrac{\sinh at}{a}$
8	$\dfrac{p}{p^2-a^2}$	$\cosh at$
9	$\dfrac{1}{(p-b)^2+a^2}$	$\dfrac{e^{bt}\sin at}{a}$
10	$\dfrac{p}{(p-b)^2+a^2}$	$e^{bt}\left(\cos at + \dfrac{b}{a}\sin at\right)$
11	$\dfrac{1}{(p^2+a^2)^2}$	$\dfrac{\sin at - at\cos at}{2a^3}$
12	$\dfrac{p}{(p^2+a^2)^2}$	$\dfrac{t\sin at}{2a}$

* If n is not an integer, replace $(n-1)!$ by $\Gamma(n)$. Convergence requires that $n > 0$.

The Subsidiary Equation

$$(a_0 p^n + a_1 p^{n-1} + \cdots + a_n)Y(p)$$
$$= y_0(a_0 p^{n-1} + a_1 p^{n-2} + \cdots + a_{n-1})$$
$$+ y_1(a_0 p^{n-2} + a_1 p^{n-3} + \cdots + a_{n-2}) + \cdots$$
$$+ y_{n-2}(a_0 p + a_1) + y_{n-1}(a_0) + R(p)$$

EXERCISES

Set up the subsidiary equation and express $Y(p)$ in terms of partial fractions. Using the table, write down the solution.

1. $y'' + 5y' + y = 2,\ y_0 = 2,\ y_1 = 0$.
2. $y'' + 3y' + 2y = 2(t^2 + t + 1),\ y_0 = 2,\ y_1 = 0$.
3. $y' + 2y = 2t - 1,\ y_0 = 3$.
4. $y'' + y = 2\sin t,\ y_0 = 0,\ y_1 = -1$.
5. $y'' + y' = 3t^2 - 6,\ y_0 = 0,\ y_1 = 1$.

6. $y'' + 4y' + 5y = 0$, $y_0 = 2$, $y_1 = 3$.
7. $y'' - 2y' + 5y = 10$, $y_0 = y_1 = 0$.
8. $y'' - 2y' + y = 2e^t$, $y_0 = 3$, $y_1 = 1$.
9. $y^{(4)} - y''' = 0$, $y_0 = y_1 = y_2 = y_3 = 1$.
10. $y^{(4)} - y''' = 0$, $y_0 = y_1 = 2$, $y_2 = y_3 = 1$.
11. $y''' + y = 3$, $y_0 = y_1 = y_2 = 0$.
12. Show that the transform of $t \cos at$ is $\dfrac{p^2 - a^2}{(p^2 + a^2)^2}$.
13. Show that the inverse of $\dfrac{1}{p^4 - a^4}$ is $\dfrac{1}{2a^3}(\sinh at - \sin at)$.
14. Derive Formula 12 of the table from Formula 5 by a differentiation with respect to p.
15. Derive Formula 12 from Formula 6 by a differentiation with respect to a.
16. Derive Formula 11 from an earlier formula.
17. Let $f(t)$ be the step function which is equal to 1 in the intervals 0 to 1, 2 to 3, 4 to 5, etc., and is zero in the intervening intervals. Show that

$$F(p) = \frac{1}{p(1 + e^{-p})}.$$

18. A particle of mass m lb at rest at the origin is set in motion at time $t = 0$ by an impulse of k units,

$$my'' = k\delta(t)$$

where $\delta(t)$ is a very large force acting for a very short time, its integral with respect to t being unity. Show that $y = kt/m$.

19. Solve the preceding exercise if there is a restoring force proportional to y, so that

$$y'' + n^2 y = k'\delta(t) \qquad k' = k/m$$

CHAPTER 5

THE METHOD OF SUCCESSIVE APPROXIMATIONS

5-1. The Method. The method of successive approximations, which is due to Picard, furnishes a mode of attack quite unlike any the student has used hitherto in solving differential equations. Furthermore it will supply us with a proof that, under suitable conditions, a differential equation has a solution. Before passing to rigorous proofs we shall give a brief account of the method.

We shall consider the differential equation of the first order,

$$\frac{dy}{dx} = f(x,y) \tag{5-1}$$

and we shall be interested in the solution $y = Y(x)$ such that $y = b$ when $x = a$. Graphically we wish to find an integral curve which passes through the point (a,b).

If the second member of (5-1) is a function of x alone, the solution can be got by an integration. If the second member contains y, we shall make a guess at the solution, $y = y_0(x)$, and substitute into the second member. The second member is now a function of x alone and the altered equation can be solved by an integration. Call its solution $y = y_1(x)$. We have then

$$\frac{dy}{dx} = f[x,y_0(x)]$$

$$y_1(x) = b + \int_a^x f[x,y_0(x)]\, dx$$

We shall now substitute $y_1(x)$ for y in the second member of (5-1) and solve again, calling the solution $y_2(x)$. Putting $y_2(x)$ into the second member we determine y_3, and so on, the solution at the nth step being

$$y_n(x) = b + \int_a^x f[x,y_{n-1}(x)]\, dx$$

We shall prove that in a suitable interval and under very general conditions as n becomes infinite $y_n(x)$ approaches a limiting function $Y(x)$ such that $y = Y(x)$ is a solution of (5-1) and that $y = b$ when $x = a$.

We shall speak of $y_n(x)$ as the *n*th *approximation* to $Y(x)$. To illustrate the method we consider the equation

$$\frac{dy}{dx} = x - y$$

with the condition $y = 1$ when $x = 0$. The equation is easily solved by other methods. The general solution is

$$y = Ce^{-x} + x - 1$$

and the particular solution desired is

$$y = 2e^{-x} + x - 1$$

Using the method of successive approximations, we have the recurrence formula

$$y_n = 1 + \int_0^x (x - y_{n-1})\, dx$$

Making the guess $y_0(x) = 1$, we have the following approximations:

$$y_1 = 1 + \int_0^x (x - 1)\, dx = 1 - x + \frac{x^2}{2}$$

$$y_2 = 1 + \int_0^x \left[x - \left(1 - x + \frac{x^2}{2}\right)\right] dx = 1 - x + x^2 - \frac{x^3}{6}$$

$$y_3 = 1 + \int_0^x \left[x - \left(1 - x + x^2 - \frac{x^3}{6}\right)\right] dx = 1 - x + x^2 - \frac{x^3}{3} + \frac{x^4}{24}$$

$$y_4 = 1 + \int_0^x \left[x - \left(1 - x + x^2 - \frac{x^3}{3} + \frac{x^4}{24}\right)\right] dx$$

$$= 1 - x + x^2 - \frac{x^3}{3} + \frac{x^4}{12} - \frac{x^5}{120}$$

If the solution, as found by earlier methods, be expanded in powers of x, it will be found to agree with y_4 up to and including the term in x^4. How well the approximations represent the actual solution y is shown for a few values in the following table. The convergence is usually most rapid near $x = a$.

x	y_0	y_1	y_2	y_3	y_4	y
0.0	1	1	1	1	1	1
0.5	1	0.625	0.729	0.711	0.713	0.713
1.0	1	0.500	0.833	0.708	0.742	0.736
1.5	1	0.625	1.187	0.836	0.984	0.946
2.0	1	1.000	1.667	1.000	1.400	1.271

THE METHOD OF SUCCESSIVE APPROXIMATIONS

EXERCISES

In Exercises 1 to 6 solve the problem in the text with the following initial guesses, carrying the work out to y_4. In each case prepare a table similar to that in the text showing the values of y_0, y_1, and y_2.

1. $y_0 = 1 - x$.
2. $y_0 = e^{-x}$.
3. $y_0 = \cos x$.
4. $y_0 = 2e^{-x}$.
5. $y_0 = 1 + x$.
6. $y_0 = 2 - x$.

In Exercises 7 to 10, get the successive approximations up to y_3, starting with a constant y_0. Solve by other methods and compare results.

7. $y' = y$ $(y = 2, x = 0)$.
8. $y' = 1 + y^2$ $(y = 0, x = 0)$.
9. $y' = y^2$ $(y = 1, x = 0)$.
10. $y' = y^2$ $(y = 2, x = 1)$.

11. Solve

$$y' = x^2 - y^2 - 1 \qquad (y = 0, x = 0)$$

with the initial guess $y_0 = x$.

12. Solve by successive approximations

$$y'' = x - y \qquad (y = 1, y' = 0, x = 0)$$

5-2. The Successive Approximations. We shall now study the approximating curves in a closed region S of the xy plane. We shall assume that $f(x,y)$ is a single-valued continuous function of x and y in S. (The function may be a many-valued function of which we consider a single-valued branch, such as $\sqrt{x^2 + y^2}$ where we take, say, only the positive square root.) The function then is bounded in S, that is, there exists a positive constant M such that for all points of S

$$|f(x,y)| < M \qquad (5-2)$$

that is,

$$-M < f(x,y) < M$$

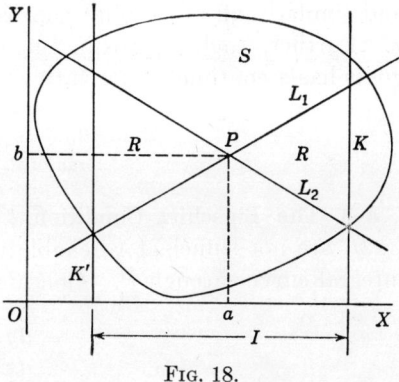

FIG. 18.

Let $P(a,b)$ be an interior point of S. We wish to investigate the existence of an integral curve through P. Through P draw straight lines L_1 and L_2 with the slopes M and $-M$, respectively (see the figure). Next draw two lines K and K' parallel to the y axis, one to the right of P and the other to the left, such that the two isosceles triangles R, formed by the lines L_1, L_2, K, and K', lie in S. K and K' cut off an interval I on the x axis, and we shall be concerned for the present only with values of x in this interval.

Consider now the recurrence formula

$$y_n(x) = b + \int_a^x f[x, y_{n-1}(x)]\, dx \qquad (5\text{-}3)$$

If we are to use this over and over, we need to know that the points on each approximating curve lie in S, so that $f(x, y_{n-1})$ is defined. This is assured by the following result.

THEOREM. *If in the interval I the curve $y = y_{n-1}(x)$ lies in S, then the following approximating curve $y = y_n(x)$ lies in R.*

We see first from (5-3) that when $x = a$, $y_n = b$; that is, $y = y_n(x)$ passes through P. If $x \neq a$, we have from (5-3)

$$|y_n - b| = \left| \int_a^x f(x, y_{n-1})\, dx \right| < M \int_a^x |dx| = M|x - a|$$

whence

$$\left| \frac{y_n - b}{x - a} \right| < M$$

We see from this that if a line be drawn from $P(a, b)$ to any point (x, y_n) on the curve $y = y_n(x)$, the slope of that line lies between $-M$ and M. It follows that the curve $y = y_n(x)$ lies in R.

If now we take for our guess $y = y_0(x)$ a continuous function of x in I such that its graph lies in S, then $y = y_1(x)$ will be a curve passing through P and lying in R and hence also in S. By induction all succeeding approximating curves will lie in R.

Since $y_0(x)$ is continuous, $f(x, y_0)$ is continuous and its integral is a continuous function of x. Hence $y_1(x)$ is a continuous function of x in I; and similarly all succeeding approximations are continuous functions of x. Further, each approximating function except possibly the initial guess has a continuous derivative, namely,

$$\frac{dy_n}{dx} = f[x, y_{n-1}(x)]$$

5-3. The Lipschitz Condition. The conditions already imposed on $f(x, y)$ are not sufficient to enable us to prove that there exists a single integral curve through P. Consider, for example, the equation

$$\frac{dy}{dx} = y^{\frac{2}{3}}$$

where P is the origin. The second member is continuous in any region enclosing the origin, yet both $y = 0$ and $y = x^3/27$ are solutions for which $y = 0$ when $x = 0$.

We shall impose a further requirement known as the *Lipschitz condition:*

THE METHOD OF SUCCESSIVE APPROXIMATIONS

There shall exist a constant A such that

$$|f(x,y) - f(x,z)| \leq A|y - z| \tag{5-4}$$

where (x,y) and (x,z) are any two points of S having the same abscissa and such that the segment of straight line joining the points lies in S.

For example, in the illustrative problem at the beginning of the chapter

$$|f(x,y) - f(x,z)| = |x - y - (x - z)| = |y - z|$$

We can take $A = 1$ in this case. In Exercise 11 we have

$$|x^2 - y^2 - 1 - (x^2 - z^2 - 1)| = |y^2 - z^2| = |y + z| \cdot |y - z|$$

If all points of S lie within a distance c of the x axis, we have $|y + z| < 2c$, and we can take $A = 2c$.

The Lipschitz condition is a weak requirement. It holds for most of the functions which the student encounters except possibly for special values of the variables. It certainly holds if $f(x,y)$ has a continuous first partial derivative with respect to y in the closed region S. For in S, $|f_y(x,y)| \leq A$, where A is some constant. Then by the law of the mean

$$f(x,y) - f(x,z) = (y - z)f_y(x,t)$$

where t lies between y and z, and hence (x,t) is in S. Then

$$|f(x,y) - f(x,z)| \leq |y - z| \cdot A$$

5-4. The Convergence. We shall now prove that, with the Lipschitz condition imposed, $y_n(x)$ approaches a limit, $Y(x)$, continuous in I, with a continuous derivative, and such that $y = Y(x)$ satisfies the differential equation (5-1) and $Y(a) = b$.

Writing $y_n(x)$ in the form

$$y_n(x) = y_1(x) + [y_2(x) - y_1(x)] + \cdots + [y_n(x) - y_{n-1}(x)]$$

we see that $y_n(x)$ approaches a limit which is equal to the sum of the infinite series

$$y_1(x) + [y_2(x) - y_1(x)] + [y_3(x) - y_2(x)] + \cdots \tag{5-5}$$

provided this series converges. We investigate its convergence.

From the recurrence formulas,

$$y_n(x) = b + \int_a^x f(x,y_{n-1}) \, dx$$

$$y_{n+1}(x) = b + \int_a^x f(x,y_n) \, dx$$

we have

$$y_{n+1}(x) - y_n(x) = \int_a^x [f(x,y_n) - f(x,y_{n-1})] \, dx$$

Then, using the Lipschitz condition,

$$|y_{n+1} - y_n| \leq \int_a^x |f(x,y_n) - f(x,y_{n-1})| \cdot |dx|$$

$$\leq A \int_a^x |y_n - y_{n-1}| \cdot |dx| \qquad (5\text{-}6)$$

Letting N be a quantity greater than $|y_2 - y_1|$ in I, we have by repeated application of (5-6)

$$|y_3 - y_2| \leq A \int_a^x |y_2 - y_1| \cdot |dx| \leq AN \int_a^x |dx| = NA|x - a|$$

$$|y_4 - y_3| \leq A \int_a^x |y_3 - y_2| \cdot |dx| \leq ANA \int_a^x |x - a| \cdot |dx|$$

$$= NA^2 \frac{|x - a|^2}{2}$$

and in general

$$|y_{n+1} - y_n| \leq NA^{n-1} \frac{|x - a|^{n-1}}{(n - 1)!}$$

the equality holding only at a. Let h be the distance from a to the more remote of the two ends of the interval I. Then

$$|x - a| \leq h$$

and

$$|y_{n+1} - y_n| < N \frac{A^{n-1}h^{n-1}}{(n - 1)!}$$

Discarding the first term of series (5-5), we see that each remaining term is less than the corresponding term of the series of constant terms

$$N + NAh + N \frac{A^2h^2}{2!} + N \frac{A^3h^3}{3!} + \cdots$$

This series is known to converge, being the series expansion of Ne^{Ah}. It follows that (5-5) converges uniformly in I.

Call the sum of (5-5) $Y(x)$. Since the terms of (5-5) are continuous functions and the series converges uniformly in I, $Y(x)$ is a continuous function of x in I.

It remains to show that $y = Y(x)$ satisfies the differential equation (5-1). It will suffice to show that

$$Y(x) = b + \int_a^x f(x,Y) \, dx \qquad (5\text{-}7)$$

Writing the terms of (5-7) in the first member and subtracting and adding the two members of (5-3), we have

$$\left| Y - b - \int_a^x f(x,Y) \, dx \right| = \left| Y - y_n + \int_a^x [f(x,y_{n-1}) - f(x,Y)] \, dx \right|$$

$$\leq |Y - y_n| + A \int_a^x |y_{n-1} - Y| \cdot |dx|$$

THE METHOD OF SUCCESSIVE APPROXIMATIONS   115

Given $\epsilon > 0$, we can find n sufficiently large that $|Y - y_n| < \epsilon$ and $|y_{n-1} - Y| < \epsilon$ in I. Then the last expression is less than

$$\epsilon + A\epsilon \int_a^x |dx| = \epsilon + A\epsilon|x - a| \leqslant \epsilon(1 + Ah)$$

Since this can be made as small as we please by taking ϵ small enough, it follows that

$$Y - b - \int_a^x f(x,Y)\, dx = 0$$

which was to be proved.

Since the second member of (5-7) has a derivative, the first member has also, and, differentiating,

$$\frac{dY}{dx} = f(x,Y)$$

The differential equation is thus satisfied. Also when $x = a$ in (5-7), $Y = b$. We note further that since $f(x,Y)$ is a continuous function of x in I, $Y(x)$ has a continuous derivative.

5-5. Uniqueness of the Solution. We shall now show that there is no solution other than $y = Y(x)$ such that $y = b$ when $x = a$. Suppose a second solution $y = Y_1(x)$; then

$$Y_1(x) = b + \int_a^x f(x,Y_1)\, dx$$

Subtracting (5-7),

$$Y_1(x) - Y(x) = \int_a^x [f(x,Y_1) - f(x,Y)]\, dx$$

and

$$|Y_1 - Y| \leqslant A \int_a^x |Y_1 - Y| \cdot |dx| \tag{5-8}$$

Let C be the maximum of the difference $|Y_1 - Y|$ in I. Then, employing (5-8) repeatedly,

$$|Y_1 - Y| \leqslant A \int_a^x C|dx| = CA|x - a|$$

$$|Y_1 - Y| \leqslant A \int_a^x CA|x - a| \cdot |dx| = CA^2 \frac{|x - a|^2}{2!}$$

and, in general,

$$|Y_1 - Y| \leqslant CA^n \frac{|x - a|^n}{n!}$$

As n increases without limit, the second member approaches zero; hence, the first member is zero, and $Y_1(x) \equiv Y(x)$. There is thus but one solution.

We can summarize the results of the preceding analysis in the following important statement:

The Existence Theorem. *In a closed region S of the xy plane let $f(x,y)$ be continuous and satisfy a Lipschitz condition*

$$|f(x,y) - f(x,z)| \leqslant A|y - z|$$

and let (a,b) be an interior point of S. Then the differential equation

$$\frac{dy}{dx} = f(x,y)$$

has a unique solution $y = Y(x)$ such that $y = b$ when $x = a$; that is to say, in a suitable interval $a - h_1 < x < a + h_2$ there exists one and only one function $Y(x)$, continuous and having a continuous derivative in the interval, such that $Y'(x) \equiv f(x,Y(x))$ in the interval and $Y(a) = b$.

5-6. Remarks. We have proved that if $f(x,y)$ is single-valued and continuous and satisfies the Lipschitz condition in S, then the method of successive approximations furnishes us with an integral curve of (5-1) through any interior point $P(a,b)$ of S, and that furthermore there is no other integral curve through P. The process converges throughout an interval I of the x axis.

The curve $y = Y(x)$ meets the line K (see Fig. 18) in a point P_1 which is in general an interior point of S, although it may lie on the boundary. If P_1 is an interior point we can apply the method of successive approximations to find an integral curve through P_1. Proceeding in this manner, the solution $y = Y(x)$ can be continued to the right so as to approach as near to the boundary of S as we wish. Similarly, the solution $y = Y(x)$ can be extended to the left of the line K'.

We see from the preceding analysis that the general solution of (5-1) involves an arbitrary constant. For a given value of x, say, $x = a$, the solution is uniquely determined by the ordinate, $y = b$; that is, the solution is a function not only of x but of the ordinate b; thus $y = Y(x,b)$, where b is a constant.

Some further facts will be stated without proof. If we assume continuity only for $f(x,y)$ there is always an integral curve through (a,b), but, as we have already shown by an example, the curve is not necessarily unique. Osgood has shown that, to the right of a, say, there are maximum and minimum solutions $y = Y(x)$ and $y = y(x) \leqslant Y(x)$, such that there is an integral curve through (a,b) and any point lying between the graphs of these two solutions.

Uniqueness of the solution can be proved under milder restrictions than the Lipschitz condition. It suffices to replace (5-4) by

$$|f(x,y) - f(x,z)| < A|y - z| \log \frac{1}{|y - z|}$$

for instance, and still other inequalities may be used.

5-7. Alteration of the Function. We shall show that a small change in the function appearing in (5-1) will result in a small change in the solution throughout the interval I. Let $Y(x)$ be the solution of (5-1) already found, and let $y(x)$ be a solution of the altered equation

$$\frac{dy}{dx} = f(x,y) + g(x,y)$$

where $g(x,y)$ is continuous and

$$|g(x,y)| < \epsilon$$

in S, both solutions having the value b at $x = a$.

Let us take $y(x)$, the solution of the second equation, for $y_0(x)$ in solving the first by successive approximations. We have

$$y_1(x) = b + \int_a^x f(x,y)\, dx$$
$$y(x) = b + \int_a^x [f(x,y) + g(x,y)]\, dx$$

whence

$$|y_1(x) - y(x)| = \left| \int_a^x g(x,y)\, dx \right| \leq \epsilon |x - a|$$

From the recurrence formula

$$y_n(x) = b + \int_a^x f(x,y_{n-1})\, dx$$

we have

$$|y_n(x) - y(x)| = \left| \int_a^x [f(x,y_{n-1}) - f(x,y) - g(x,y)]\, dx \right|$$
$$\leq A \int_a^x |y_{n-1} - y| \cdot |dx| + \epsilon |x - a|$$

Taking the inequality already found for $|y_1 - y|$ and putting $n = 2, 3, \ldots$, we find

$$|y_2 - y| \leq \epsilon |x - a| + A\epsilon \int_a^x |x - a| \cdot |dx| = \epsilon |x - a| + A\epsilon \frac{|x - a|^2}{2!}$$
$$|y_3 - y| \leq \epsilon |x - a| + \frac{A\epsilon |x - a|^2}{2!} + \frac{A^2 \epsilon |x - a|^3}{3!}$$

and, in general,

$$|y_n - y| \leq \epsilon |x - a| + \frac{A\epsilon |x - a|^2}{2!} + \cdots + \frac{A^{n-1} \epsilon |x - a|^n}{n!}$$

In the limit we have, summing the resulting series in the second member,

$$|Y(x) - y(x)| \leqslant \frac{\epsilon}{A} (e^{A|x-a|} - 1) \leqslant \frac{\epsilon}{A} (e^{Ah} - 1) \qquad (5\text{-}9)$$

We see from this that by taking ϵ small enough we can make the difference in the two solutions throughout I as small as we wish.

It will be observed that we have not imposed a Lipschitz condition on the function appearing in the altered equation. The assumed solution $y(x)$ need not be unique. If there are other solutions, they also satisfy (5-9).

5-8. Change of Initial Conditions. Consider the solution $z(x)$ of (5-1) where $z(a) = b + \eta$, given that $|\eta| < \epsilon$. We have

$$z(x) = b + \eta + \int_a^x f(x,z)\, dx$$

This can be reduced to the preceding case by putting

$$z(x) = y(x) + \eta$$

so that

$$y(x) = b + \int_a^x f(x, y + \eta)\, dx$$

From the Lipschitz condition we have

$$|f(x, y + \eta) - f(x,y)| \leqslant A|\eta| < A\epsilon$$

so that we may write

$$f(x, y + \eta) = f(x,y) + g(x,y)$$

where

$$|g(x,y)| < A\epsilon$$

The inequality (5-9) is then satisfied with ϵ replaced by $A\epsilon$:

$$|Y(x) - y(x)| \leqslant \epsilon(e^{A|x-a|} - 1)$$

and

$$|Y(x) - z(x)| < \epsilon e^{A|x-a|} \leqslant \epsilon e^{Ah}$$

By taking ϵ small enough, the two solutions can be made to differ by as little as we wish.

Again, if b is fixed but a is replaced by $a + \zeta$, where $|\zeta| < \epsilon$, the solution satisfies

$$z(x) = b + \int_{a+\zeta}^x f(x,z)\, dx = b + \eta + \int_a^x f(x,z)\, dx$$

where

$$\eta = -\int_a^{a+\zeta} f(x,z)\, dx$$

and

$$|\eta| < M|\zeta| < M\epsilon$$

THE METHOD OF SUCCESSIVE APPROXIMATIONS

We thus have the preceding case with ϵ replaced by $M\epsilon$, whence

$$|Y(x) - z(x)| < M\epsilon e^{A|x-a|} \leqslant M\epsilon e^{Ah}$$

We may combine the results of this section and the preceding into the following statement:

THEOREM. *If the function in (5-1) and the initial point (a,b) vary continuously, then the solution varies continuously.*

Thus, if the function and the quantities a and b are continuous functions of a parameter λ then the solution $y(x,\lambda)$ is also a continuous function of λ.

EXERCISES

1. Let the graph of $y_0(x)$ lie entirely above or entirely below that of $Y(x)$. Prove that the approximating curves to the right of P are all on the same side of the integral curve if $\partial f/\partial y > 0$, and alternately above and below if $\partial f/\partial y < 0$. What is the situation to the left of P?

2. If $|y_0 - Y| < C$, show that

$$|y_n - Y| \leqslant CA^n \frac{|x-a|^n}{n!}$$

3. Derive a bound for $|y_n - Y|$ in terms of N, where $|y_1 - y_0| < N$ in I.

4. Illustrating Osgood's theorem, show how to draw an integral curve of $y' = y^{\frac{2}{3}}$ through the origin and a point lying between the integral curves $y = x^3/27$ and $y = 0$.

CHAPTER 6

SYSTEMS OF ORDINARY EQUATIONS

6-1. Equations of the First Order. We shall consider first a system of n equations involving one independent variable, u dependent variables, and the n first derivatives of the latter. We suppose the equations solved for these derivatives, so that we have a system of the form

$$\frac{dy_1}{dx} = f_1(x, y_1, \ldots, y_n)$$
$$\frac{dy_2}{dx} = f_2(x, y_1, \ldots, y_n)$$
$$\cdots\cdots\cdots\cdots\cdots$$
$$\frac{dy_n}{dx} = f_n(x, y_1, \ldots, y_n)$$
(6-1)

If we have n functions

$$y_1 = y_1(x), \ldots, y_n = y_n(x) \quad (6\text{-}2)$$

which satisfy (6-1) identically in x, they constitute a solution of the system. The solution often appears, however, as a system of n equations in $x, y_1, \ldots, y_n$, which give the functions (6-2) in implicit form.

For $x = a$, let values $y_1 = b_1, \ldots, y_n = b_n$ be prescribed. Under suitable conditions we prove the existence of functions satisfying the equations and taking on the prescribed values when $x = a$.

THEOREM. *Let the functions f_i be continuous in the region S, defined as follows,*

$$|x - a| \leqslant k \quad |y_1 - b_1| \leqslant k_1, \ldots \quad |y_n - b_n| \leqslant k_n \quad (6\text{-}3)$$

and in S let the functions satisfy a Lipschitz condition of the form

$$|f_i(x, y_1, \ldots, y_n) - f_i(x, z_1, \ldots, z_n)| \leqslant A_1 |y_1 - z_1|$$
$$+ \cdots + A_n |y_n - z_n| \quad (6\text{-}4)$$

then in a suitable interval I, $|x - a| \leqslant h$, there exists a unique set of n functions

$$y_1(x), y_2(x), \ldots, y_n(x) \quad (6\text{-}5)$$

continuous and having continuous derivatives in I, which satisfy the differential equations (6-1) *identically and which have the property that*

$$y_1(a) = b_1, \ldots, y_n(a) = b_n \tag{6-6}$$

The proof of this theorem follows the lines of the existence theorem for the differential equation of the first order and will be sketched very briefly here.

Let M be the maximum absolute value of the n functions f_i in S, and let h be the least of the quantities $k, k_1/M, \ldots, k_n/M$.

We proceed by the method of successive approximations. Setting $y_1 = b_1, \ldots, y_n = b_n$ (or other suitable functions might be used) in the second members of (6-1), we determine values $y_{11}(x), \ldots, y_{n1}(x)$ taking on the prescribed values at a by integration. We put these new values into the second members of (6-1) and determine $y_{12}(x), \ldots, y_{n2}(x)$ by integration; and so on. We use in this process the recurrence formulas

$$y_{1m}(x) = b_1 + \int_a^x f_1(x, y_{1\,m-1}, \ldots, y_{n\,m-1})\, dx$$
$$\cdots\cdots\cdots\cdots\cdots\cdots\cdots\cdots\cdots\cdots\cdots\cdots \tag{6-7}$$
$$y_{nm}(x) = b_n + \int_a^x f_n(x, y_{1\,m-1}, \ldots, y_{n\,m-1})\, dx$$

We note first that for x in I the set $(x, y_{1m}, \ldots, y_{nm})$ lies in S. Assume that $(x, y_{1\,m-1}, \ldots, y_{n\,m-1})$ lies in S, then for any equation of (6-7)

$$|y_{im} - b_i| \leqslant \left| \int_a^x |f_i(x, y_{1\,m-1}, \ldots, y_{n\,m-1})\, dx \right|$$
$$\leqslant M|x - a| \leqslant Mh \leqslant k_i \tag{6-8}$$

Hence $(x, y_{1m}, \ldots, y_{nm})$ is in S. Since the first values $(x, b_1, \ldots, b_n)$ are in S, the result follows by induction.

For any equation of (6-7) we have, using m and $m + 1$ and subtracting,

$$|y_{i\,m+1} - y_{im}| \leqslant \left| \int_a^x [f_i(x, y_{1m}, \ldots, y_{nm}) - f_i(x, y_{1\,m-1}, \ldots, y_{n\,m-1})]\, dx \right|$$
$$\leqslant A \left| \int_a^x [|y_{1m} - y_{1\,m-1}| + \cdots + |y_{nm} - y_{n\,m-1}|]\, |dx| \right| \tag{6-9}$$

where A is the greatest of the quantities $A_1, \ldots, A_n$ in (6-4).

Now let

$$|y_{i1} - b_i| \leqslant C$$

in I. Using this for each difference under the integral sign we have from (9), with $m = 1$,

$$|y_{i2} - y_{i1}| \leqslant CnA \int_a^x |dx| = CnA|x - a|$$

Using these values in (6-9), we have

$$|y_{i3} - y_{i2}| \leqslant Cn^2A^2 \int_a^x |x - a| \cdot |dx| = \frac{Cn^2A^2}{2!} |x - a|^2$$

and, in general,

$$|y_{i\,m+1} - y_{im}| \leqslant \frac{Cn^mA^m}{m!} |x - a|^m \leqslant \frac{Cn^mA^mh^m}{m!} \tag{6-10}$$

From the convergence of the series

$$C + CnAh + C\frac{n^2A^2h^2}{2!} + \cdots = Ce^{nAh}$$

(6-10) establishes the uniform convergence of the series

$$y_i(x) = b_i + (y_{i1} - b_i) + (y_{i2} - y_{i1}) + \cdots \tag{6-11}$$

in I and the continuity of its sum $y_i(x)$.

The facts that the resulting functions $y_1(x), \ldots, y_n(x)$ are differentiable, that they satisfy Eqs. (6-1), that they take on the prescribed values $b_1, \ldots, b_n$ at $x = a$, and that they are unique, are established by repeating the reasoning of the previous chapter. There is no gain to the reader in repeating it here and it will be left to him to check the statements.

Since $b_1, \ldots, b_n$ are arbitrary there is an n-parameter family of solutions.

The Lipschitz condition is a weak restriction. It holds if the functions f_i have continuous first partial derivatives with respect to the variables y_k.

It can be shown also that slight changes in the functions appearing in (6-1) or slight changes in the initial values b_k or in a produce slight alterations in the solutions (6-5).

6-2. Linear Systems. If the second members of (6-1) are linear in the dependent variables,

$$\frac{dy_1}{dx} = a_{11}(x)y_1 + a_{12}(x)y_2 + \cdots + a_{1n}(x)y_n + b_1(x)$$

$$\cdots \cdots \cdots \cdots \cdots \cdots \cdots \cdots \cdots \cdots \cdots \cdots \tag{6-12}$$

$$\frac{dy_n}{dx} = a_{n1}(x)y_1 + a_{n2}(x)y_2 + \cdots + a_{nn}(x)y_n + b_n(x)$$

more precise results may be stated.

THEOREM. *In an interval $A < x < B$, in which the coefficients*

$$a_{11}(x), \ldots, a_{nn}(x), b_1(x), \ldots, b_n(x)$$

are continuous, there is a unique set of functions

$$y_1(x), \ldots, y_n(x)$$

continuous and having continuous derivatives in AB which satisfy (6-12) *and have prescribed values*

$$y_1(a) = b_1, \ldots, y_n(a) = b_n$$

at a point a of AB.

It suffices to prove the theorem in a slightly smaller closed interval $A < A' \leqslant x \leqslant B' < B$, in which the coefficients are bounded. We observe that the Lipschitz condition (6-4) is satisfied if A_k is an upper bound of $|a_{1k}(x)|, \ldots, |a_{nk}(x)|$ in $A'B'$.

The second members of (6-12) exist for x in $A'B'$ whatever the y's are. Thus the successive approximations exist, and the problem is one of convergence only. The inequality (6-10) holds for x in $A'B'$, where h is the distance from a to the more remote of the points A' and B'; and the proof is complete.

The linear system thus has the important property that its solutions can have singularities only at the fixed points at which the coefficients have singularities. In the general case (6-1), the position of a singularity of $y_i(x)$ is usually quite unpredictable. For example, in a very simple case,

$$\frac{dy}{dx} = y^2 \qquad y(x) = \frac{b}{1 - bx}$$

we have written the solution with the condition $y(0) = b$. Here, although y^2 is continuous everywhere, the solution has a singularity $x = 1/b$ which varies with the initial conditions.

EXERCISES

1. Call the system *reduced* if $b_1(x), \ldots, b_n(x)$ are canceled from (6-12). Show that if $z_1(x), \ldots, z_n(x)$ satisfy the complete equation (6-12) and $y_1(x), \ldots, y_n(x)$ satisfy the reduced equation, then $y_1(x) + z_1(x), \ldots, y_n(x) + z_n(x)$ satisfy the complete equation.

2. Given k sets of solutions of the reduced equation, show that a suitable linear combination with arbitrary constants is also a solution.

3. Show that if $y_1(x), \ldots, y_n(x)$, solutions of (6-12) reduced, all vanish at a point of AB, they are all identically zero.

4. If the coefficients in (6-12) are constants show how to find solutions of the reduced equation of the form

$$y_1 = c_1 e^{rx}, \ldots, y_n = c_n e^{rx}$$

5. Show that $y_2, \ldots, y_n$ may be eliminated from (6-12) giving a linear equation which y_1 must satisfy.

6. Solve by first eliminating one of the variables:

$$\frac{dy}{dx} = y - \frac{2}{3}z + x$$
$$\frac{dz}{dx} = \frac{4}{3}y - z + 1$$

7. Solve the preceding by successive approximations out to y_3 and z_3 with the initial conditions $x = 0$, $y = 0$, $z = 0$.

8. Find in series form to terms in x^3 the solutions of

$$\frac{dy}{dx} = x^2 + y + z$$
$$\frac{dz}{dx} = y + u + 2z$$
$$\frac{du}{dx} = 2x + y^2$$

such that $y = 0$, $z = 1$, $u = 1$, when $x = 0$.

6-3. Geometrical Interpretations. The set of functions (6-5) which satisfy (6-1) may be presented in geometric guise in various ways. The most obvious is the following. In the xy plane let the points $y = b_1, \ldots, b_n$ be marked off on the line $x = a$ (Fig. 19). Then through such a set of selected points, if the conditions of the theorem are satisfied for these points, pass the graphs of a set of solutions (6-5).

Equations (6-1) give us, when we set $x = a$, $y_1 = b_1$, etc., the slopes of the graphs at the points selected. On differentiating (6-1), if the functions f_i are differentiable, we get the second and higher derivatives of the solution at $x = a$.

An instructive interpretation results from considering the quantities $x, y_1, \ldots, y_n$ as the coordinates of a point in $(n + 1)$-dimensional space. For simplicity we illustrate with two equations:

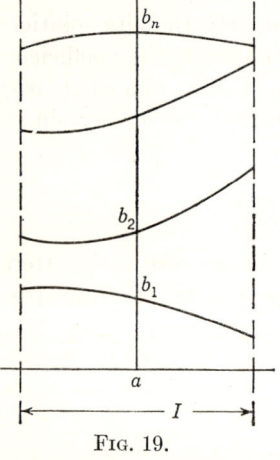

Fig. 19.

$$\frac{dy}{dx} = f_1(x,y,z)$$
$$\frac{dz}{dx} = f_2(x,y,z) \tag{6-13}$$

with the conditions $y = b$, $z = c$, at $x = a$. The solutions $y = y(x)$, $z = z(x)$ are a pair of cylinders (Fig. 20), the first with elements parallel to the z axis, the second with elements parallel to the y axis. These cylinders intersect in a curve C through (a,b,c). C is the geometrical picture of the solution.

Writing (6-13) in the form

$$\frac{dx}{1} = \frac{dy}{f_1} = \frac{dz}{f_2}$$

there is determined, at each point of space at which the functions are defined a lineal element with direction ratios

$$1, f_1, f_2$$

We thus have a three-dimensional direction field. An integral curve, such as C, is a curve tangent at each of its points to the lineal element at that point.

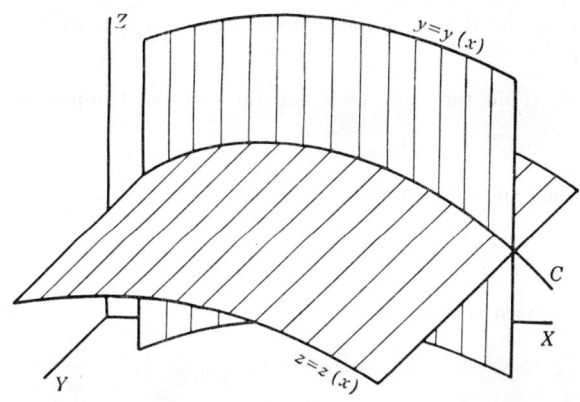

Fig. 20.

A third geometrical interpretation results if we consider $y_1, \ldots, y_n$ as the coordinates of a point in n-dimensional space and consider x as a parameter. We illustrate with (6-13), giving it, however, a more familiar appearance by using t for the independent variable and x and y for the dependent variables:

$$\frac{dx}{dt} = f_1(t,x,y)$$
$$\frac{dy}{dt} = f_2(t,x,y) \tag{6-14}$$

A solution $x = x(t)$, $y = y(t)$ is a curve in the xy plane in parametric form. Through the point (b,c) pass in general an infinite number of integral curves, one for each initial value of t. An exception occurs if the slope $dy/dx = f_2/f_1$ is independent of t.

We get a mechanical interpretation of (6-14) if we consider t as the time. Then f_1 and f_2 are the components of the velocity in the x and y directions, respectively. The integral curve is then the path of a particle moving in the plane. The solutions of (6-1) for $n = 3$ may be interpreted as the paths of particles moving in three-dimensional space; and so on.

It is often found convenient to introduce a parameter into the work. Thus the equation of the first order,

$$M\,dx + N\,dy = 0$$

may be written

$$\frac{dx}{N} = \frac{dy}{-M} = dt$$

or

$$\frac{dx}{dt} = N(x,y) \qquad \frac{dy}{dt} = -M(x,y)$$

Here the slope is independent of t. Much of the fundamental work on the equation of the first order has been based on this method of writing the equation.

Example 6-1. Find the differential equations of the two-parameter family of parabolas

$$y = ax^2 \qquad x + y + z = b$$

We differentiate and eliminate the parameters,

$$\frac{dy}{dx} = 2ax = \frac{2y}{x} \qquad 1 + \frac{dy}{dx} + \frac{dz}{dx} = 0$$

These may be written in the form

$$\frac{dx}{x} = \frac{dy}{2y} = \frac{dz}{-(x + 2y)}$$

Curves Orthogonal to Surfaces. The curves at right angles to a one-parameter family of surfaces appear in many applications. The lines of force are orthogonal to the equipotential surfaces in electric, magnetic, and gravitational fields; the lines of heat flow are orthogonal to the isothermal surfaces; and so on.

The differential equations of the orthogonal curves are readily found. Let

$$F(x,y,z) = C$$

be the family of surfaces. The normal at each point has as direction ratios the partial derivatives F_x, F_y, F_z. The curves we seek, with ratios dx, dy, dz, must have these same directions; hence

$$\frac{dx}{F_x} = \frac{dy}{F_y} = \frac{dz}{F_z}$$

If the constant has not been isolated so that it disappears in differentiating, it must be eliminated from the partial derivatives.

We shall find in a later section that if, conversely, a two-parameter family of curves is given, there is usually no one-parameter family of surfaces orthogonal to them.

Example 6-2. Given that the temperature u in a body is

$$u = k(e^{-y} \sin x + z)$$

find the differential equations of the lines of heat flow.

The isothermal surfaces are $u = C$. We have at once

$$\frac{dx}{e^{-y}\cos x} = \frac{dy}{-e^{-y}\sin x} = \frac{dz}{1}$$

EXERCISES

Find the differential equations of the following families of curves:
1. $z = ax^2$, $z = by^2$.
2. $x^2 + y^2 + z^2 = a^2$, $2x + 3y + z = b$.
3. All straight lines through the point (a,b,c).
4. All circles through the origin, orthogonal to the xy plane.
5. The family $y = ae^x + 5be^{4x}$, $z = ae^x + 2be^{4x}$.

Find the differential equations of the curves orthogonal to
6. All spheres with center at the origin.
7. The paraboloids $z = x^2 + Cy^2$.
8. All unit spheres with centers on the x axis.

9. Show that the equations of the lines meeting the x axis and the line $x = 0$, $z = 1$ are

$$\frac{dx}{xz} = \frac{dy}{y(z-1)} = \frac{dz}{z^2 - z}$$

10. If the equipotential surfaces are the cylinders

$$x^2 + y^2 + Cy = 1$$

find the differential equations of the lines of force.

6-4. Elementary Methods of Solution. In simple cases we can often solve a system of the form (6-1) by combining the equations in various ways. We illustrate with Eq. (6-13), the extension of the methods to a greater number of equations being easily made. For greater symmetry we write

$$\frac{dx}{P} = \frac{dy}{Q} = \frac{dz}{R} \qquad (6\text{-}15)$$

where P, Q, R are functions of x, y, z.

The integral curves of (6-15) are a two-parameter family of curves in three-dimensional space. If we can get from (6-15) two relations connecting the variables, we shall have two surfaces whose curve of intersection is an integral curve. If these relations contain two arbitrary constants, there results, by varying these constants, a two-parameter family of integral curves.

First Method. It may happen that one variable is missing, or can be canceled, from one equation of (6-15). Suppose, for instance, that z does not appear in, or can be canceled from, P and Q. We have an equation in two variables $dy/dx = Q/P$, whose solution, $g(x,y,a) = 0$, gives one of the desired relations.

If a second variable can be removed from another equation we get a second relation, and the system is solved; or the relation already found may be used to eliminate one of the variables, yielding an equation in two variables to be solved.

Example. Find the lines of heat flow in Example 6-2.
We have to solve
$$\frac{dx}{e^{-y}\cos x} = \frac{dy}{-e^{-y}\sin x} = dz$$

From the first and second members,
$$-\tan x\, dx = dy$$
whence, integrating,
$$\log \cos x = y + C \quad \text{or} \quad e^{-y}\cos x = a$$

From the first and third members we now have
$$dx = a\, dz \qquad x = az + b$$

The lines of flow are the plane curves
$$e^{-y}\cos x = a \qquad x = az + b$$

Second Method. Useful equations resulting from (6-15) may be written down as follows. Let λ_i, μ_i, ν_i be any functions; then
$$\frac{dx}{P} = \frac{dy}{Q} = \frac{dz}{R} = \frac{\lambda_1\, dx + \mu_1\, dy + \nu_1\, dz}{\lambda_1 P + \mu_1 Q + \nu_1 R}$$
$$= \frac{\lambda_2\, dx + \mu_2\, dy + \nu_2\, dz}{\lambda_2 P + \mu_2 Q + \nu_2 R} = \cdots \qquad (6\text{-}16)$$

The multipliers may be chosen in infinitely many ways, so that
$$\lambda_i P + \mu_i Q + \nu_i R = 0$$
then
$$\lambda_i\, dx + \mu_i\, dy + \nu_i\, dz = 0$$

If it happens that the first member of this equation is the exact differential of some function $u(x,y,z)$, then we have at once one relation
$$u(x,y,z) = a$$

Example.
$$\frac{dx}{mz - ny} = \frac{dy}{nx - lz} = \frac{dz}{ly - mx}$$

Using first the multipliers l, m, n and then x, y, z, the members above are equal to
$$\frac{l\, dx + m\, dy + n\, dz}{0} = \frac{x\, dx + y\, dy + z\, dz}{0}$$

Setting the numerators equal to zero and integrating, we have
$$lx + my + nz = a$$
$$x^2 + y^2 + z^2 = b$$

The first equation represents a family of parallel planes; the second, a family of spheres with center at the origin. These families intersect in the integral curves. The integral curves then consist of all circles with centers on a certain line L through the origin (with direction ratios l, m, n) which lie in planes perpendicular to L. The circles are like the pieces got by slicing an onion.

Third Method. Equations of the form (6-16) may be used in many ways. We mention one further use. We may be able to choose the multipliers so that two members of (6-16) are exact differentials:

$$\frac{\lambda_1\, dx + \mu_1\, dy + \nu_1\, dz}{\lambda_1 P + \mu_1 Q + \nu_1 R} = d\Omega_1$$

$$\frac{\lambda_2\, dx + \mu_2\, dy + \nu_2\, dz}{\lambda_2 P + \mu_2 Q + \nu_2 R} = d\Omega_2$$

If so, we have the relation

$$\Omega_1 = \Omega_2 + C_1$$

Example. The Linear Case.

$$\frac{dx}{a_1 x + b_1 y + c_1 z + d_1} = \frac{dy}{a_2 x + b_2 y + c_2 z + d_2} = \frac{dz}{a_3 x + b_3 y + c_3 z + d_3}$$

The preceding are equal to

$$\frac{\lambda\, dx + \mu\, dy + \nu\, dz}{(\lambda a_1 + \mu a_2 + \nu a_3)x + (\lambda b_1 + \mu b_2 + \nu b_3)y + (\lambda c_1 + \mu c_2 + \nu c_3)z + \lambda d_1 + \mu d_2 + \nu d_3}$$

Using constant multipliers, we shall have an exact differential if the numerator is a constant multiple of the differential of the denominator:

$$\lambda a_1 + \mu a_2 + \nu a_3 = \rho\lambda$$
$$\lambda b_1 + \mu b_2 + \nu b_3 = \rho\mu$$
$$\lambda c_1 + \lambda c_2 + \nu c_3 = \rho\nu$$

These homogeneous linear equations in the multipliers may be solved if and only if the determinant vanishes,

$$\begin{vmatrix} a_1 - \rho & a_2 & a_3 \\ b_1 & b_2 - \rho & b_3 \\ c_1 & c_2 & c_3 - \rho \end{vmatrix} = 0$$

For each of the roots ρ_1, ρ_2, ρ_3 of this cubic in ρ the three equations for the multipliers may be solved. Calling the resulting denominators Ω_1, Ω_2, Ω_3, we have the equations

$$\frac{1}{\rho_1} d\log\Omega_1 = \frac{1}{\rho_2} d\log\Omega_2 = \frac{1}{\rho_3} d\log\Omega_3$$

and the solutions
$$\Omega_1^{1/\rho_1} = C_1 \Omega_2^{1/\rho_2} = C_2 \Omega_3^{1/\rho_3}$$

This treatment requires modification if two or more roots of the determinantal equation are equal. The solutions here derived involve imaginaries if there are complex roots.

A solution in parametric form results if we set each of the members equal to dt We have

$$\frac{1}{\rho_1} d \log \Omega_1 = dt$$

$$\Omega_1 = C_1 e^{\rho_1 t}$$

with similar expressions for Ω_2 and Ω_3.

EXERCISES

Solve:

1. $\dfrac{dx}{x} = \dfrac{dy}{y} = \dfrac{dz}{z}$.

2. $\dfrac{dx}{y} = \dfrac{dy}{-x} = \dfrac{dz}{k}$.

3. $\dfrac{dx}{y+z} = \dfrac{dy}{z+x} = \dfrac{dz}{x+y}$.

4. $\dfrac{dx}{xz} = \dfrac{dy}{yz} = \dfrac{dz}{4xy}$.

5. $\dfrac{dx}{x^2} = \dfrac{dy}{z^2} = \dfrac{dz}{-yz}$.

6. $\dfrac{dx}{z(y-x)} = \dfrac{dy}{z(y+x)} = \dfrac{dz}{x^2+y^2}$.

7. $\dfrac{dy}{dx} + 1 = 0, \quad \dfrac{dz}{dx} = x \sin(x+y)$.

8. $z\,dx + x\,dy + y\,dz = 0, \quad dx + dy - 2\,dz = 0$.

9. By the use of suitable multipliers solve

$$\frac{dx}{x^2 + y^2} = \frac{dy}{2xy}$$

Find the curves orthogonal to the following one-parameter families of surfaces:

10. The planes $y = Cx$.
11. The planes $z = ax + by + C$.
12. The paraboloids $z = C(x^2 + y^2)$.
13. All spheres touching the xy plane at the origin.

14. If the temperature is
$$u = k(2xy + z^2)$$
find the lines of flow.

15. Find the two most general functions of x such that each is the derivative of the other.

16. Find the two most general functions such that the derivative of each is the reciprocal of the other.

17. Find the equation of the surface formed by integral curves of Exercise 1 through the circle $x^2 + y^2 = 1, z = 1$.

18. Find the surface formed by integral curves of

$$\frac{dx}{yz} = \frac{dy}{xz} = \frac{dz}{4xy}$$

through the ellipse $4y^2 + z^2 = 1, x = 0$.

SYSTEMS OF ORDINARY EQUATIONS   131

19. Show that the members of Eqs. (6-16) are equal to $F(dx,dy,dz)/F(P,Q,R)$ if F is homogeneous of degree 1 in its three arguments.

20. Show that the members of (6-16) are equal to

(a) $\dfrac{ds}{\sqrt{P^2 + Q^2 + R^2}}$ (b) $\left(\dfrac{dx\,dy\,dz}{PQR}\right)^{1/3}$ (c) $\dfrac{(ds)^2}{P\,dx + Q\,dy + R\,dz}$

6-5. Two Applications. *A. Problem of the Flexible Cord.* Let a cord, loaded in some way, be hanging from two points of support. We suppose the cord is horizontal at A and we take the y axis vertically there. Let $P(x,y)$ be a point on the cord. The portion AP is acted on by three forces: the horizontal tension H at A; the tension T along the cord at P;

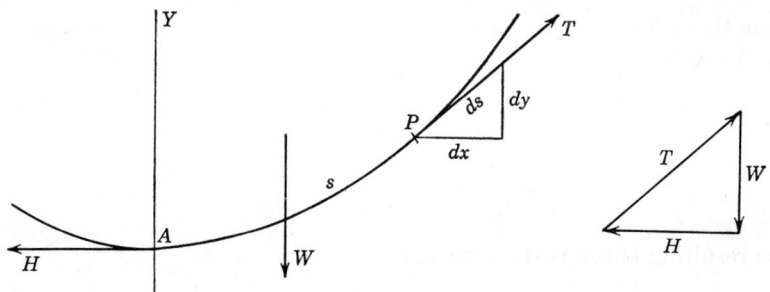

Fig. 21.

and the weight W borne by AP. Since there is equilibrium, these forces form a closed triangle and $T = \sqrt{W^2 + H^2}$. From the similarity of the force triangle and the differential triangle, we have

$$\frac{dx}{H} = \frac{dy}{W} = \frac{ds}{T}$$

where s is the arc length AP. The solution of the equations requires a knowledge of the character of the loading.

Cord Hanging under Own Weight. Let w equal the weight per unit length, then $W = ws$. We put $H = wc$, and take the origin c units below A. The equations are

$$\frac{dx}{c} = \frac{dy}{s} = \frac{ds}{\sqrt{s^2 + c^2}}$$

From the second and third members we find

$$y = \sqrt{s^2 + c^2} + K$$

and, since $y = c$ when $s = 0$, we have $K = 0$.

The system may be written—adding two members by the use of suitable multipliers—

$$\frac{dx}{c} = \frac{dy}{s} = \frac{ds}{y} = \frac{dy+ds}{y+s} = \frac{dy-ds}{-y+s}$$

From the first and fourth members,

$$x/c = \log(y+s) + C_1 \qquad y+s = k_1 e^{x/c}$$

and from the first and fifth,

$$x/c = -\log(y-s) + C_2 \qquad y-s = k_2 e^{-x/c}$$

From the values $x = 0$, $s = 0$, $y = c$ we find $k_1 = k_2 = c$. We then have y and s in terms of x,

$$y = \frac{c}{2}[e^{x/c} + e^{-x/c}] = c \cosh \frac{x}{c}$$

$$s = \frac{c}{2}[e^{x/c} - e^{-x/c}] = c \sinh \frac{x}{c}$$

The resulting curve is the catenary.

EXERCISES

Find the curve in which the cord hangs in the following cases:
1. The load is uniformly distributed along the horizontal, $W = kx$. This is the suspension-bridge problem.
2. The cord is of varying thickness so that AP has the weight $k\sqrt{s}$.
3. The cord, of negligible weight, supports a series of vertical rods forming a portiere. The rods are close together, equally spaced horizontally, and their lower ends lie in a horizontal line.

B. A Parasite Problem. The following illustrates the use of differential equations in biological problems.[1]

A certain parasite is born from an egg deposited in the host, the latter being thereby killed. It is required to find the way in which the host and parasite populations vary with the time.

We make our assumptions as simple as possible. Let x and y be the numbers of hosts and parasites, respectively, at the time t. The number of eggs deposited in unit time depends on the probability of parasites and hosts coming together and is taken to be proportional to the product of the populations, that is, kxy. Let b be the birth rate per head of the host species and c its death rate when no parasites are present, and let s be the death rate of the parasites.

[1] See Lotka, "Elements of Physical Biology."

The changes in the populations in a small time Δt are approximately

$$\Delta x = bx\,\Delta t - cx\,\Delta t - kxy\,\Delta t$$
$$\Delta y = kxy\,\Delta t - sy\,\Delta t$$

The terms in the second member of the first equation may be identified as the number born, the number dying from causes other than the birth of parasites, and the number dying from the hatching of parasites' eggs during the period. The terms in the second equation give the number of parasites born and dying.

Dividing by Δt and letting Δt approach zero, we have, putting $b - c = r$,

$$\frac{dx}{dt} = rx - kxy$$
$$\frac{dy}{dt} = kxy - sy$$

It is assumed that $b > c$, so that $r > 0$. The study of the equations is left for the exercises below. It may be stated that actual cases agree fairly well with theoretical results.

EXERCISES

1. Find populations x and y which remain stationary.

Show that if the origin be translated to this equilibrium point, the equations have the form

$$\frac{dX}{dt} = -Y(s + kX)$$
$$\frac{dY}{dt} = X(r + kY)$$

2. If, near the equilibrium point, the terms in XY be discarded, show that the point (X,Y) moves about the point in an ellipse. Show that the time to pass through a cycle is the same for all the ellipses.

3. Solve the problem without discarding terms.

Discuss the problem with the following variations in the conditions:

4. The host population is stationary if no parasites are present, that is, the birth rate is equal to the death rate.

5. The death rate of the host exceeds the birth rate when no parasites are present.

6. A certain fraction of the parasites' eggs fail to hatch, although they kill the host.

7. None of the eggs hatches.

8. The host population is stationary when no parasites are present and, for a limited time, the death rate of the parasites is negligible.

6-6. The Equation of the nth Order. By a simple device we can prove the existence theorem for the equation of the nth order,

$$\frac{d^n y}{dx^n} = f\left(x, y, \frac{dy}{dx}, \frac{d^2 y}{dx^2}, \ldots, \frac{d^{n-1} y}{dx^{n-1}}\right) \qquad (6\text{-}17)$$

and determine the number of arbitrary constants involved in its solution. Here the nth derivative is expressed as a function of x, y and the derivatives of y of order less than n.

THEOREM. *Let f be a continuous function of its arguments when they range over the set of values*

$$|x - a| \leqslant k, \ |y - b| \leqslant k_0, \ |y' - b_1| \leqslant k_1, \ \ldots, \ |y^{(n-1)} - b_{n-1}| \leqslant k_{n-1}$$

and for this range of values let f satisfy a Lipschitz condition of the form

$$|f(x,y_1,y_1', \ldots, y_1^{(n-1)}) - f(x,y_2,y_2', \ldots, y_2^{(n-1)})|$$
$$\leqslant A|y_1 - y_2| + A_1|y_1' - y_2'| + \cdots + A_{n-1}|y_1^{(n-1)} - y_2^{(n-1)}|$$

then in a suitable interval I, $|x - a| \leqslant h$, there exists a unique function $y(x)$, having a continuous derivative of the nth order, which satisfies the differential equation (6-17) *and takes on at $x = a$ the values*

$$y = b \quad \frac{dy}{dx} = b_1 \quad \frac{d^2y}{dx^2} = b_2 \quad \cdots \quad \frac{d^{n-1}y}{dx^{n-1}} = b_{n-1}$$

By introducing auxiliary variables $z_1, z_2, \ldots, z_{n-1}$ for the first $n - 1$ derivatives we reduce the problem directly to the solution of a system of equations of the form (6-1):

$$\frac{dy}{dx} = z_1 \quad \frac{dz_1}{dx} = z_2 \ \cdots$$

$$\frac{dz_{n-2}}{dx} = z_{n-1} \quad \frac{dz_{n-1}}{dx} = f(x,y,z_1, \ldots, z_{n-1})$$

The functions $z_1, \ldots, z_{n-1}$ appearing in the second members obviously satisfy a Lipschitz condition. The existence theorem for the system (6-1) then applies to prove the existence of a unique set of solutions

$$y(x), z_1(x), \ldots, z_{n-1}(x)$$

in an interval about $x = a$ which take on the values

$$b, b_1, \ldots, b_{n-1}$$

at a. Here $z_k(a) = y^{(k)}(a) = b_k$, and the theorem is established.

Since $b, b_1, \ldots, b_{n-1}$ may be chosen arbitrarily, the solution involves n arbitrary constants.

If the equation is linear,

$$\frac{d^ny}{dx^n} + P_1(x)\frac{d^{n-1}y}{dx^{n-1}} + \cdots + P_n(x)y = R(x) \tag{6-18}$$

the second members of the preceding set of equations are linear in y, $z_1, \ldots, z_{n-1}$. The treatment of Sec. 6-2 applies, and we have the following existence theorem:

THEOREM. *In an interval $A < x < B$ in which $P_1(x)$, ..., $P_n(x)$, $R(x)$ are continuous, there exists one and only one solution $y = y(x)$ of (6-18), continuous together with its first n derivatives and such that $y(x)$ and its first $n - 1$ derivatives take on prescribed values at a point a of the interval.*

6-7. Systems of Equations of Higher Order. By the method of the preceding section we can readily prove the existence theorem for systems of ordinary equations. We omit the statement of the theorem, which would be somewhat long, but illustrate the method by an example.

Given the system

$$\frac{d^3y}{dx^3} = f_1\left(x, y, \frac{dy}{dx}, \frac{d^2y}{dx^2}, z, \frac{dz}{dx}\right)$$

$$\frac{d^2z}{dx^2} = f_2\left(x, y, \frac{dy}{dx}, \frac{d^2y}{dx^2}, z, \frac{dz}{dx}\right)$$

This may be written as a system of five equations of the first order:

$$\frac{dy}{dx} = s_1 \qquad \frac{ds_1}{dx} = s_2 \qquad \frac{dz}{dx} = t$$

$$\frac{ds_2}{dx} = f_1(x,y,s_1,s_2,z,t) \qquad \frac{dt}{dx} = f_2(x,y,s_1,s_2,z,t)$$

If now f_1 and f_2 are continuous and satisfy a Lipschitz condition in a range of values about $x = a$, $y = b$, $s_1 = b_1$, $s_2 = b_2$, $z = c$, $t = c_1$, the existence theorem for the system (6-1) demonstrates the existence of a unique set of solutions in an interval about $x = a$, which take on the prescribed values.

This proves for the original pair of equations the existence of unique solutions $y(x)$, $z(x)$ in an interval about $x = a$ such that at $x = a$ the values

$$y = b \qquad \frac{dy}{dx} = b_1 \qquad \frac{d^2y}{dx^2} = b_2 \qquad z = c \qquad \frac{dz}{dx} = c_1$$

are taken on. The solution of the system thus involves five arbitrary constants—the sum of the orders of the highest derivatives of y and z that appear.

6-8. Total Differential Equations. If the number of dependent variables is greater than the number of equations, there will be ordinarily a great variety of solutions. Thus, if we are given only the first equation of the preceding pair,

$$\frac{d^3y}{dx^3} = f_1\left(x, y, \frac{dy}{dx}, \frac{d^2y}{dx^2}, z, \frac{dz}{dx}\right)$$

we can take $z(x)$ arbitrarily and determine $y(x)$ therefrom.

Of this character is the *total differential equation*, or *Pfaffian*, in three variables,

$$P(x,y,z) + Q(x,y,z)\frac{dy}{dx} + R(x,y,z)\frac{dz}{dx} = 0$$

or, in symmetric form,

$$P\,dx + Q\,dy + R\,dz = 0 \qquad (6\text{-}19)$$

The functions $y = y(x)$, $z = z(x)$ constitute a solution if they reduce the equation to an identity in x.

Geometrically the solution is a curve in three-dimensional space, and the solution is usually obtained as two equations in the variables. Equation (6-19) shows that the tangent to the integral curve through a point

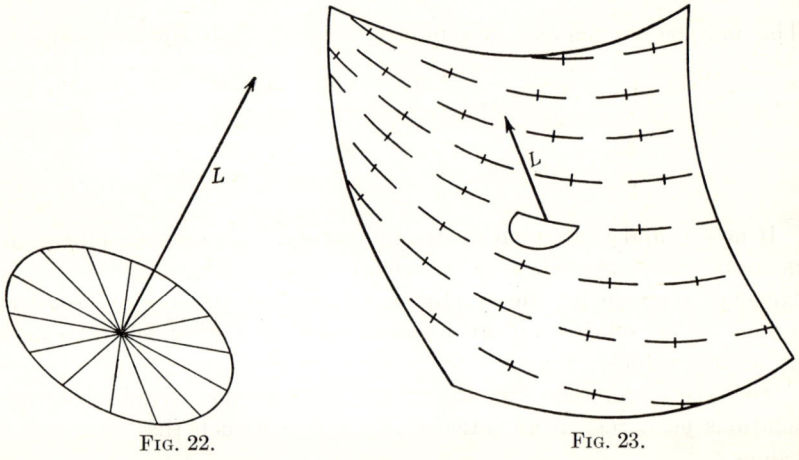

FIG. 22.   FIG. 23.

$A(x,y,z)$ is orthogonal to the line L through A with direction ratios P, Q, R. The integral curves thus cut at right angles the integral curves of (6-15). If we draw at A a disk perpendicular to L (Fig. 22), the tangent to the integral curve lies in this disk. Conversely, a curve is an integral curve if it is tangent at each point to the disk there.

There is, in general, a one-parameter family of integral curves lying on a given surface, $z = f(x,y)$, say. Using this equation to eliminate z and dz from (6-19), we have an equation of the first order in x and y. Its solution, $\varphi(x,y,C) = 0$, is a one-parameter family of cylinders with rulings parallel to the z axis cutting the given surface in integral curves.

We have another picture of the situation in Fig. 23. The disks cut the surface in a family of lineal elements forming a direction field on the surface. The lineal elements combine into a one-parameter family of integral curves.

6-9. The Integrable Case.

It may happen that (6-19) is the exact differential of a function or can be made so by the introduction of a suitable factor. The Pfaffian is then called *integrable*. Thus

(i) $$y\,dx + x\,dy + 2z\,dz = 0$$

is recognized to be

$$d(xy + z^2) = 0$$

whence we have

(ii) $$xy + z^2 = C$$

In what sense is (ii) the solution of (i)? Any pair of functions $y(x)$, $z(x)$ satisfying (i) satisfies an equation of form (ii), and conversely. Thus all solutions can be derived from (ii). Taking $z = \sin x$, $C = 1$, for example, we have the solution

$$y = \frac{1}{x}\cos^2 x \qquad z = \sin x$$

The geometric meaning of integrability may be expressed in several ways: (1) The integral curves lie on a one-parameter family of surfaces, and any curve on any of the surfaces is an integral curve. (2) The disks of the preceding section fit together to form a one-parameter family of surfaces. (3) There is a one-parameter family of surfaces orthogonal to the two-parameter family of curves defined by Eqs. (6-15).

We shall now derive the condition that the Pfaffian be integrable. If integrable, multiplication of (6-19) by a suitable factor $\lambda(x,y,z) \neq 0$ will make the first member the exact differential of some function $u(x,y,z)$,

$$\lambda(P\,dx + Q\,dy + R\,dz) \equiv du \equiv \frac{\partial u}{\partial x}dx + \frac{\partial u}{\partial y}dy + \frac{\partial u}{\partial z}dz$$

From this we have

$$\frac{\partial u}{\partial x} \equiv \lambda P \qquad \frac{\partial u}{\partial y} \equiv \lambda Q \qquad \frac{\partial u}{\partial z} \equiv \lambda R$$

From the condition

$$\frac{\partial^2 u}{\partial y\,\partial x} \equiv \frac{\partial^2 u}{\partial x\,\partial y}$$

or

$$\frac{\partial}{\partial y}(\lambda P) \equiv \frac{\partial}{\partial x}(\lambda Q)$$

we have (representing partial derivatives by subscripts)

$$Q\lambda_x - P\lambda_y + \lambda(Q_x - P_y) \equiv 0$$

In a similar manner, using the other cross derivatives,

$$R\lambda_y - Q\lambda_z + \lambda(R_y - Q_z) \equiv 0$$
$$P\lambda_z - R\lambda_x + \lambda(P_z - R_x) \equiv 0$$

On multiplying these three equations by R, P, Q, respectively, and adding, the derivatives of λ drop out, and there results, after canceling λ, the necessary condition

$$P(R_y - Q_z) + Q(P_z - R_x) + R(Q_x - P_y) \equiv 0 \qquad (6\text{-}20)$$

We shall show that, conversely, if (6-20) holds, the equation is integrable. We first give a process for finding the integral and illustrate its use; then we show that (6-20) insures the success of the method.

The rule is as follows:

Hold z constant and integrate the resulting equation in x and y, replacing the constant of integration by a function $f(z)$. Differentiate the result with respect to all the variables, and compare with the original equation. There will result an equation for the determination of $f(z)$.

Obviously the process may be altered by holding x or y constant.

Example. Solve

(iii) $\qquad\qquad 2yz\,dx + zx\,dy + 3xy\,dz = 0$

It is readily found that (6-20) is satisfied. Holding z constant,

$$2y\,dx + x\,dy = 0$$

and integrating, we get

$$x^2 y = f(z)$$

Differentiating,

$$2xy\,dx + x^2\,dy - f'(z)\,dz = 0$$

and comparing with (iii) multiplied by x/z,

$$2xy\,dx + x^2\,dy + \frac{3x^2 y}{z}\,dz = 0$$

we require that

$$-\frac{df}{dz} = \frac{3x^2 y}{z} = \frac{3f}{z}$$

From this we get

$$f(z) = \frac{C}{z^3}$$

The solution is then

$$x^2 y = \frac{C}{z^3}$$

or

$$x^2 y z^3 = C$$

We turn now to the theory. Holding z constant and integrating

$$P\,dx + Q\,dy = 0$$

SYSTEMS OF ORDINARY EQUATIONS 139

we have a solution $u = C$, or more generally

$$u = f(z)$$

such that

$$u_x\,dx + u_y\,dy = \mu(P\,dx + Q\,dy)$$

where μ is some factor. The differential of the solution with all three quantities varying is

$$u_x\,dx + u_y\,dy + [u_z - f'(z)]\,dz = 0$$

We have a solution if $f(z)$ can be determined so that

$$u_z - f'(z) = \mu R$$

This is possible if $u_z - \mu R$ is a function of z and f (that is, u) alone.

It suffices for this that the Jacobian of u and $u_z - \mu R$ with respect to x and y be zero,

$$u_x \frac{\partial}{\partial y}(u_z - \mu R) - u_y \frac{\partial}{\partial x}(u_z - \mu R) \equiv 0 \qquad (6\text{-}21)$$

Now we find readily—as we expect—that if (6-19) be multiplied by a factor μ, Eq. (6-20) still holds for the new coefficients,

$$\mu P[(\mu R)_y - (\mu Q)_z] + \mu Q[(\mu P)_z - (\mu R)_x] + \mu R[(\mu Q)_x - (\mu P)_y] \equiv 0$$

Noting that $\mu P = u_x$, $\mu Q = u_y$, this last equation reduces at once to (6-21).

It follows from the preceding that $f(z)$ satisfies a differential equation of the first order in f and z, from which f can be determined. This establishes the sufficiency of condition (6-20).

Mayer's Method. If an integral surface S be cut by any other surface, the curve of section will be an integral curve. A family of surfaces through a point of S yields a family of integral curves on S. From this family the equation of S may be found. This idea is the genesis of a method of solution due to Mayer—a method which requires but a single integration.

We take a family of planes parallel to the z axis through the point (x_0, y_0, z_0),

$$y = m(x - x_0) + y_0 \qquad (6\text{-}22)$$

Here $dy = m\,dx$, and we eliminate y from (6-19). We solve the resulting differential equation of the first order in z and x, and get the particular solution such that $z = z_0$ when $x = x_0$. This solution together with (6-22) gives, with varying m, a family of integral curves through the point. By eliminating m we have the equation of the integral surface through the point.

By taking z_0, say, as arbitrary, we get a family of surfaces. We would choose x_0 and y_0 so as to simplify the work as much as possible ($x_0 = y_0 = 0$, for example), provided the resulting line on which the point moves is not objectionable.

Example. Solve

(iv) $\qquad (yz - 1)\,dx + (z - x)x\,dy + (1 - xy)\,dz = 0$

It is readily found that the equation is integrable. Put

$$y = mx \qquad dy = m\,dx$$

We have, on rearranging terms,

$$\frac{dz}{dx} + \frac{2mx}{1 - mx^2}\,z = \frac{1 + mx^2}{1 - mx^2}$$

A particular solution of this linear equation is $z = x$; a solution of the reduced equation is $z = 1 - mx^2$. The general solution is

$$z = x + C(1 - mx^2)$$

Here C takes the place of z_0 in the preceding treatment. Eliminating m by putting $m = y/x$, we have the final result

$$z = x + C(1 - xy)$$

Line Integrals. If the equation is already exact, so that $\lambda = 1$ is an integrating factor, and hence

$$P_y = Q_x \qquad P_z = R_x \qquad Q_z = R_y$$

we can write the solution as a line integral. The reasoning here is analogous to that in the treatment of the exact equation in two variables. We have the solution $u = C$, where

$$u(x,y,z) = \int_{a,b,c}^{x,y,z} [P(X,Y,Z)\,dX + Q(X,Y,Z)\,dY + R(X,Y,Z)\,dZ]$$

the integral, which is independent of the path, being taken along some curve connecting (a,b,c) and (x,y,z).

Example. Solve

(v) $\qquad (2xz^2 + y)\,dx + x\,dy + (2x^2z + 3z^2)\,dz = 0$

This equation is exact and could be solved by inspection. However, we integrate along the straight line from the origin to (x,y,z),

$$X = xt \qquad Y = yt \qquad Z = zt$$

getting the solution

$$u = \int_0^1 [(2xz^2t^3 + yt)x\,dt + xty\,dt + (2x^2zt^3 + 3z^2t^2)z\,dt]$$
$$= x^2z^2 + xy + z^3 = C$$

SYSTEMS OF ORDINARY EQUATIONS 141

EXERCISES

1. Solve (iii) by inspection, after division by xyz.
2. Solve (v) by integrating from the origin along a broken line whose parts are parallel to the axes.
3. Find the two most general functions of x whose sum is equal to the sum of their derivatives.
4. If relation (6-20) holds, show that, as stated in the text, the relation still holds for the new coefficients after (6-19) has been multiplied by some function.
5. Solve by Mayer's method

$$x^2 \, dz = (x^2 - y) \, dx + x \, dy$$

putting $y = m(x - 1)$ and getting, before the elimination of m,

$$z = x - m/x + m - 1 + z_0$$

Show integrable and solve:

6. $ky^2 \, dx + z \, dy - y \, dz = 0$.
7. $(y^2 + yz) \, dx + (xz + z^2) \, dy + (y^2 - xy) \, dz = 0$.
8. $2xyz \, dx + z(1 - yz^2) \, dy + y(3 - 2yz^2) \, dz = 0$.
9. Show that if (6-19) is integrable the curves defined by the equations

$$\frac{dx}{\dfrac{\partial R}{\partial y} - \dfrac{\partial Q}{\partial z}} = \frac{dy}{\dfrac{\partial P}{\partial z} - \dfrac{\partial R}{\partial x}} = \frac{dz}{\dfrac{\partial Q}{\partial x} - \dfrac{\partial P}{\partial y}}$$

lie on the integral surfaces.

Find the integral curves of $y \, dx - x \, dy + dz = 0$
10. Lying in the plane $z = 2x + 3y + a$.
11. On the quadratic surface $y = xz$. 12. On the paraboloid $2z = x^2 + y^2$.
13. Orthogonal to the circles $x^2 + z^2 = a$, $y = b$.
14. Find the simultaneous solutions of the Pfaffians,

$$(x^2 + y^2) \, dx + 2xy \, dy + (xz + y) \, dz = 0$$
$$(x + y)(dx + dy) + (1 + z) \, dz = 0$$

15. Show that there is a family of planes orthogonal to the curves defined by

$$\frac{dx}{mz - ny} = \frac{dy}{nx - lz} = \frac{dz}{ly - mx}$$

16. Show that there is no family of surfaces orthogonal to the lines meeting the two skew lines of Exercise 9, Sec. 6-3. Show that the curves orthogonal to the lines of the family and lying in a plane perpendicular to either of the skew base lines are circles.
17. Show that the twisted cubics, $z = ax^2$, $z^2 = bxy$, are orthogonal to a family of ellipsoids.
18. Find the two most general functions of x whose difference is equal to the sum of their derivatives.

6-10. Systems of Linear Equations. If we have a system of n equations which are linear in n dependent variables and their derivatives, we can ordinarily eliminate all the dependent variables but one. We can differentiate the equations, getting always linear equations, until we have equations enough to perform the elimination.

We shall study here only systems with constant coefficients. We shall treat three equations in three dependent variables. The formulas will be less complicated than in the general case, whereas the method for more equations will be obvious. The system has the form

$$f_1(D)y + f_2(D)z + f_3(D)w = R_1(x)$$
$$g_1(D)y + g_2(D)z + g_3(D)w = R_2(x)$$
$$h_1(D)y + h_2(D)z + h_3(D)w = R_3(x)$$

where $f_1(D), \ldots, h_3(D)$ are polynomials in the operator D.

To eliminate z and w, say, we operate with D, D^2, etc., and combine, so that z, w, Dz, Dw, etc., drop out; that is, we manipulate the operators algebraically so that those operating on z and w vanish. The result of this elimination is most readily furnished by determinantal methods,

$$\Delta(D)y = \begin{vmatrix} f_1 & f_2 & f_3 \\ g_1 & g_2 & g_3 \\ h_1 & h_2 & h_3 \end{vmatrix} y = \begin{vmatrix} R_1 & f_2 & f_3 \\ R_2 & g_2 & g_3 \\ R_3 & h_2 & h_3 \end{vmatrix}$$

Here $\Delta(D)$ is used to represent briefly the first determinant. The second determinant, with one column of functions, has the following meaning: Expand according to the column in which R_1, R_2, R_3 lie, using the minors as operators on these functions.

If $\Delta(D) \neq 0$, we have thus an equation in y. If $\Delta(D)$ is a constant, not zero, we have y immediately; otherwise we can solve the differential equation for y. We get similarly equations for z and w:

$$\Delta(D)z = \begin{vmatrix} f_1 & R_1 & f_3 \\ g_1 & R_2 & g_3 \\ h_1 & R_3 & h_3 \end{vmatrix} \qquad \Delta(D)w = \begin{vmatrix} f_1 & f_2 & R_1 \\ g_1 & g_2 & R_2 \\ h_1 & h_2 & R_3 \end{vmatrix}$$

The preceding equations must be satisfied by the solutions. They are *necessary* conditions. Not all their solutions need satisfy the original system. By substituting the general solutions of these derived equations into the original system we find what further restrictions are necessary. In fact, the number of arbitrary constants is equal to the degree of $\Delta(D)$.

If $\Delta(D) = 0$, the first members of the derived equations vanish. If any of the second members fails to vanish, the equations cannot be satisfied and the original system has no solutions.

Example. We illustrate with a system of two equations:

$$y'' + z' - y - z = 1$$
$$z'' + y' + y + z = x$$

These may be written
$$(D^2 - 1)y + (D - 1)z = 1$$
$$(D + 1)y + (D^2 + 1)z = x$$
Eliminating z,
$$\begin{vmatrix} D^2 - 1 & D - 1 \\ D + 1 & D^2 + 1 \end{vmatrix} y = \begin{vmatrix} 1 & D - 1 \\ x & D^2 + 1 \end{vmatrix}$$
$$(D^4 - D^2)y = (D^2 + 1)1 - (D - 1)x = x$$

Similarly
$$(D^4 - D^2)z = \begin{vmatrix} D^2 - 1 & 1 \\ D + 1 & x \end{vmatrix} = -(x + 1)$$

The general solutions of these equations are
$$y = a + bx + ce^x + de^{-x} - \frac{x^3}{6}$$
$$z = k + fx + ge^x + he^{-x} + \frac{x^3}{6} + \frac{x^2}{2}$$

On substituting these values into the original equations and collecting terms, we have
$$(f - a - k - 1) - (b + f)x - 2he^{-x} \equiv 0$$
$$(a + b + k + 1) + (b + f)x + 2(c + g)e^x + 2he^{-x} \equiv 0$$

Since the functions 1, x, e^x, e^{-x} are linearly independent, their coefficients (in parentheses) must be zero. From them we have
$$k = -(a + b + 1) \qquad f = -b \qquad g = -c \qquad h = 0$$
and the general solution is
$$y = a + bx + ce^x + de^{-x} - \frac{x^3}{6}$$
$$z = -(a + b + 1) - bx - ce^x + \frac{x^3}{6} + \frac{x^2}{2}$$
which contains four arbitrary constants.

EXERCISES

1. Solve the problem in the text after simplifying the first equation as follows:
$$(D + 1)y + z = \frac{1}{D - 1}1 = ke^x - 1$$

2. Solve:
$$y'' - z'' + w = 1$$
$$z'' - w'' + y = x$$
$$w'' - y'' - y = x^2$$

3. Suppose a radioactive substance is changing into a second which, in turn, is changing into a third product which is stable. Given that the rate at which one substance is being changed into the next is proportional to its mass, derive the equations
$$\frac{dx}{dt} = -ax \qquad \frac{dy}{dt} = ax - by \qquad \frac{dz}{dt} = by$$
where x, y, z are the respective masses.

Find the masses at any time if initially (*a*) only the first substance is present, (*b*) all three substances are present.

4. Suppose there is a chain of n substances involved in the radioactive series. Derive a formula for the mass of the kth substance.

6-11. The Motion of a Particle. The behavior of a particle moving in a plane or in space under the influence of given forces is governed by a system of differential equations of the second order. Taking units so that $\lambda = 1$ (Sec. 1-17) Newton's second law of motion is

$$m\mathbf{a} = \mathbf{F} \qquad (6\text{-}23)$$

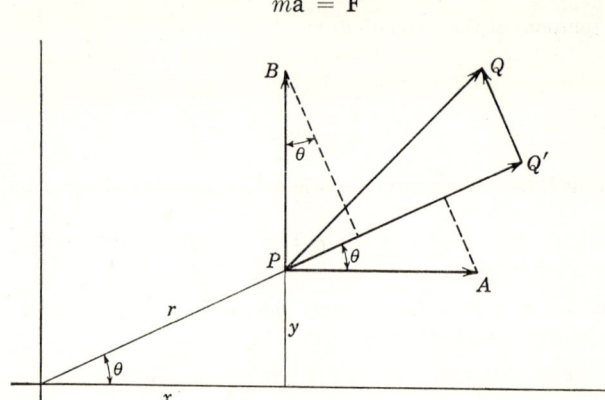

FIG. 24.

where m is the mass of the particle, $\mathbf{a}$ is its vector acceleration, and $\mathbf{F}$ is the resultant force acting upon it.

We may equate projections of the vectors in (6-23) in any direction. Locating the particle by rectangular coordinates (x,y,z), and taking components in the direction of the axes, we have

$$mx'' = X \qquad my'' = Y \qquad mz'' = Z \qquad (6\text{-}24)$$

where X, Y, Z are components of the force in the direction of the axes. We shall use primes throughout to denote derivatives with respect to the time t.

If the motion takes place in a plane, we can select this as the xy plane. There is no force in the z direction. The equations of motion now are

$$mx'' = X \qquad my'' = Y \qquad (6\text{-}25)$$

We shall be interested in the use of polar coordinates. From Fig. 24 we can read off the formulas for the magnitudes of components of vectors along the radius vector and perpendicular thereto in terms of components parallel to the x and y axes,

$$\begin{aligned} PQ' &= PA \cos \theta + PB \sin \theta \\ Q'Q &= -PA \sin \theta + PB \cos \theta \end{aligned} \qquad (6\text{-}26)$$

SYSTEMS OF ORDINARY EQUATIONS 145

If **PQ** is a force, we have $PA = X$, $PB = Y$; if **PQ** is an acceleration, $PA = x''$, $PB = y''$; if **PQ** is a velocity, $PA = x'$, $PB = y'$.
The relation between the coordinates is

$$x = r \cos \theta \qquad y = r \sin \theta \qquad (6\text{-}27)$$

Differentiating with respect to the time,

$$x' = -r \sin \theta \, \theta' + \cos \theta \, r' \qquad y' = r \cos \theta \, \theta' + \sin \theta \, r'$$

Substituting x' and y' for PA and PB in (6-26), we have

$$v_r = r' \qquad v_\theta = r\theta' \qquad (6\text{-}28)$$

then

$$v^2 = x'^2 + y'^2 = r'^2 + r^2 \theta'^2 \qquad (6\text{-}29)$$

A second differentiation gives

$$x'' = -r \sin \theta \, \theta'' - r \cos \theta \, \theta'^2 - 2 \sin \theta \, \theta' r' + \cos \theta \, r''$$
$$y'' = r \cos \theta \, \theta'' - r \sin \theta \, \theta'^2 + 2 \cos \theta \, \theta' r' + \sin \theta \, r''$$

Putting $PA = x''$, $PB = y''$, we have from (6-26)

$$a_r = r'' - r\theta'^2 \qquad a_\theta = r\theta'' + 2\theta' r' \qquad (6\text{-}30)$$

The equations of motion now take the form

$$m(r'' - r\theta'^2) = F_r \qquad m(r\theta'' + 2\theta' r') = F_\theta \qquad (6\text{-}31)$$

Here F_r and F_θ are the components of the resulting force along the radius vector and perpendicular thereto. We count them as positive in the direction of increasing r and θ, and negative in the opposite direction.

Central Forces. If the force is directed always toward, or directly away from, a fixed point, it is convenient to take this point at the origin. Then $F_\theta = 0$, and the second equation of (6-31) is

$$r\theta'' + 2\theta' r' = 0$$

On multiplying by r,

$$r^2 \theta'' + 2rr'\theta' = 0$$

the first member becomes exact. Integrating, we have

$$r^2 \theta' = h \qquad (6\text{-}32)$$

This last has a geometric meaning. If A is the area swept over by the radius vector from some starting position, we have

$$dA = \tfrac{1}{2} r^2 \, d\theta = \tfrac{1}{2} h \, dt$$

Integrating and taking $t = 0$ when $A = 0$,

$$A = \tfrac{1}{2} ht \qquad (6\text{-}33)$$

We have established the *law of areas:*

If the force is central, the radius vector (joining the center of force to the moving particle) sweeps out equal areas in equal times.

6-12. Planetary Motion. Attraction according to the inverse-square law is so common a phenomenon of nature that it is a matter of general culture to have some knowledge of its consequences. It explains the orbits of the planets, the behavior of comets and meteors, and the paths of the electrons.

We shall discuss the motion of a small body under the attraction of a large one. We consider the latter as fixed at the origin. The force of attraction between the bodies is GmM/r^2, where m and M are the masses, and G is a constant. The equations of motion of the small body are

$$m(r'' - r\theta'^2) = -\frac{GmM}{r^2} \qquad m(r\theta'' + 2\theta'r') = 0$$

$$r'' - r\theta'^2 = -\frac{k}{r^2} \qquad r\theta'' + 2\theta'r' = 0 \qquad (6\text{-}34)$$

Here $k = GM$, a constant proportional to the mass of the large body but independent of the mass of the particle. Thus k would be the same, for example, for all planets in their motion about the sun.

Since the force is central we have the first integral (6-32), and the law of areas holds. We shall use (6-32) to eliminate t from the first equation of (6-34), thus getting the differential equation of the orbit.

It turns out that we are led to a familiar equation if we use the reciprocal of r as a new variable:

$$u = \frac{1}{r} \qquad \theta' = hu^2 \qquad (6\text{-}35)$$

We have

$$r' = -\frac{u'}{u^2} = -\frac{1}{u^2}\frac{du}{d\theta}\theta' = -h\frac{du}{d\theta}$$

$$r'' = -h\frac{d^2u}{d\theta^2}\theta' = -h^2u^2\frac{d^2u}{d\theta^2}$$

On substituting these values into the first equation of (6-34) and canceling $-h^2u^2$, we have

$$\frac{d^2u}{d\theta^2} + u = \frac{k}{h^2} \qquad (6\text{-}36)$$

The solution of this linear equation can be written down at once

$$u = \frac{k}{h^2} + K \cos(\theta - \beta)$$

where K and β are arbitrary constants. We have the equation of the orbit in the form

$$r = \frac{h^2/k}{1 + e \cos (\theta - \beta)} \quad (6\text{-}37)$$

This will be recognized as the equation of a conic with focus at the origin and with eccentricity e, supposed non-negative. (Adding π to β would change the sign of this constant, if necessary.)

We recall the facts from Fig. 25. A conic is defined as the locus of a point whose distance from a fixed point (the focus) is e times its distance from a fixed line (the directrix), where e is the eccentricity:

$$OP = e \cdot PM = e(ON - OQ)$$

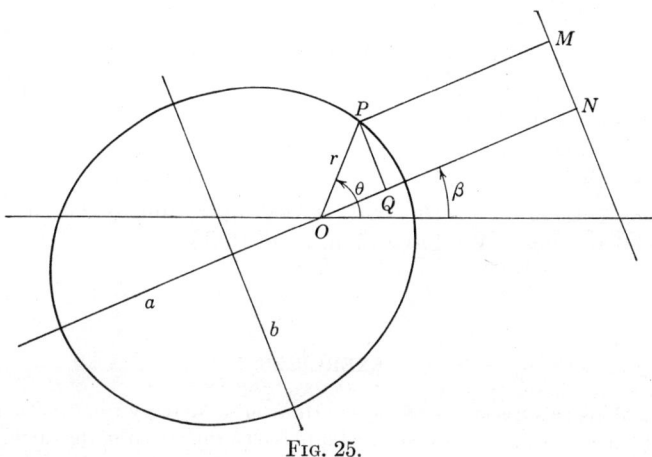

Fig. 25.

If the distance from focus to directrix is p,

$$r = e[p - r \cos (\theta - \beta)]$$

which reduces to (6-37) with $ep = h^2/k$. The conic is an ellipse, a parabola, or a hyperbola according as $e < 1$, $e = 1$, or $e > 1$.

Consider the case of a closed orbit, with semiaxes a and b. The major axis is the sum of greatest and least distances from O to the orbit,

$$2a = \frac{h^2/k}{1 - e} + \frac{h^2/k}{1 + e} = \frac{2h^2/k}{1 - e^2}$$

Making use of the well-known relation $b^2 = a^2(1 - e^2)$, we find

$$b^2 = \frac{h^2 a}{k}$$

We use these results to get a relation between a (the so-called mean distance) and the period of revolution T. We have from (6-33) for a complete revolution

$$A = \pi ab = \frac{hT}{2}$$

Squaring and eliminating b,

$$\pi^2 a^2 b^2 = \frac{\pi^2 h^2 a^3}{k} = \frac{h^2 T^2}{4}$$

or

$$T^2 = \frac{4\pi^2}{k} a^3 \qquad (6\text{-}38)$$

We have established Kepler's third law:

The squares of the periods of the planets are proportional to the cubes of their mean distances.

The first and second laws have emerged in the preceding discussion; namely, (1) that the orbit is an ellipse with the sun at one focus, and (2) that the law of areas is valid.

In considering motions about an attracting body it is often convenient to take as units of time and distance the period and mean distance of some body revolving about it—the year, for example, and the earth's distance of 93,000,000 miles. We have then, from (6-38),

$$k = 4\pi^2 \qquad T^2 = a^3 \qquad (6\text{-}39)$$

EXERCISES

1. (*a*) Mercury's "year" is 88 days. How far is Mercury from the sun?

(*b*) The mean distance of the planet Pluto is 39.5 times that of the earth. What is its period of revolution?

2. By differentiating (6-29) and using (6-34) get $vv' = -kr'/r^2$. Hence derive the energy integral

$$\frac{v^2}{2} - \frac{k}{r} = C$$

3. By considering the energy integral at the perihelion point, show that

$$C = \frac{k^2(e^2 - 1)}{2h^2}$$

Show that the path is elliptic, parabolic, or hyperbolic according as v^2 is less than, equal to, or greater than $2k/r$.

4. Show that the velocity at any point in an elliptic path is

$$v^2 = k\left(\frac{2}{r} - \frac{1}{a}\right)$$

5. If the orbit is circular, show that $v^2 = k/a$.

By what factor must the preceding velocity be multiplied in order that the body will escape?

6. In his space ship high above the earth Buck Rogers fires a blast of gunfire, the bullets having the same velocity (less than the velocity of escape) and varying directions. Show that all bullets which do not strike the earth will reassemble later at the point of departure.

7. Solve the equations of motion,

$$x'' = 0 \qquad y'' = -g$$

to find the path of a projectile moving in a vacuum.

8. Discuss the motion of the projectile in air, given that the resistance (directed along the path) is proportional to the velocity.

9. If a particle is attracted toward the origin with a force proportional to its distance, show that the equations of motion are

$$x'' = -a^2 x \qquad y'' = -a^2 y$$

Show that the path is an ellipse, and find the period.

10. Find the motion of the preceding particle if the force is repulsive.

11. Consider the motion in Exercise 9, given that there is a small resistance proportional to the velocity.

CHAPTER 7

CERTAIN CLASSICAL EQUATIONS

There are several differential equations of the second order whose investigation has produced an extensive literature. They have arisen, as a rule, in connection with physical problems. Their solutions have varied and interesting properties. Some of the best known of these equations will be discussed in this chapter.

7-1. Analytic Solutions. An adequate knowledge of solutions in series requires a grasp of the theory of functions of a complex variable. We shall here merely state certain of the results. A function $f(x)$ is called *analytic* at a if it can be expanded in a Taylor's series in powers of $x - a$ which is valid near a. A function of two or more variables, say $f(x,y,z)$, is analytic at a point (a,b,c) if it can be expanded in a Taylor's series in powers of $x - a$, $y - b$, $z - c$ valid about the point.

If the functions which enter into an equation are analytic we can state, under suitable conditions, that the solution is analytic. The following theorem can be proved.

THEOREM. *If the second members of the set of equations,*

$$\frac{dy_1}{dx} = f_1(x,y_1, \ldots ,y_n)$$
$$\ldots \ldots \ldots \ldots \ldots \ldots \quad (7\text{-}1)$$
$$\frac{dy_n}{dx} = f_n(x,y_1, \ldots ,y_n)$$

are analytic at the point $(a,b_1, \ldots ,b_n)$, *the unique set of solutions,*

$$y_1 = y_1(x) \quad \ldots \quad y_n = y_n(x)$$

which at $x = a$ *take on the values* $b_1, \ldots , b_n$, *respectively, are functions analytic at* a.

Accepting this theorem, we may find the solution in series form by substituting for the unknowns power series with undetermined coefficients and equating like powers of $x - a$.

We shall be particularly concerned with the linear system

CERTAIN CLASSICAL EQUATIONS

$$\frac{dy_1}{dx} = a_{11}(x)y_1 + \cdots + a_{1n}(x)y_n + b_1(x)$$
$$\cdots\cdots\cdots\cdots\cdots\cdots\cdots\cdots\cdots\cdots\cdots\cdots\cdots \quad (7\text{-}2)$$
$$\frac{dy_n}{dx} = a_{n1}(x)y_1 + \cdots + a_{nn}(x)y_n + b_n(x)$$

in which the functions of x appearing in the second members are analytic at a. Very precise results can be stated concerning the region of convergence of the sequence of functions furnished by the method of successive approximations and the circle of convergence of the solutions in power series.

In the complex x plane (Fig. 26) let us plot a and mark the various singularities of the functions of x in (7-2). By the *star* with center a we shall mean the region each of whose points can be joined to a by a line

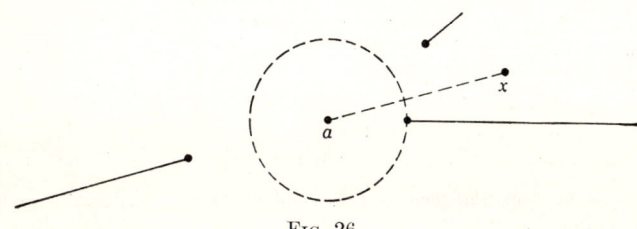

Fig. 26.

segment not meeting a singular point. The star thus consists of the whole plane except the "shadows" of the singular points cast by a light at a.

Armed with a knowledge of complex variables, we can carry through the convergence proofs of the preceding chapter, making use of bounds of the functions along rectilinear paths of integration. We have the following theorem:

THEOREM. *The successive approximations of Sec. 6-2 converge in the star with center at a and the resultant solution is analytic in the star.*

The expansion of the solution in a series of powers of $x - a$ converges in a circle with a as center and with radius at least as great as the distance from a to the nearest singularity.

The circle of convergence of the power series will usually pass through the nearest singularity. It may happen, however, that the solution is analytic there and that there is convergence in a circle through a more distant singularity, or indeed in the whole plane.

The preceding theorem applies to the equation which we shall be interested in,

$$\frac{d^2y}{dx^2} + P(x)\frac{dy}{dx} + Q(x)y = 0 \quad (7\text{-}3)$$

since, as we have observed before, it can be written as the linear system

$$\frac{dy}{dx} = v$$

$$\frac{dv}{dx} = -P(x)v - Q(x)y$$

The singularities of $P(x)$ and $Q(x)$ are marked in the x plane to determine the facts of convergence. A point at which $P(x)$ and $Q(x)$ are analytic is called an *ordinary point;* a point at which one or both of the functions have singularities is called a *singular point*.

Example. Find a solution of

$$(x - 1)(x - k)y'' + (x - 2k + 1)y' - y = 0$$

in powers of x, with the conditions $y(0) = y'(0) = 1$.
We substitute a series of the form

$$y = 1 + x + c_2 x^2 + c_3 x^3 + \cdots$$

into the differential equation:

$$[k - (k+1)x + x^2][2c_2 + 6c_3 x + \cdots + n(n-1)c_n x^{n-2} + \cdots]$$
$$+ (1 - 2k + x)(1 + 2c_2 x + \cdots + nc_n x^{n-1} + \cdots)$$
$$- (1 + x + c_2 x^2 + \cdots + c_n x^n + \cdots) = 0$$

Equating to zero the constant term and the coefficients of x, ..., x^{n-2}, we have after some simplification,

$$2kc_2 = 2k$$
$$3kc_3 = 3kc_2$$
$$\cdots\cdots\cdots\cdots\cdots\cdots\cdots\cdots\cdots$$
$$nkc_n = (kn + n - 3)c_{n-1} - (n - 3)c_{n-2}$$

We have $c_2 = c_3 = 1$; and if $c_{n-1} = c_{n-2} = 1$, then also $c_n = 1$. Hence $c_n = 1$ for all n, and the series solution is

$$y = 1 + x + x^2 + x^3 + \cdots = \frac{1}{1 - x}$$

The region of convergence is $|x| < 1$. The circle of convergence thus passes through the singular point $x = 1$, which may or may not be the singularity nearest to $x = 0$, depending upon the value of k. As a matter of fact, all solutions are analytic at $x = k$, if $k \neq 1$. For we have by inspection the solution $y = x - 2k + 1$; whence the general solution is

$$y = \frac{A}{1 - x} + B(x - 2k + 1)$$

7-2. Regular Singular Points. It often happens that a function fails to be expansible in a Taylor's series at a because it becomes infinite; for example, $x^2/(x - a)^2$, $\cot(x - a)$, etc. If the function be multiplied by a suitable integral power of $x - a$, the resulting function is analytic.

A singular point a of (7-3) is called a *regular singular point* if the functions $(x - a)P(x)$ and $(x - a)^2 Q(x)$ are analytic. About such a point

CERTAIN CLASSICAL EQUATIONS

the equation has, as we shall see, a special type of series solution. We may write the equation, on multiplying by $(x - a)^2$, in the form

$$(x - a)^2 \frac{d^2y}{dx^2} + (x - a)p(x)\frac{dy}{dx} + q(x)y = 0 \qquad (7\text{-}4)$$

where $p(x)[= (x - a) P(x)]$ and $q(x)[= (x - a)^2 Q(x)]$ are expansible in Taylor's series:

$$p(x) = p_0 + p_1(x - a) + p_2(x - a)^2 + \cdots$$
$$q(x) = q_0 + q_1(x - a) + q_2(x - a)^2 + \cdots \qquad (7\text{-}5)$$

We try a solution of the form

$$y = (x - a)^\alpha[1 + c_1(x - a) + c_2(x - a)^2 + \cdots]$$
$$= (x - a)^\alpha + c_1(x - a)^{\alpha+1} + c_2(x - a)^{\alpha+2} + \cdots$$

where the exponent α is to be presently determined. Substituting into the equation, we have

$$(x - a)^2[\alpha(\alpha - 1)(x - a)^{\alpha-2} + c_1(\alpha + 1)\alpha(x - a)^{\alpha-1}$$
$$+ c_2(\alpha + 2)(\alpha + 1)(x - a)^\alpha + \cdots]$$
$$+ (x - a)[p_0 + p_1(x - a) + \cdots][\alpha(x - a)^{\alpha-1}$$
$$+ c_1(\alpha + 1)(x - a)^\alpha + \cdots]$$
$$+ [q_0 + q_1(x - a) + \cdots][(x - a)^\alpha + c_1(x - a)^{\alpha+1} + \cdots] = 0$$

We first equate to zero the coefficient of $(x - a)^\alpha$, the lowest power of $x - a$ to appear. We have

$$\alpha(\alpha - 1) + p_0\alpha + q_0 = 0 \qquad (7\text{-}6)$$

This quadratic equation in α is called the *indicial equation*. Its roots α_1, α_2 determine the possible exponents in the series we have assumed.

Having chosen α, we next equate to zero the coefficients of $(x - a)^{\alpha+1}$, $(x - a)^{\alpha+2}$, etc.

$$[(\alpha + 1)\alpha + p_0(\alpha + 1) + q_0]c_1 + p_1\alpha + q_1 = 0$$
$$\cdots\cdots\cdots\cdots\cdots\cdots\cdots\cdots\cdots\cdots\cdots\cdots\cdots$$
$$[(\alpha + n)(\alpha + n - 1) + p_0(\alpha + n) + q_0]c_n \qquad (7\text{-}7)$$
$$+ \text{ terms in } c_1, \ldots, c_{n-1} = 0$$
$$\cdots\cdots\cdots\cdots\cdots\cdots\cdots\cdots\cdots\cdots\cdots\cdots\cdots$$

We proceed to get $c_1, c_2, \ldots,$ in order from these equations. From the two roots α_1, α_2 of the indicial equation we get usually two series solutions.

This method fails only if the coefficient of some c_n is zero. Representing the first member of the indicial equation by $f(\alpha)$, we observe that the coefficient of c_n is $f(\alpha + n)$. If for the root α_1 we have $f(\alpha_1 + n) = 0$, we see that the second root is $\alpha_2 = \alpha_1 + n$. The method can break down only if the roots of the indicial equation differ by an integer.

In general, then, we get two formal series solutions. If the roots of the indicial equation are unequal but differ by an integer, only one root leads to a formal solution. If the roots are equal, we have, of course, only one series.

Except for the initial multiplier $(x - a)^\alpha$, we have a power series. Its region of convergence is a circle with center at a. It can be shown that this circle reaches at least to the nearest of the other singularities, and that within this domain the series provides a solution of the differential equation.

If the preceding method gives us two solutions, the general solution of the equation is a linear combination of the two. If it gives but one, a second solution can be found as explained in Sec. 3-9.

Solutions valid for large values of x may be found by changing the variable to $t = 1/x$, and getting series valid near $t = 0$. Infinity is said to be an ordinary point or a regular singular point, if $t = 0$ is an ordinary point or a regular singular point of the transformed equation. A solution in powers of t valid near the origin is then a solution of the original equation in powers of $1/x$ valid for large values of x. The region of convergence is the *exterior* of a circle with center at $x = 0$; and this circle is at least as small as the circle through the farthest finite singularity.

EXERCISES

1. Solve in a series of powers of x

$$y'' - xy = 0$$

with the conditions (a) $y(0) = 1$, $y'(0) = 0$; (b) $y(0) = 0$, $y'(0) = 1$.

2. Solve the Legendre equation

$$(1 - x^2)y'' - 2xy' + 20y = 0$$

in powers of x, given $y(0) = 1$, $y'(0) = 0$.

3. Solve the problem in the text as an Euler equation in the case $k = 1$.

4. The equation

$$(x^3 + 4x^2 + 13x)y'' - x^5y' + y = 0$$

has certain solutions whose power series start as follows:

(a) $y = 1 + 2(x - 1) + a_2(x - 1)^2 + \cdots$
(b) $y = 5(x + 2) + b_2(x + 2)^2 + \cdots$
(c) $y = 1 + c_2(x + 4)^2 + \cdots$
(d) $y = x + d_2x^2 + \cdots$
(e) $y = e_2(x - 2)^2 + e_3(x - 2)^3 + \cdots$

In each case, and without finding the series, state what you can about the radius of convergence.

CERTAIN CLASSICAL EQUATIONS 155

5. Solve in powers of x
$$y'' - xy' - ky = 0$$
given (a) $y(0) = 1$; $y'(0) = 0$; (b) $y(0) = 0$; $y'(0) = 1$.

6. Solve
$$xy'' + y = 0$$
in a series of the form
$$y = x + c_2 x^2 + \cdots$$

Find the exponents at the origin and solve in series form:
7. $2xy'' + y' + y = 0$.
8. $4x(1 - x)y'' + 2(1 + x^2)y' - (1 + x)y = 0$.
9. $x^2 y'' - 2xy' + (2 + x^2)y = 0$.   **10.** $xy'' - y' - 4x^3 y = 0$.

11. Show that the change of variable $t = 1/x$ transforms Eq. (7-3) into the form
$$\frac{d^2 y}{dt^2} + \left(\frac{2}{t} - \frac{P}{t^2}\right)\frac{dy}{dt} + \frac{Q}{t^4} y = 0$$

12. Find the exponents at all the singular points of
$$\frac{d^2 y}{dx^2} + \left[\frac{1 - \lambda}{x} + \frac{1 - \mu}{x - 1}\right]\frac{dy}{dx} + \frac{(1 - \lambda - \mu)^2 - \nu^2}{4x(x - 1)} y = 0$$

7-3. The Hypergeometric Differential Equation. This equation has the form
$$x(1 - x)y'' + [\gamma - (\alpha + \beta + 1)x]y' - \alpha\beta y = 0 \tag{7-8}$$
We observe that $x = 0$ and $x = 1$ are regular singular points; all other finite points are ordinary points.

To obtain the exponents at the origin we get the initial terms in the expansions (7-5):
$$p(x) = \frac{\gamma - (\alpha + \beta + 1)x}{1 - x} = \gamma + [\gamma - (\alpha + \beta + 1)]x + \cdots$$
$$q(x) = \frac{-\alpha\beta x}{1 - x} = -\alpha\beta x - \alpha\beta x^2 - \cdots$$

Here $p_0 = \gamma$, $q_0 = 0$; and the indicial equation (using primes to avoid confusion with the α of the equation),
$$\alpha'(\alpha' - 1) + \gamma\alpha' = 0$$
has the roots
$$\alpha_1 = 0 \qquad \alpha_2 = 1 - \gamma$$

If α_2 is not a positive integer, that is, if γ is not zero or a negative integer, we have a solution of the form
$$y = 1 + c_1 x + c_2 x^2 + c_3 x^3 + \cdots$$
Substituting this into (7-8) and equating to zero the coefficient of x^n, we find
$$(n + 1)(\gamma + n)c_{n+1} = (\alpha + n)(\beta + n)c_n$$

for $n \geq 1$. This holds also for $n = 0$ if we take $c_0 = 1$. From this relation we can calculate the coefficients in order. The resulting series, which is called the *hypergeometric series* and is commonly designated by $F(\alpha,\beta,\gamma,x)$, is

$$F(\alpha,\beta,\gamma,x) = 1 + \frac{\alpha\beta}{\gamma} x + \frac{\alpha(\alpha+1)\beta(\beta+1)}{2!\gamma(\gamma+1)} x^2$$
$$+ \frac{\alpha(\alpha+1)(\alpha+2)\beta(\beta+1)(\beta+2)}{3!\gamma(\gamma+1)(\gamma+2)} x^3 + \cdots \quad (7\text{-}9)$$

That this series converges out to the singularity at $x = 1$ may be established directly. If α or β is a negative integer, the series terminates; otherwise we apply the test-ratio test. For the ratio of successive terms we have

$$\left| \frac{c_{n+1} x^{n+1}}{c_n x^n} \right| = \left| \frac{(\alpha+n)(\beta+n)}{(n+1)(\gamma+n)} x \right| \to |x|$$

The series converges when $|x| < 1$ and diverges when $|x| > 1$.

There is a second solution with the exponent $1 - \gamma$, provided γ is not a positive integer. We can substitute

$$y = x^{1-\gamma}(1 + c_1 x + c_2 x^2 + \cdots)$$

into (7-8) and determine the coefficients. We may, however, proceed as follows. Putting

$$y = x^{1-\gamma} w$$

we get

$$x(1 - x)w'' + [\gamma_1 - (\alpha_1 + \beta_1 + 1)x]w' - \alpha_1 \beta_1 w = 0$$

where

$$\alpha_1 = \alpha - \gamma + 1 \qquad \beta_1 = \beta - \gamma + 1 \qquad \gamma_1 = 2 - \gamma$$

This has the series solution

$$w = F(\alpha_1,\beta_1,\gamma_1,x)$$

whence we have the second solution

$$y = x^{1-\gamma} F(\alpha - \gamma + 1, \beta - \gamma + 1, 2 - \gamma, x) \quad (7\text{-}10)$$

The general solution of the equation is a linear combination of the two solutions found above.

Other solutions, which are, of course, linear combinations of the preceding, may be found at the singular point $x = 1$. The series may be derived directly or may be deduced from the preceding by a change of variables. Putting

$$t = 1 - x$$

we find

$$t(1 - t)y'' + [\gamma_1 - (\alpha + \beta + 1)t]y' - \alpha\beta y = 0$$

primes indicating derivatives with respect to t, where
$$\gamma_1 = \alpha + \beta - \gamma + 1$$
We have then the solutions
$$y = F(\alpha, \beta, \alpha + \beta - \gamma + 1, 1 - x)$$
$$y = (1 - x)^{\gamma-\alpha-\beta} F(\gamma - \alpha, \gamma - \beta, \gamma - \alpha - \beta + 1, 1 - x) \quad (7\text{-}11)$$

The hypergeometric equation has the interesting property that the derivative of a solution satisfies an associated hypergeometric equation. If we differentiate (7-8), we find
$$x(1 - x)y''' + [\gamma + 1 - (\alpha + 1 + \beta + 1 + 1)x]y''$$
$$- (\alpha + 1)(\beta + 1)y' = 0$$
a hypergeometric equation in y' in which α, β, γ have been replaced by $\alpha + 1, \beta + 1, \gamma + 1$. Hence the nth derivative of a solution satisfies the hypergeometric equation with the constants $\alpha + n, \beta + n, \gamma + n$.

Various equations can be reduced to the hypergeometric form by a suitable change of variable. The following class of equations is an important example:
$$(x^2 + Ax + B)y'' + (Cx + D)y' + Ey = 0$$
where $A, \ldots, E$ are constants, and the roots of the quadratic coefficient of y'' are unequal. This coefficient may be written in the form $(x - a)(x - b)$. If we put
$$x - a = (b - a)t$$
we have
$$x - b = (a - b)(1 - t)$$
and the equation after the change of variable to t has the form
$$t(1 - t)y'' + (lt + m)y' + qy = 0$$
This is a hypergeometric equation in which $\gamma = m$, and α and β are defined by the equations
$$\alpha + \beta + 1 = -l \qquad \alpha\beta = -q$$

EXERCISES

From a consideration of the series verify the following:
1. $F(\alpha,\beta,\beta,x) = (1 - x)^{-\alpha}$.
2. $xF(1,1,2,x) = -\log(1 - x)$.
3. $xF(\frac{1}{2},\frac{1}{2},\frac{3}{2},x^2) = \arcsin x$.
4. $xF(\frac{1}{2},1,\frac{3}{2},-x^2) = \arctan x$.

Get a solution of each of the equations in Exercises 5 to 8 in terms of the hypergeometric function F.

158   DIFFERENTIAL EQUATIONS

5. $(x^2 - x)y'' + (3x + \frac{1}{2})y' + y = 0$.
6. $(x^2 - x)y'' + (4x - 3)y' + 2y = 0$.
7. $(x^2 - 1)y'' + (2x + 3)y' + \frac{2}{9}y = 0$.
8. $(x^2 - 4x + 3)y'' + (3x - 4)y' + y = 0$.

9. Solve in finite form

$$x(1 - x)y'' + [\beta - (\beta + 2)x]y' - \beta y = 0$$

10. Show that

$$F'(\alpha,\beta,\gamma,x) = \frac{\alpha\beta}{\gamma} F(\alpha + 1, \beta + 1, \gamma + 1, x)$$

and get an analogous expression for the nth derivative.

11. Show that in Eq. (7-8) the change of the dependent variable

$$y = (1 - x)^{\gamma-\alpha-\beta} u$$

leads to a hypergeometric equation, and get the solutions

$$y = (1 - x)^{\gamma-\alpha-\beta} F(\gamma - \alpha, \gamma - \beta, \gamma, x)$$
$$y = x^{1-\gamma}(1 - x)^{\gamma-\alpha-\beta} F(1 - \alpha, 1 - \beta, 2 - \gamma, x)$$

Show that these are identical with (7-9) and (7-10), respectively.

12. Setting $x = 1/t$, find the exponents at infinity. Derive the solutions

$$\frac{1}{x^\alpha} F\left(\alpha, \alpha - \gamma + 1, \alpha - \beta + 1, \frac{1}{x}\right)$$
$$\frac{1}{x^\beta} F\left(\beta, \beta - \gamma + 1, \beta - \alpha + 1, \frac{1}{x}\right)$$

7-4. The Legendre Differential Equation. An example of the preceding is the equation of Legendre,

$$(1 - x^2)y'' - 2xy' + n(n + 1)y = 0 \qquad (7\text{-}12)$$

which is important for certain physical problems. Taking $a = 1$, $b = -1$, we make the change of variable

$$x - 1 = -2t$$

The resulting equation is

$$t(1 - t)y'' + (1 - 2t)y' + n(n + 1)y = 0 \qquad (7\text{-}13)$$

a hypergeometric equation with the constants $\gamma = 1$, $\alpha = n + 1$, $\beta = -n$. We have, from this, the solution

$$y_1 = F[n + 1, -n, 1, \tfrac{1}{2}(1 - x)] \qquad (7\text{-}14)$$

Since the roots of the indicial equation are equal ($\gamma - 1 = 0$), we get no new solution from formula (7-10). We can get a second solution of (7-13), valid near $t = 0$, by quadratures. Put

$$y = y_1 w$$

where y_1 is the solution just found, and (7-13) becomes

CERTAIN CLASSICAL EQUATIONS

$$\frac{w''}{w'} = \frac{-2y_1'}{y_1} - \frac{1-2t}{t-t^2}$$

Integrating to get a particular solution,

$$\log w' = -2 \log y_1 - \log (t - t^2)$$

$$w' = \frac{1}{y_1{}^2(t-t^2)} = \frac{1}{[1 - 2(n+1)nt + \cdots](t-t^2)}$$

$$= \frac{1}{t} + (2n^2 + 2n + 1) + \cdots$$

$$w = \log t + k_0 + k_1 t + k_2 t^2 + \cdots$$

where k_0 is arbitrary, and k_1, k_2 are suitable coefficients. We have then a second solution

$$y_2 = y_1[\log (1 - x) + c_1(1 - x) + c_2(1 - x)^2 + \cdots]$$

The general solution is

$$y = C_1 y_1 + C_2 y_2$$

If y_2 is present, the solution has a logarithmic term. Hence, the most general solution analytic at $x = 1$ is a constant multiple of y_1.

7-5. Legendre Polynomials. Henceforth we suppose that n is a positive integer or zero. The hypergeometric series indicated in (7-14) breaks off and y_1 is a polynomial. It is called the *Legendre polynomial* of degree n and is represented by $P_n(x)$. Substituting in the series (7-9), we find

$$y_1 = P_n(x) = F[n+1, -n, 1, \tfrac{1}{2}(1-x)]$$

$$= 1 + \frac{(n+1)n}{2}(x-1) + \frac{(n+2)(n+1)n(n-1)}{(2!)^2 2^2}(x-1)^2$$

$$+ \cdots + \frac{2n(2n-1)(2n-2) \cdots 2 \cdot 1}{(n!)^2 2^n}(x-1)^n \quad (7\text{-}15)$$

We shall obtain an alternative expression for $P_n(x)$. We shall show first that

$$y = \frac{d^n}{dx^n}(x^2 - 1)^n \quad (7\text{-}16)$$

satisfies (7-12). It may be verified by direct substitution that

$$y = (t - t^2)^n$$

satisfies the hypergeometric equation

$$t(1-t)y'' + (1-n)(1-2t)y' + 2ny = 0$$

Here $\alpha = 1$, $\beta = -2n$, $\gamma = 1 - n$. Then the nth derivative

$$y = \frac{d^n}{dt^n}(t - t^2)^n$$

satisfies the hypergeometric equation with α, β, γ increased by n; that is, $n+1$, $-n$, 1; in other words Eq. (7-13). Replacing t in terms of x and discarding certain constant factors, we get (7-16) as a solution of (7-12).

Now (7-16) is a polynomial, and it follows from the remark at the close of the preceding section that it is a constant factor times $P_n(x)$. To get this factor it suffices to evaluate both polynomials at some point. Put

$$u = (x-1)^n \qquad v = (x+1)^n$$

and differentiate n times the product uv,

$$y = (uv)^{(n)} = u^{(n)}v + nu^{(n-1)}v' + \cdots + nu'v^{(n-1)} + uv^{(n)} \qquad (7\text{-}17)$$

For $x = 1$, u and its first $n-1$ derivatives vanish; hence

$$y(1) = u^{(n)}(1)v(1) = n!2^n$$

Since from (7-15), $P_n(1) = 1$, we have

$$P_n(x) = \frac{1}{n!2^n} \frac{d^n}{dx^n} (x^2 - 1)^n \qquad (7\text{-}18)$$

This is known as Rodrigues's formula.

At $x = -1$ all terms in (7-17) except the last vanish, and we have

$$y(-1) = u(-1)v^{(n)}(-1) = (-2)^n n!$$

whence

$$P_n(-1) = (-1)^n$$

If we expand $(x^2 - 1)^n$ by the binomial theorem and then differentiate n times, we get the following:

$$P_n(x) = \frac{(2n)!}{2^n(n!)^2} \left[x^n - \frac{n(n-1)}{2(2n-1)} x^{n-2} \right.$$
$$\left. + \frac{n(n-1)(n-2)(n-3)}{2 \cdot 4 \cdot (2n-1)(2n-3)} x^{n-4} + \cdots \right] \qquad (7\text{-}19)$$

We observe that $P_n(x)$ contains only even powers of x or only odd powers of x. Hence $P_n(x)$ is an even function of x, $P_n(-x) = P_n(x)$, if n is even; it is an odd function of x, $P_n(-x) = -P_n(x)$, if n is odd.

Following are the first few Legendre polynomials:

$$P_0(x) = 1 \qquad P_1(x) = x \qquad P_2(x) = \tfrac{1}{2}(3x^2 - 1)$$
$$P_3(x) = \tfrac{1}{2}(5x^3 - 3x) \qquad P_4(x) = \tfrac{1}{8}(35x^4 - 30x^2 + 3)$$

Their graphs are shown in Fig. 27.

7-6. Integral Properties of Legendre Polynomials. Among the striking properties of the Legendre polynomials are those connected with their integrals over the interval from -1 to 1. We prove first

CERTAIN CLASSICAL EQUATIONS

$$\int_{-1}^{1} P_m(x) P_n(x)\, dx = 0 \qquad (m \neq n) \qquad (7\text{-}20)$$

$P_n(x)$ satisfies Legendre's equation, which may be written in the form

$$\frac{d}{dx}\left[(1 - x^2) P_n'(x)\right] + n(n + 1) P_n(x) = 0$$

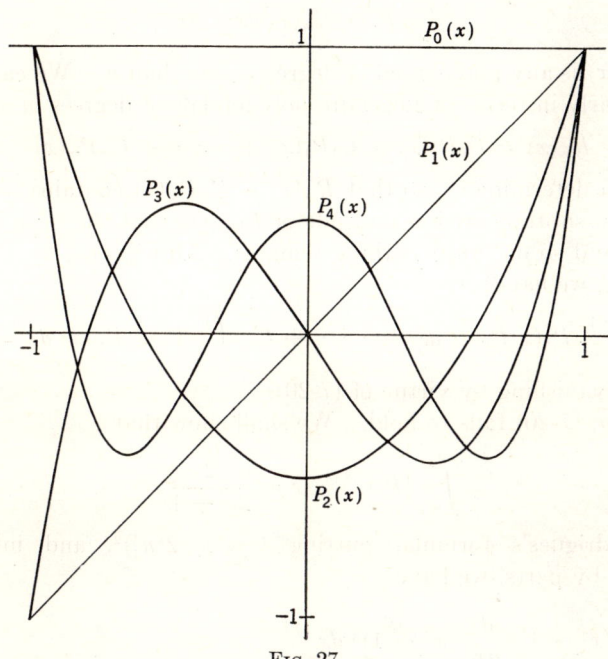

Fig. 27.

Multiplying by $P_m(x)$ and integrating,

$$\int_{-1}^{1} P_m(x) \frac{d}{dx}\left[(1 - x^2) P_n'(x)\right] dx + n(n + 1) \int_{-1}^{1} P_m(x) P_n(x)\, dx = 0$$

The first term may be integrated by parts, giving

$$\{P_m(x)[(1 - x^2) P_n'(x)]\}_{-1}^{1} - \int_{-1}^{1} (1 - x^2) P_n'(x) P_m'(x)\, dx$$

of which the first term vanishes. We have then

$$-\int_{-1}^{1} (1 - x^2) P_n'(x) P_m'(x)\, dx + n(n + 1) \int_{-1}^{1} P_m(x) P_n(x)\, dx = 0$$

Interchanging m and n, we have

$$-\int_{-1}^{1} (1 - x^2) P_m'(x) P_n'(x)\, dx + m(m + 1) \int_{-1}^{1} P_n(x) P_m(x)\, dx = 0$$

Subtracting, we get

$$(n - m)(n + m + 1) \int_{-1}^{1} P_m(x) P_n(x) \, dx = 0$$

which establishes (7-20).

From (7-20) we can show easily that

$$\int_{-1}^{1} P_n(x) R_m(x) \, dx = 0 \tag{7-21}$$

where $R_m(x)$ is any polynomial of degree m less than n. We can express $R_m(x)$ linearly in terms of Legendre polynomials of degrees m and less:

$$R_m(x) = C_0 P_0(x) + C_1 P_1(x) + \cdots + C_m P_m(x)$$

For we can determine C_m so that $R_m(x) - C_m P_m(x)$ contains no term in x^m; we then subtract such a multiple of $P_{m-1}(x)$ that the term in x^{m-1} is canceled, and so on, until nothing remains. Multiplying by $P_m(x)$ and integrating, we have

$$C_0 \int_{-1}^{1} P_n(x) P_0(x) \, dx + \cdots + C_m \int_{-1}^{1} P_n(x) P_m(x) \, dx = 0$$

each term vanishing by virtue of (7-20).

If $m = n$, (7-20) fails to hold. We shall show that

$$\int_{-1}^{1} [P_n(x)]^2 \, dx = \frac{2}{2n + 1} \tag{7-22}$$

Using Rodrigues's formula, putting $k = 1/(2^n n!)^2$, and integrating repeatedly by parts, we have

$$k \int_{-1}^{1} \frac{d^n}{dx^n} (x^2 - 1)^n \frac{d^n}{dx^n} (x^2 - 1)^n \, dx$$

$$= -k \int_{-1}^{1} \frac{d^{n-1}}{dx^{n-1}} (x^2 - 1)^n \frac{d^{n+1}}{dx^{n+1}} (x^2 - 1)^n \, dx$$

$$\cdots \cdots \cdots \cdots \cdots \cdots \cdots \cdots \cdots$$

$$= (-1)^n k \int_{-1}^{1} (x^2 - 1)^n \frac{d^{2n}}{dx^{2n}} (x^2 - 1)^n \, dx$$

$$= (2n)! k \int_{-1}^{1} (1 - x^2)^n \, dx$$

$$= 2k(2n)! \int_{0}^{1} (1 - x^2)^n \, dx$$

Putting $x = \cos \theta$, this becomes

$$2k(2n)! \int_{0}^{\frac{1}{2}\pi} \sin^{2n+1} \theta \, d\theta = 2k(2n)! \frac{2 \cdot 4 \cdots 2n}{3 \cdot 5 \cdots (2n + 1)}$$

$$= \frac{2(2n)!}{(2^n n!)^2} \cdot \frac{(2^n n!)^2}{(2n + 1)!} = \frac{2}{2n + 1}$$

CERTAIN CLASSICAL EQUATIONS 163

7-7. Expansion of an Arbitrary Function. It has been proved—by methods which are not at all simple—that many functions can be expanded in series of Legendre polynomials. If $f(x)$ has a continuous first derivative in the interval $-1 \leq x \leq 1$; or if $f(x)$ is continuous with a finite number of maxima and minima in the interval; even if $f(x)$ has a finite number of finite jumps or breaks between which it has the qualities of smoothness just named, it can be expanded in a series

$$f(x) = a_0 P_0(x) + a_1 P_1(x) + \cdots$$

valid, except where the jumps occur, in the interval $-1 < x < 1$.

Where such an expansion is possible the coefficients are readily found. Multiplying by $P_n(x)$ and integrating,

$$\int_{-1}^{1} f(x) P_n(x) \, dx = a_n \int_{-1}^{1} [P_n(x)]^2 \, dx = \frac{2 a_n}{2n + 1} \quad (7\text{-}23)$$

the remaining terms on the right vanishing by virtue of (7-20). From this we have a_n in terms of the integral on the left.

This expansion has the property that if the series be broken off after the term in $P_n(x)$, the resulting polynomial is the "best" approximation to the function in the interval by a polynomial of degree not exceeding n according to the requirement of least squares; that is, it is the polynomial $R_n(x)$ which minimizes the integral

$$U = \int_{-1}^{1} [f(x) - R_n(x)]^2 \, dx$$

Writing $R_n(x)$ in terms of Legendre polynomials,

$$R_n(x) = a_0 P_0(x) + \cdots + a_n P_n(x)$$

U is a function of $a_0, \ldots, a_n$ which are to be so chosen as to make U a minimum. Setting the various derivatives equal to zero, we have

$$\frac{1}{2} \frac{\partial U}{\partial a_k} = - \int_{-1}^{1} [f(x) - R_n(x)] P_k(x) \, dx$$

$$= - \int_{-1}^{1} f(x) P_k(x) \, dx + a_k \int_{-1}^{1} [P_k(x)]^2 \, dx = 0$$

whence a_k has the value found in (7-23). It is easily shown that we have an actual minimum, and not a maximum or minimax.

It can be proved that if $f(x)$ is *analytic* in a region enclosing the interval $-1 \leq x \leq 1$, the series converges to the function in the largest ellipse with ± 1 as foci within which $f(x)$ is analytic.

7-8. Roots of Legendre Polynomials. We now show that:
All the roots of $P_n(x)$ lie in the interval $-1 < x < 1$ and are distinct.

That the roots in the interval are distinct is immediate. At a multiple root both $P_n(x)$ and $P_n'(x)$ would be zero, and the only solution of the differential equation with this property is identically zero.

We put aside $P_0(x)$, which has no roots. Let $a_1, \ldots, a_m$ be the roots of $P_n(x)$ in the interval. At these points the graph of $P_n(x)$ crosses the x axis. If we assume that the theorem is not true, then $m < n$. We now form the function

$$R_m(x) = (x - a_1)(x - a_2) \cdots (x - a_m)$$

a polynomial of degree less than n. This, like $P_n(x)$, is positive at $x = 1$, and it has the same sign throughout the interval as $P_n(x)$ since its graph crosses the axis at the same points. [If $P_n(x)$ does not cross the x axis we use the function $R_0(x) \equiv 1$ which has this property.] Then in the interval

$$P_n(x) R_m(x) \geqslant 0 \qquad \int_{-1}^{1} P_n(x) R_m(x)\, dx > 0$$

This contradicts Eq. (7-21), which establishes the proposition.

EXERCISES

1. Compute $P_5(x)$.
2. Show that

$$\frac{d}{dx}\left[(x - a)(x - b) y'\right] - cy = 0$$

is reduced to a Legendre equation by the change of variable

$$x = \tfrac{1}{2}[(a - b)t + a + b]$$

3. Express x^4 in a series of Legendre polynomials.
4. Prove that the roots of $P_n(x)$ lie in the interval $-1 < x < 1$ from Rodrigues's formula, by applying Rolle's theorem to the successive derivatives of $(x^2 - 1)^n$.
5. Show that the change of variable $t = x^2$ reduces Legendre's equation to a hypergeometric equation, and express $P_n(x)$ in terms of a hypergeometric function whose fourth element is x^2.
6. Show that

$$\int_{-1}^{1} P_n(x) P'_{n+1}(x)\, dx = 2$$

7. Show that

$$\int_{-1}^{1} x P_n(x) P_{n-1}(x)\, dx = \frac{2n}{4n^2 - 1}$$

8. Show that the minimum value of U in Sec. 7-7 is

$$\int_{-1}^{1} [f(x)]^2\, dx - 2\left(a_0^2 + \frac{a_1^2}{3} + \cdots + \frac{a_n^2}{2n + 1}\right)$$

where the a's have the values found in the text.

9. Find the best polynomial of the second degree to approximate to $\cos\dfrac{\pi x}{2}$ in the interval $-1 \leqslant x \leqslant 1$.

10. Show that if $f(x)$ is an even function of x, then in its expansion in Legendre polynomials,

$$a_{2k} = (4k + 1) \int_0^1 f(x) P_{2k}(x)\, dx \qquad a_{2k+1} = 0$$

11. Show that $f(x)$ defined in $0 \leqslant x \leqslant 1$ may be expressed in a series of even Legendre polynomials. (In $-1 \leqslant x < 0$, put $f(x) = f(-x)$ and apply the preceding results.)

12. State results analogous to the preceding two problems for odd functions.

13. Expand x in the interval $0 \leqslant x \leqslant 1$ in a series of even Legendre polynomials. Sketch the first three approximations.

Derive the following recurrence formulas:

14. $P'_{n+1}(x) - xP'_n(x) = (n + 1)P_n(x)$.
15. $(n + 1)P_{n+1}(x) - (2n + 1)xP_n(x) + nP_{n-1}(x) = 0$.
16. $xP'_n(x) - P'_{n-1}(x) = nP_n(x)$.
17. $P'_{n+1}(x) - P'_{n-1}(x) = (2n + 1)P_n(x)$.
18. $(x^2 - 1)P'_n(x) = nxP_n(x) - nP_{n-1}(x)$.

7-9. Bessel's Differential Equation.

The equation

$$x^2 y'' + xy' + (x^2 - n^2)y = 0 \tag{7-24}$$

is called Bessel's equation. Like Legendre's equation, it arose from physical problems. It also has a very extensive literature.[1] We shall here find the customary solution in series form and derive a few of its properties.

We observe that $x = 0$ is a regular singular point and that all other finite points are ordinary points. It will be found that infinity is an irregular singular point. To find the exponents at the origin we write the expressions (7-5):

$$p(x) = 1 \qquad q(x) = -n^2 + x^2$$

The indicial equation then is

$$\alpha(\alpha - 1) + \alpha - n^2 = 0$$

and has the roots n and $-n$. Suppose $n \geqslant 0$.

Substituting the series

$$y = x^n + c_1 x^{n+1} + c_2 x^{n+2} + \cdots$$

into (7-24) and equating the coefficient of each power to zero, we find from the terms in x^{n+1}

$$(2n + 1)c_1 = 0$$

and from the terms in x^{n+r}, $r > 1$

$$r(2n + r)c_r + c_{r-2} = 0$$

---

[1] The 800-page "Treatise on the Theory of Bessel Functions," by G. N. Watson, has 36 pages of bibliography.

Each coefficient is a multiple of the second preceding. Since $c_1 = 0$, all coefficients with odd subscripts vanish. Those with even subscripts are readily built up, starting from $c_0 = 1$ and using the relation

$$c_r = -\frac{c_{r-2}}{r(2n+r)}$$

We find the series

$$y = x^n \left[1 - \frac{x^2}{2(2n+2)} + \frac{x^4}{2 \cdot 4(2n+2)(2n+4)} - \cdots \right]$$

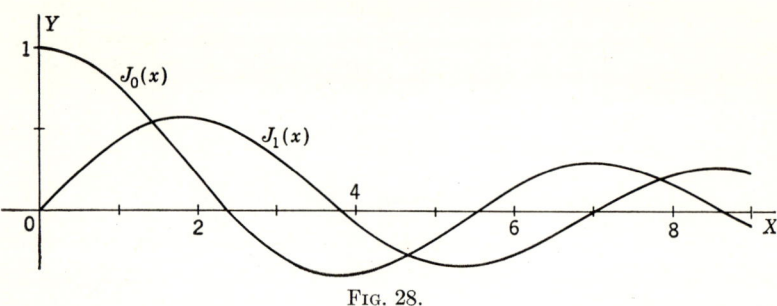

Fig. 28.

We have a like solution, replacing n by $-n$ in this series, for the other root of the indicial equation, provided n is not an integer.

The test-ratio test readily establishes the convergence of the series. As $r \to \infty$, we have

$$\left| \frac{c_r x^r}{c_{r-2} x^{r-2}} \right| = \left| \frac{x^2}{r(2n+r)} \right| \to 0$$

and the series converges for all values of x. This agrees with the theory, since the nearest singular point to $x = 0$ is at infinity.

If n is an integer, it is customary to introduce a factor $1/2^n n!$ and designate the function by $J_n(x)$:

$$J_n(x) = \frac{x^n}{2^n n!} \left[1 - \frac{x^2}{2^2(n+1)} + \frac{x^4}{2^4 \cdot 2!(n+1)(n+2)} \right.$$
$$\left. - \frac{x^6}{2^6 \cdot 3!(n+1)(n+2)(n+3)} + \cdots \right] \quad (7\text{-}25)$$

$J_n(x)$ is defined for all values of n except negative integers by putting $\Gamma(n+1)$ in place of $n!$.

The case $n = 0$ is of particular importance. The series then takes the form

$$J_0(x) = 1 - \frac{x^2}{2^2} + \frac{x^4}{2^2 \cdot 4^2} - \frac{x^6}{2^2 \cdot 4^2 \cdot 6^2} + \cdots \quad (7\text{-}26)$$

The graphs of $J_0(x)$ and $J_1(x)$ are shown in Fig. 28.

CERTAIN CLASSICAL EQUATIONS 167

If n is not integral, the general solution of (7-24) is

$$y = AJ_n(x) + BJ_{-n}(x) \tag{7-27}$$

If we use a nonreal value of n, this general solution also holds.
If n is a positive integer, let

$$y = J_n(x)w$$

Then (7-24) becomes

$$\frac{w''}{w'} = -\frac{2J_n'}{J_n} - \frac{1}{x}$$

whence, integrating,

$$\log w' = -2 \log J_n - \log x + k \qquad w' = \frac{C_1}{xJ_n^2}$$

$$y = J_n(x)w = C_1 J_n(x) \int \frac{dx}{xJ_n(x)^2} + C_2 J_n(x)$$

The expansion of the integrand in ascending powers of x will contain a term in $1/x$, so the integration will produce a logarithmic term. Many combinations have been used for the second solution, and the complicated expansions will be found in books of tables. If $n = 0$, one such, taking $C_1 = 1$, $C_2 = 0$, is the Neumann solution

$$y = K_0(x) = J_0(x) \log x + \frac{x^2}{2^2} - \frac{x^4}{2^2 \cdot 4^2}\left(1 + \frac{1}{2}\right)$$

$$+ \frac{x^6}{2^2 \cdot 4^2 \cdot 6^2}\left(1 + \frac{1}{2} + \frac{1}{3}\right) - \cdots$$

By a change of variable many practical equations can be reduced to Bessel's form. The following type is useful:

$$x^2 y'' + Axy' + (Bx^m + C)y = 0 \tag{7-28}$$

where A, B, C are constants. To solve this let us, for convenience, write Bessel's equation (7-24) with the variables X and Y and put

$$Y = x^{-p}y \qquad X = ax^q$$

We get

$$x^2 y'' + (1 - 2p)xy' + (a^2 q^2 x^{2q} + p^2 - q^2 n^2)y = 0 \tag{7-29}$$

This is (7-28) if

$$p = \frac{1 - A}{2} \qquad q = \frac{m}{2} \qquad a = \frac{2\sqrt{B}}{m} \qquad n = \frac{\sqrt{(1 - A)^2 - 4C}}{m}$$

We then have as a solution $y = x^p J_n(ax^q)$

Example. Solve

$$x^2 y'' - xy' + (x + 1)y = 0$$

168   DIFFERENTIAL EQUATIONS

Here $p = 1$, $q = \frac{1}{2}$, $a = 2$, $n = 0$. The general solution is

$$y = x[C_1 J_0(2\sqrt{x}) + C_2 K_0(2\sqrt{x})]$$

7-10. Roots of Bessel Functions. For n a nonnegative integer, as we continue to assume, $J_n(x)$ is an even or an odd function. Hence the negative of a root is a root, and we shall consider only positive roots. We shall prove the following theorem:

THEOREM. *Between two positive roots of $J_n(x)$ lie a root of $J_{n+1}(x)$ and a root of $J_{n-1}(x)$.*

We make use of certain relations between successive Bessel functions. From the series (7-25) we have

$$\frac{d}{dx}[x^{-n} J_n(x)] = \frac{1}{2^n n!}\left[-\frac{x}{2(n+1)} + \frac{x^3}{2^3(n+1)(n+2)}\right.$$

$$\left. - \frac{x^5}{2^5 2!(n+1)(n+2)(n+3)} + \cdots \right] = -\frac{x}{2^{n+1}(n+1)!}$$

$$\left[1 - \frac{x^2}{2^2(n+2)} + \frac{x^4}{2^4 2!(n+2)(n+3)} - \cdots \right] = -x^{-n} J_{n+1}(x) \quad (7\text{-}30)$$

Between two positive roots of $J_n(x)$, and hence also of $x^{-n} J_n(x)$, the derivative of this latter function must vanish at least once. Hence, from the relation just established, $J_{n+1}(x) = 0$.

Similarly, for $n > 0$, we get from (7-25):

$$\frac{d}{dx}[x^n J_n(x)] = \frac{1}{2^n n!}\left[2nx^{2n-1} - \frac{(2n+2)x^{2n+1}}{2^2(n+1)}\right.$$

$$\left. + \frac{(2n+4)x^{2n+3}}{2^4 2!(n+1)(n+2)} - \cdots \right] = \frac{x^{2n-1}}{2^{n-1}(n-1)!}\left[1 - \frac{x^2}{2^2 n}\right.$$

$$\left. + \frac{x^4}{2^4 2! n(n+1)} - \cdots \right] = x^n J_{n-1}(x) \quad (7\text{-}31)$$

From this, between two roots of $x^n J_n(x)$ lies a root of its derivative and hence of $J_{n-1}(x)$.

The oscillatory character of the Bessel functions is best shown after an alteration of the differential equation. If we put

$$y = \frac{u}{\sqrt{x}}$$

into (7-24), we get

$$u'' + \left(1 - \frac{4n^2 - 1}{4x^2}\right) u = 0 \quad (7\text{-}32)$$

This has the solution $u = \sqrt{x} J_n(x)$, which has the same positive roots as $J_n(x)$.

We see from the preceding that for large x the equation differs little from $u'' + u = 0$, whose solutions are sine functions of period 2π. So

CERTAIN CLASSICAL EQUATIONS

for large x a solution y of (7-24) oscillates roughly in the same way but, owing to the factor $1/\sqrt{x}$, with steadily decreasing amplitude.

The more precise study of the roots is probably best made by the use of the following proposition, which is one of a series of oscillation theorems due to Sturm.

THEOREM. *Let $u(x)$, $v(x)$, be solutions of two differential equations*

$$u'' + g(x)u = 0$$
$$v'' + h(x)v = 0$$

in an interval in which the coefficients of the equations are continuous. If $\alpha < \beta$ are consecutive roots of $u(x)$ and if

$$h(x) \geqslant g(x) \qquad h(x) \not\equiv g(x)$$

in the interval $\alpha\beta$, then there is a root of $v(x)$ between α and β.

In proving this, we have

$$\frac{d}{dx}(vu' - uv') = vu'' - uv'' = (h - g)uv$$

and, integrating,

$$[vu' - uv']_\alpha^\beta = \int_\alpha^\beta (h - g)uv \, dx \qquad (7\text{-}33)$$

Suppose $v(x)$ has no root between α and β. We may suppose without loss of generality that u and v are both positive for $\alpha < x < \beta$ (replacing either by its negative, if necessary). Then $u'(\alpha) > 0$, $u'(\beta) < 0$. The first member of (7-33) is negative or zero,

$$v(\beta)u'(\beta) - v(\alpha)u'(\alpha) \leqslant 0$$

whereas the second member is positive. This contradiction proves the theorem.

We investigate first the zeros of $J_0(x)$. We compare

$$u = \sqrt{x}J_0(x)$$
$$v = \sin(x - a)$$

which satisfy

$$u'' + \left(1 + \frac{1}{4x^2}\right)u = 0$$
$$v'' + v = 0$$

Since

$$\left(1 + \frac{1}{4x^2}\right) - 1 = \frac{1}{4x^2} > 0$$

there is a root of u between the roots a and $a + \pi$ of v. We have the following result:

There is a root of $J_0(x)$ in each interval of length π of the positive x axis.

It follows from this result and the interlacing of the roots that $J_1(x)$, $J_2(x)$, etc., have each infinitely many positive roots.

Let α, β be consecutive roots of $J_n(x)$. We compare

$$u = \sqrt{x}\, J_n(x)$$
$$v = \sin(x - \alpha)$$

which satisfy

$$u'' + \left(1 - \frac{4n^2 - 1}{4x^2}\right) u = 0$$
$$v'' + v = 0$$

Here

$$1 - \left(1 - \frac{4n^2 - 1}{4x^2}\right) > 0 \qquad n > \frac{1}{2}$$

hence there is a root of v between α and β. Since the root of v following α is $\alpha + \pi$, we have $\beta > \alpha + \pi$. Hence:

Consecutive roots of $J_n(x)$ ($n > \frac{1}{2}$) differ by more than π.

If we compare u, as just defined, with $v = \sin p(x - \alpha)$ which satisfies the equation

$$v'' + p^2 v = 0$$

and if we take p less than 1 but as near 1 as we wish, we have

$$\left(1 - \frac{4n^2 - 1}{4x^2}\right) - p^2 > 0$$

for $x > \xi$, sufficiently large. For $\alpha > \xi$ the root β occurs before the next root $\alpha + \dfrac{\pi}{p}$ of v. The nearer p is to 1 the larger the corresponding ξ.

The difference between sufficiently large consecutive roots of $J_n(x)$ ($n > \frac{1}{2}$) exceeds π by an arbitrarily small amount.

We can show directly from the differential equation,

$$x^2 J_n'' + x J_n' + (x^2 - n^2) J_n = 0$$

that the first positive root α_1 exceeds n. At some point x_0, between 0 and α_1, J_n has a maximum. There

$$J_n(x_0) > 0 \qquad J_n'(x_0) = 0 \qquad J_n''(x_0) \leqslant 0$$

and the equation becomes

$$x_0^2 J_n''(x_0) + (x_0^2 - n^2) J_n(x_0) = 0$$

The first member is negative for $x_0 < n$; hence we have $\alpha_1 > x_0 \geqslant n$. That $J_n(x)$ has no complex roots will appear in the next section.

CERTAIN CLASSICAL EQUATIONS

7-11. Integral Properties. Expansions. Setting αx in place of x, Eq. (7-32) takes the form

$$u'' + \left(\alpha^2 - \frac{4n^2 - 1}{4x^2}\right)u = 0$$

which has the solution

$$u = \sqrt{x}\, J_n(\alpha x)$$

Similarly

$$v = \sqrt{x}\, J_n(\beta x)$$

satisfies

$$v'' + \left(\beta^2 - \frac{4n^2 - 1}{4x^2}\right)v = 0$$

Applying Eq. (7-33) to these equations and integrating from 0 to x, we have

$$(\beta^2 - \alpha^2)\int_0^x uv\, dx = [vu' - uv']_0^x$$

that is,

$$(\beta^2 - \alpha^2)\int_0^x xJ_n(\alpha x)J_n(\beta x)\, dx$$
$$= x[\alpha J_n(\beta x)J_n'(\alpha x) - \beta J_n(\alpha x)J_n'(\beta x)] \quad (7\text{-}34)$$

If we differentiate the last equation with respect to β and then set $\beta = \alpha$, we find

$$2\alpha \int_0^x xJ_n^{\,2}(\alpha x)\, dx$$
$$= x[\alpha xJ_n'^{\,2}(\alpha x) - J_n(\alpha x)J_n'(\alpha x) - \alpha xJ_n(\alpha x)J_n''(\alpha x)] \quad (7\text{-}35)$$

From (7-34) we have

$$(\beta^2 - \alpha^2)\int_0^1 xJ_n(\alpha x)J_n(\beta x)\, dx = \alpha J_n(\beta)J_n'(\alpha) - \beta J_n(\alpha)J_n'(\beta)$$

The second member vanishes if α and β are one of the following:

(i)   Roots of $J_n(x)$
(ii)   Roots of $J_n'(x)$
(iii)   Roots of $xJ_n'(x) + hJ_n(x)$
(iv)   Roots of $J_{n+1}(x)$
(v)   Roots of $J_{n-1}(x)$

Here (i) and (ii) are immediate; (iii) is easily established; and (iv) and (v) are special cases of (iii), using the results of Exercises 11 and 12 of the following group. If α and β are distinct positive roots, or in general if $\beta^2 - \alpha^2 \neq 0$, we have for any of the five cases mentioned above the important integral property

$$\int_0^1 xJ_n(\alpha x)J_n(\beta x)\, dx = 0 \quad (7\text{-}36)$$

From this equation we can show that there are no imaginary roots. Owing to the reality of $J_n(x)$, if there is a complex root $\alpha = a + ib$, then $\beta = a - ib$ is a root. Then $J_n(\alpha x)$ and $J_n(\beta x)$ are conjugate imaginaries and their product is positive, so (7-36) fails to hold. This shows that there are no complex roots except for the case of a pure imaginary $\alpha = ib$, when $\beta^2 = \alpha^2$. But we see directly from the series (7-25) that if x is a pure imaginary, the terms are all positive and their sum cannot be zero.

If we assume the expansion of an arbitrary function in the interval 01 in a series of the form

$$f(x) = c_1 J_n(\alpha_1 x) + c_2 J_n(\alpha_2 x) + \cdots$$

where $\alpha_1, \alpha_2, \ldots$ are the successive positive roots of any one of the functions (i), ..., (v), the coefficients are readily found, assuming that we can integrate term by term. Multiplying by $xJ_n(\alpha_k x)$ and integrating, we have, by virtue of (7-36),

$$\int_0^1 xJ_n(\alpha_k x)f(x)\,dx = c_k \int_0^1 xJ_n^2(\alpha_k x)\,dx$$

This determines c_k. The second integral, which is independent of $f(x)$, may be evaluated by means of (7-35). In case (i), we have

$$\int_0^1 xJ_n^2(\alpha_k x)\,dx = \tfrac{1}{2}J_n'^2(\alpha_k) = \tfrac{1}{2}J_{n+1}^2(\alpha_k) = \tfrac{1}{2}J_{n-1}^2(\alpha_k)$$

the last two expressions coming from Exercises 11 and 12.

With the values of c_k here found it is readily shown that for each of the five cases the first m terms of the series define the function of the form

$$R(x) = c_1 J_n(\alpha_1 x) + \cdots + c_m J_n(\alpha_m x)$$

which minimizes the integral

$$U = \int_0^1 x[f(x) - R(x)]^2\,dx$$

EXERCISES

1. Work out the series for $J_0(x)$ directly from the differential equation.
2. Use the series to compute the following values:

$$J_0(1) = .7652 \quad J_0(2) = .2239 \quad J_0(3) = -.2601$$

3. Carry through the change of variables resulting in Eq. (7-29) in the text.
4. Find, using a Bessel function, a solution of $xy'' + y = 0$ which is analytic at $x = 0$.

Find the general solutions of the following:
5. $xy'' + (2n+1)y' + xy = 0$.
6. $x^2 y'' + xy' - (x^2 + 2)y = 0$.
7. $y'' + 9k^2 xy = 0$.
8. $x^2 y'' + 3xy' + (b^2 x^2 + 1)y = 0$.

CERTAIN CLASSICAL EQUATIONS

9. Solve by putting $y = e^x u$,
$$xy'' + (1 - 2x)(y' - y) = 0$$

10. Show that
$$xy'' + y' - ixy = 0$$
has the solution
$$y = \operatorname{ber} x + i \operatorname{bei} x$$
where
$$\operatorname{ber} x = 1 - \frac{x^4}{2^2 \cdot 4^2} + \frac{x^8}{2^2 \cdot 4^2 \cdot 6^2 \cdot 8^2} - \cdots$$
$$\operatorname{bei} x = \frac{x^2}{2^2} - \frac{x^6}{2^2 \cdot 4^2 \cdot 6^2} + \frac{x^{10}}{2^2 \cdot 4^2 \cdot 6^2 \cdot 8^2 \cdot 10^2} - \cdots$$

By the use of Eqs. (7-30) and (7-31), or otherwise, establish the following:

11. $J_n'(x) = \frac{n}{x} J_n(x) - J_{n+1}(x)$. **12.** $J_n'(x) = J_{n-1}(x) - \frac{n}{x} J_n(x)$.

13. $J_{n+1}(x) + J_{n-1}(x) = \frac{2n}{x} J_n(x)$. **14.** $J_2(x) - J_0(x) = 2J_0''(x)$.

15. $J_2(x) = J_0''(x) - \frac{1}{x} J_0'(x)$. **16.** $J_0'(x) = -J_1(x)$.

Given $\Gamma(\frac{1}{2}) = \sqrt{\pi}$ and recalling that $\Gamma(n + 1) = n\Gamma(n)$, show that

17. $J_{1/2}(x) = \sqrt{\frac{2}{\pi x}} \sin x$. **18.** $J_{3/2}(x) = \sqrt{\frac{2}{\pi x}} \left(\frac{\sin x}{x} - \cos x \right)$.

19. Verify that 17 satisfies Bessel's equation with $n = \frac{1}{2}$.
20. Show that Eq. (7-36) holds for cases (iii), (iv), and (v) of the text.
21. Carry through the work of minimizing the integral U above.
22. Show that $J_n(x)$ is the coefficient of t^n in the expansion in powers of t of
$$e^{\frac{x}{2}\left(t - \frac{1}{t}\right)}$$

CHAPTER 8

INTERPOLATION AND NUMERICAL INTEGRATION

In the present chapter the reader will be made acquainted with certain fundamental numerical processes. The methods to be investigated are of the greatest importance in all sorts of problems in which definite numerical results are sought and should form a part of the mathematical equipment of every scientific worker.

8-1. Interpolation. Interpolation is the process of replacing a function by another function which can, for practical purposes, be used in its stead. If the difference between the two functions is sufficiently small, the values of the interpolating function can be computed and used in lieu of those of the original function, or its integral may be used instead of the integral of the original function. If the difference of their slopes is sufficiently small, the slope of the interpolating function may be employed in place of the slope of the original function, or the length of one graph be used for the length of the other; and so on.

The reader has been interpolating since his high-school days. From the entries in a table of common logarithms,

$$\log 1.81 = 0.2577 \qquad \log 1.82 = 0.2601$$

one computes intermediate values,

$$\log 1.815 = 0.2577 + \tfrac{1}{2}(0.2601 - 0.2577) = 0.2589$$
$$\log 1.813 = 0.2577 + \tfrac{3}{10}(0.2601 - 0.2577) = 0.2584$$

In this computation, no properties of the logarithm are used. It is assumed merely that the change in the function is proportional to the change in the independent variable in the interval. This assumption is strictly true only if the graph of the function is a straight line. What has actually been done has been to replace the curved graph of $\log x$ in the interval $1.81 < x < 1.82$ by its chord (Fig. 29). If the entries of the table are sufficiently close together, the errors of this proceeding are negligible.

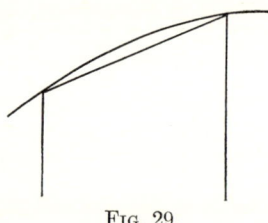

FIG. 29.

The use of an interpolating function in the preceding example was a matter of convenience. We could, at the expense of some labor, compute the logarithm of any number in the interval to any desired degree of accuracy. There are other cases, however, in which the use of an interpolating function is forced upon us. For example, the values of the function may be found experimentally at isolated points and an analytic formula for the function be quite unknown. We should thus have isolated points on the graph of the function. Plotting these points and drawing a smooth curve through them, we make the hypothesis that the curve so drawn is a reasonable representation of the graph of the desired function. The curve thus constructed is the graph of an interpolating function.

8-2. Parabolic Interpolation. Lagrange's Formula. Various types of functions have been employed as interpolating functions, certain

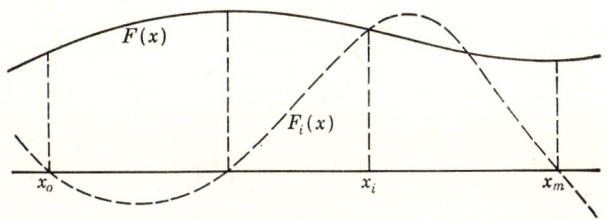

Fig. 30.

classes of functions being particularly well represented by special kinds of interpolating functions. For general use it is especially important that the interpolating functions be easy to manipulate—to compute values, to integrate, to differentiate, and the like. The simplest functions from all these points of view are the polynomials. The graph of a polynomial of the nth degree is called a *parabola of the nth degree*. Hence, interpolation in which the interpolating function is a polynomial is called *parabolic interpolation*.

In parabolic interpolation goodness of fit is secured by making the parabola pass through specified points on the graph of the function. The possibility of accomplishing this results from the following theorem.

THEOREM. *There is one and only one polynomial $F(x)$ of degree not exceeding m which takes on prescribed values $f_0, f_1, \ldots, f_m$ for distinct values $x = x_0, x_1, \ldots, x_m$, respectively.*

We shall show first that there is one such polynomial. We begin by setting up a polynomial which vanishes at all the points $x_0, \ldots, x_m$, except the point x_i at which it has the value f_i (Fig. 30). A polynomial vanishing in the desired way is

$$F_i(x) = C_i(x - x_0)(x - x_1) \cdots (x - x_{i-1})(x - x_{i+1}) \cdots (x - x_m)$$

We can now determine C_i so that the value f_i is taken on at x_i,

$$F_i(x_i) = C_i(x_i - x_0) \cdots (x_i - x_{i-1})(x_i - x_{i+1}) \cdots (x_i - x_m)$$
$$= f_i$$

None of the factors by which C_i is multiplied is zero, so a division gives C_i. The polynomial is then explicitly

$$F_i(x) = \frac{(x - x_0)(x - x_1) \cdots (x - x_{i-1})(x - x_{i+1}) \cdots (x - x_m)}{(x_i - x_0)(x_i - x_1) \cdots (x_i - x_{i-1})(x_i - x_{i+1}) \cdots (x_i - x_m)} f_i \quad (8\text{-}1)$$

Forming the polynomials for all the points and adding,

$$F(x) = F_0(x) + F_1(x) + \cdots + F_m(x) \quad (8\text{-}2)$$

we have a polynomial taking on the prescribed values for the given values of x. The final polynomial is

$$F(x) = \frac{(x - x_1)(x - x_2) \cdots (x - x_m)}{(x_0 - x_1)(x_0 - x_2) \cdots (x_0 - x_m)} f_0$$
$$+ \frac{(x - x_0)(x - x_2) \cdots (x - x_m)}{(x_1 - x_0)(x_1 - x_2) \cdots (x_1 - x_m)} f_1 + \cdots$$
$$+ \frac{(x - x_0)(x - x_1) \cdots (x - x_{i-1})(x - x_{i+1}) \cdots (x - x_m)}{(x_i - x_0)(x_i - x_1) \cdots (x_i - x_{i-1})(x_i - x_{i+1}) \cdots (x_i - x_m)} f_i$$
$$+ \cdots + \frac{(x - x_0)(x - x_1) \cdots (x - x_{m-1})}{(x_m - x_0)(x_m - x_1) \cdots (x_m - x_{m-1})} f_m \quad (8\text{-}3)$$

It is clear that $F(x)$ is of degree m at most, for none of the polynomials which are added together is of degree greater than m. It may happen, of course, that terms of the highest degree cancel, so that the degree of $F(x)$ is less than m.

Equation (8-3) is known as *Lagrange's formula*.

Suppose, now, that there is a second polynomial $G(x)$ of degree m or less, taking on the values $f_0, \ldots, f_m$ at $x_0, \ldots, x_m$, respectively. Then the polynomial, $G(x) - F(x)$, which is of degree m at most, is zero for the $m + 1$ values $x_0, \ldots, x_m$. But a polynomial of degree m at most cannot have $m + 1$ roots except it vanish identically:

$$G(x) - F(x) \equiv 0 \qquad G(x) \equiv F(x)$$

Hence, there is but one polynomial whose degree does not exceed m which takes on the prescribed values at the given points.

It follows from the theorem, taking $m = 1, 2, 3, \ldots$, that a straight line can be drawn through two points on the graph, a parabola of the second degree at most can be drawn through three points, a parabola of the third degree at most can be drawn through four points, etc. In each case there is only one such parabola.

8-3. Finite Differences. It often happens, as in the case of a prepared table, that the original function is known for values of x which are equally spaced along the x axis, say at $x = a, a + h, a + 2h, \ldots, a + mh$. In this case, the interpolating polynomial can be expressed in a useful alternative form by the formation of a table of differences.

The first difference, $\Delta f(x)$, of a function $f(x)$ is defined as follows:

$$\Delta f(x) = f(x + h) - f(x)$$

We note that if $f(x + h) \equiv f(x)$, then the first difference vanishes identically. This is true if $f(x)$ is a constant, or, more generally, if $f(x)$ is a periodic function with the period h.

The second difference is the first difference of the first difference,

$$\Delta^2 f(x) = \Delta[\Delta f(x)]$$
$$= [f(x + 2h) - f(x + h)] - [f(x + h) - f(x)]$$

etc., the rth difference being defined as the first difference of the $(r-1)$th difference:

$$\Delta^r f(x) = \Delta[\Delta^{r-1} f(x)]$$

We now prove a theorem concerning the differences of a polynomial.

THEOREM. *The rth difference ($r \leq m$) of a polynomial of the mth degree is a polynomial of degree $m - r$. The mth difference is a constant, and all higher differences are zero.*

Let

$$f(x) = a_0 x^m + a_1 x^{m-1} + \cdots + a_m \qquad a_0 \neq 0$$

Then

$$f(x + h) = a_0(x + h)^m + a_1(x + h)^{m-1} + \cdots + a_m$$
$$= a_0 x^m + (a_1 + mha_0) x^{m-1} + \cdots$$

Forming the difference, $f(x + h) - f(x)$, we have

$$\Delta f(x) = mha_0 x^{m-1} + \text{terms of lower degree}$$

The first difference is thus a polynomial of degree $m - 1$. Applying the result just found to the new polynomial $\Delta f(x)$, we have

$$\Delta^2 f(x) = m(m - 1) h^2 a_0 x^{m-2} + \text{terms of lower degree}$$

Continuing, we have, for $r \leq m$,

$$\Delta^r f(x) = m(m - 1) \cdots (m - r + 1) h^r a_0 x^{m-r} + \text{terms of lower degree}$$

This is a polynomial of degree $m - r$. When $r = m$ the difference is constant,

$$\Delta^m f(x) = m(m - 1) \cdots 1 h^m a_0 = m! h^m a_0$$

and the next and all succeeding differences are zero.

EXERCISES

1. Find the nth difference of the function m^x.
2. Writing
$$x^{(n)} = x(x - h) \cdots [x - (n - 1)h]$$
$$x^{(-n)} = \frac{1}{x(x + h) \cdots [x + (n - 1)h]}$$
show that
$$\Delta x^{(n)} = hnx^{(n-1)}$$
$$\Delta x^{(-n)} = -hnx^{(-n-1)}$$
and find the rth difference of each function.

3. Derive a formula for the first difference of a product; of a quotient.
4. Show that
$$y = A3^x + B2^x$$
satisfies the difference equation ($h = 1$)
$$\Delta^2 y - 3\Delta y + 2y = 0$$

5. Find solutions of the difference equations ($h = 1$)

(a) $\quad \Delta^2 y - 6\Delta y + 8y = 0$
(b) $\quad \Delta^2 y - \Delta y = 0$

6. Show that Lagrange's formula, (8-3), may be written in the alternative form
$$F(x) = c_0 + c_1(x - x_0) + c_2(x - x_0)(x - x_1) + \cdots$$
and derive the relation
$$c_i = \frac{f_0}{(x_0 - x_1) \cdots (x_0 - x_i)} + \cdots + \frac{f_i}{(x_i - x_0) \cdots (x_i - x_{i-1})}$$

8-4. Difference Tables. When a function is known for equally spaced values of the argument, the differences for the various entries may be got immediately by subtractions. It is customary to write these differences in tabular form. We illustrate the arrangement in the following table. The function used is $f(x) = x^3$, and the tabular difference is $h = 2$.

x	x^3	Δx^3	$\Delta^2 x^3$	$\Delta^3 x^3$	$\Delta^4 x^3$
10	1000	728	288	48	0
12	1728	1016	336	48	0
14	2744	1352	384	48	
16	4096	1736	432		
18	5832	2168			
20	8000				

In this arrangement, the function and its differences for a given value of the argument appear in a horizontal line. The column headed x is supposed known, and the remaining entries are formed by subtractions

INTERPOLATION AND NUMERICAL INTEGRATION

Thus,
$$\Delta f(14) = f(16) - f(14) = 4096 - 2744 = 1352$$
$$\Delta^2 f(14) = \Delta f(16) - \Delta f(14) = 1736 - 1352 = 384$$

and so on. Each entry in the difference columns is found by subtracting the entry immediately to the left from the entry just below the latter.

In general, for a function $f(x)$ given for the values $x = a, a + h, \ldots, a + mh$, the table of differences has the arrangement:

x	$f(x)$	$\Delta f(x)$	$\Delta^2 f(x)$	$\ldots$	$\Delta^m f(x)$
a	$f(a)$	$\Delta f(a)$	$\Delta^2 f(a)$		$\Delta^m f(a)$
$a + h$	$f(a + h)$	$\Delta f(a + h)$	$\Delta^2 f(a + h)$		
$a + 2h$	$f(a + 2h)$	$\Delta f(a + 2h)$	$\Delta^2 f(a + 2h)$		
$\ldots$	$\ldots$	$\ldots$	$\ldots$		
$a + (m - 1)h$	$f[a + (m - 1)h]$	$\Delta f[a + (m - 1)h]$			
$a + mh$	$f(a + mh)$				

The $m + 1$ entries supply finally one entry in the column of mth differences.

8-5. Newton's Interpolation Formula. We shall now derive formulas for the entries in the table in terms of the entries in the first row. For the entries in the second row we have, in view of the way the table was constructed,

$$f(a + h) = f(a) + \Delta f(a)$$
$$\Delta f(a + h) = \Delta f(a) + \Delta^2 f(a)$$
$$\Delta^2 f(a + h) = \Delta^2 f(a) + \Delta^3 f(a), \text{ etc.}$$

Using these equations, we get the entries in the third row:

$$f(a + 2h) = f(a + h) + \Delta f(a + h)$$
$$= f(a) + 2\Delta f(a) + \Delta^2 f(a)$$
$$\Delta f(a + 2h) = \Delta f(a) + 2\Delta^2 f(a) + \Delta^3 f(a), \text{ etc.}$$

From these we get the entries in the fourth row:

$$f(a + 3h) = f(a + 2h) + \Delta f(a + 2h)$$
$$= f(a) + 3\Delta f(a) + 3\Delta^2 f(a) + \Delta^3 f(a), \text{ etc.}$$

The coefficients thus far are those of the binomial expansion. This suggests the general formulas involving the binomial coefficients:

$$f(a + nh) = f(a) + n\Delta f(a) + \frac{n(n - 1)}{2!} \Delta^2 f(a)$$
$$+ \cdots + \Delta^n f(a) \quad (8\text{-}4)$$

$$\Delta^r f(a + nh) = \Delta^r f(a) + n\Delta^{r+1} f(a) + \frac{n(n - 1)}{2!} \Delta^{r+2} f(a)$$
$$+ \cdots + \Delta^{r+n} f(a) \quad (8\text{-}5)$$

These hold, as we have just found, for $n = 1$ and $n = 2$. We shall prove them by induction for any entry in the table.

We assume the formulas to hold for $n - 1$ and prove them for n. We have, by hypothesis,

$$f[a + (n - 1)h] = f(a) + (n - 1)\Delta f(a)$$
$$+ \frac{(n - 1)(n - 2)}{2!} \Delta^2 f(a) + \cdots + \Delta^{n-1} f(a)$$
$$\Delta f[a + (n - 1)h] = \Delta f(a) + (n - 1)\Delta^2 f(a)$$
$$+ \cdots + (n - 1)\Delta^{n-1} f(a) + \Delta^n f(a)$$

Adding these, the first member is $f(a + nh)$ and the general term on the right is

$$\frac{(n - 1) \cdots (n - r)}{r!} \Delta^r f(a) + \frac{(n - 1) \cdots (n - r + 1)}{(r - 1)!} \Delta^r f(a)$$
$$= \frac{(n - 1) \cdots (n - r + 1)}{r!} \Delta^r f(a)[(n - r) + r]$$
$$= \frac{n(n - 1) \cdots (n - r + 1)}{r!} \Delta^r f(a)$$

This is the general term of (8-4), and that formula is established.

Equation (8-5) is also established, for we have merely to apply (8-4) to the table of values of the function $\Delta^r f(x)$.

We write (8-4) in the form

$$F(x) = f(a) + n\Delta f(a) + \frac{n(n - 1)}{2!} \Delta^2 f(a)$$
$$+ \cdots + \frac{n(n - 1) \cdots (n - m + 1)}{m!} \Delta^m f(a) \quad (8\text{-}6)$$

where

$$x = a + nh \text{ or } n = \frac{x - a}{h} \quad (8\text{-}7)$$

For $n = 0, 1, \ldots, m$, that is, for $x = a, a + h, \ldots, a + mh$, the second number of (8-6) has the values $f(a), f(a + h), \ldots, f(a + mh)$, respectively.

Now $F(x)$ is a polynomial of degree m or less, as we see on replacing n by its value in terms of x from (8-7). It is the unique polynomial which has the same values as $f(x)$ at the points $x = a, a + h, \ldots, a + mh$.

Equation (8-6) is *Newton's interpolation formula*. The polynomial which it defines is necessarily the same as that given by Lagrange's formula.

In applying Newton's formula, n is the number of intervals of length h separating x and a.

INTERPOLATION AND NUMERICAL INTEGRATION 181

Example. Given that sin 30° = 0.50000, sin 35° = 0.57358, sin 40° = 0.64279, sin 45° = 0.70711, sin 50° = 0.76604. Find sin 37°.
We form a table of differences:

x	sin x	Δ	Δ^2	Δ^3	Δ^4
30	50000	7358	−437	−52	2
35	57358	6921	−489	−50	
40	64279	6432	−539		
45	70711	5893			
50	76604				

Since 7° is 1.4 times the tabular difference of 5°, we take $n = 1.4$ in Newton's formula. We neglect the fourth difference on account of its insignificance.

sin 37° = 0.50000 + 1.4(0.07358)
$$+ \frac{(1.4)(0.4)}{2}(-0.00437) + \frac{(1.4)(0.4)(-0.6)}{6}(-0.00052)$$
= 0.50000 + 0.103012 − 0.001224 + 0.000029 = 0.60182

On checking by consulting a table, we find that this value is correct.
By ordinary linear interpolation between 35° and 40° we have

$$0.57358 + \tfrac{2}{5}(0.06921) = 0.60126$$

which is seriously in error.

EXERCISES

1. Extend the table of cubes in the text backward by subtractions to $x = 0$.
2. Sunrise and sunset occurred at 31-day intervals as follows:

	Oct. 31	Dec. 1	Jan. 1	Feb. 1
Sunrise..........	6:15	6:44	7:03	6:56
Sunset..........	5:11	4:54	5:04	5:32

Find
  a. The time of sunrise and sunset on Dec. 15.
  b. The date and time of the latest sunrise.
  c. The date and time of the earliest sunset.
  d. The date of the shortest day and its length.

3. The heights of a projectile above the point from which it was fired 10, 20, and 30 sec after leaving the gun were respectively 6,000, 8,200, and 6,400 ft. Find its greatest height.
4. Derive Newton's formula by writing

$$f(a + nh) = c_0 + c_1 n + c_2 n^{(2)} + \cdots + c_m n^{(m)}$$

(see problem 2 of the preceding exercises) and determining the constants by differencing and setting $n = 0$.

182   DIFFERENTIAL EQUATIONS

5. Form a difference table from the entries

A	$\log \sin A$
23°	9.5918 780
24°	9.6093 133
25°	9.6259 483
26°	9.6418 420
27°	9.6570 468
28°	9.6716 093

and interpolate for $\log \sin 24°10'$. Compute the errors due to using only two terms, only three terms, etc., of Newton's formula, given that the correct value is 9.6121 397.

6. In like manner compute the errors in finding the common logarithm of 1.0001 by interpolation in the table

x	$\log x$
1.000	0.00000 00000 00000 0
1.001	0.00043 40774 79318 6
1.002	0.00086 77215 31226 9
1.003	0.00130 09330 20418 1
1.004	0.00173 37128 09000 5
1.005	0.00216 60617 56507 7

the correct value being 0.00004 34272 76862 7.

8-6. The Error in Parabolic Interpolation. Let $F(x)$ be the polynomial of degree m at most, which takes on the same values as a function $f(x)$ at the $m + 1$ distinct points $x_0, x_1, \ldots, x_m$. It is clear that at other points the two functions may differ widely. We can, for example, draw the graph of a function $f(x)$ through the given points which differs as much as we please from the graph of $F(x)$ at any other point. It is only when $f(x)$ has certain qualities of smoothness that $F(x)$ is a good substitute function for most purposes. We shall now derive a formula for the difference between the two functions.

We shall assume that $f(x)$ has a continuous derivative of order $m + 1$ in an interval ab in which $x_0, \ldots, x_m$ lie. Let x lie in ab. We write

$$f(x) = F(x) + R(x)$$

where $R(x)$ is the error, or, as it is often called, the *remainder*.

At $x_0, \ldots, x_m$, $f(x)$ and $F(x)$ have the same values and $R(x) = 0$. For any other value of x, we write $R(x)$ in the form

$$R(x) = (x - x_0)(x - x_1) \cdots (x - x_m)P(x)$$

This defines $P(x)$ in terms of the error $R(x)$.

We now hold x fast and consider the function

$$\Phi(z) = f(z) - F(z) - (z - x_0) \cdots (z - x_m)P(x)$$

The function $\Phi(z)$ vanishes when $z = x_0, x_1, \ldots, x_m$, since $f(z)$ and $F(z)$ are then equal and the final term of the second member is zero

INTERPOLATION AND NUMERICAL INTEGRATION   183

Also $\Phi(x) = 0$, since, when $z = x$, the final term is then the remainder at x. The function $\Phi(z)$ is thus zero at the $m + 2$ points $x_0, x_1, \ldots, x_m, x$.

We now apply Rolle's theorem. Between points at which $\Phi(z)$ vanishes there is a point at which the derivative $\Phi'(z)$ vanishes. There are $m + 1$ intervals in each of which $\Phi'(z)$ vanishes.

We now repeat the reasoning: $\Phi'(z)$ vanishes at $m + 1$ distinct points; hence $\Phi''(z)$ vanishes at m distinct points; and so on. We have finally that $\Phi^{(m+1)}(z)$ vanishes at at least one point ξ.

Now the $(m + 1)$th derivative of $F(z)$ is zero, and we have

$$\Phi^{(m+1)}(z) = f^{(m+1)}(z) - (m + 1)!P(x)$$

At $z = \xi$,

$$f^{(m+1)}(\xi) - (m + 1)!P(x) = 0$$

$$P(x) = \frac{f^{(m+1)}(\xi)}{(m + 1)!}$$

We have finally the following result:

THEOREM. *Let $F(x)$ be the polynomial of degree m at most, which has the same values as $f(x)$ at $x_0, x_1, \ldots, x_m$. Then*

$$f(x) = F(x) + \frac{(x - x_0)(x - x_1) \cdots (x - x_m)}{(m + 1)!} f^{(m+1)}(\xi) \qquad (8\text{-}8)$$

where ξ lies between the least and the greatest of the values $x_0, x_1, \ldots, x_m$, and x.

We see from this formula, what we know already, that if $f(x)$ is a polynomial of degree m or less, then $f(x)$ and $F(x)$ are identical, for $f^{(m+1)}(x) \equiv 0$.

We can write (8-8) in a somewhat different form for Newton's formula (8-6). We have

$$x = a + nh$$
$$x_r = a + rh$$
$$x - x_r = h(n - r)$$

and (8-8) takes the form

$$f(x) = F(x) + \frac{n(n - 1) \cdots (n - m)}{(m + 1)!} h^{m+1} f^{(m+1)}(\xi) \qquad (8\text{-}9)$$

The point ξ is unknown, but if we know bounds for the $(m + 1)$th derivative in ab, $|f^{(m+1)}(x)| < M$, then (8-8) and (8-9) give us limits for the error. We observe from both formulas that by taking the points $x_0, \ldots, x_m$ close enough together, and letting x lie near these points, the factor $(x - x_0) \cdots (x - x_m)$ in (8-8) or the factor h^{m+1} in (8-9) can be made arbitrarily small, and hence the error in interpolating can be made as small as is desired.

Example. Find the error due to interpolation in the example worked out in the preceding section.

Here third differences are used, the entries being from 30° to 45°. The error is

$$R = \frac{n(n-1)(n-2)(n-3)}{4!} h^4 f^{(4)}(\xi)$$

Using radians,

$$f(x) = \sin x$$
$$f^{(4)}(x) = \sin x$$
$$h = 5° = 0.08727 \text{ radian}$$

Below 45°, we have $\sin x < 0.70711$. We have, finally, since $n = 1.4$,

$$|R| < \frac{(1.4)(0.4)(0.6)(1.6)}{24} (0.08727)^4 (0.70711)$$

or

$$|R| < 0.00000092$$

The process is accurate enough for a six-place table.

8-7. Linear Interpolation. For ordinary linear interpolation, $m = 1$ and we have

$$R = \frac{n(n-1)}{2} h^2 f''(\xi)$$

Let $|f''(x)| \leq M$ between the entries, and let us find a limit for the error for all points between the two entries, $0 < n < 1$. The quantity $n(n-1)$ has its maximum absolute value between 0 and 1 at $n = \frac{1}{2}$; then $n(n-1) = -\frac{1}{4}$. Between the entries, then, we have

$$|R| \leq \frac{h^2}{8} M$$

In constructing a table, it is usually desired to make the entries close enough together that linear interpolation suffices for the determination of additional values. The formula just derived is useful for the determination of an appropriate value of h.

Problem. In a table of natural sines, how large an interval may be used if the errors due to linear interpolation are not to exceed $\frac{1}{2}$ in the fourth decimal place?

Here

$$|f''(x)| = \left|\frac{d^2 \sin x}{dx^2}\right| = |\sin x| \leq 1$$

and the error will be within the prescribed limits if

$$\frac{h^2}{8} \leq 0.00005$$

or

$$h \leq 0.02 \text{ radian} = 1.15°$$

An interval of 1° would thus be satisfactory for a four-place table.

EXERCISES

1. In a table of cube roots of numbers from 1 to 10 to four decimals to be used with linear interpolation, what interval would you use?

2. How large an interval may be used in a table of natural sines in order that the error due to interpolation with second differences shall not exceed 5 in the fifth decimal place?

3. Generalize Lagrange's formula so that the values of the polynomial and of certain first, second, etc., derivatives agree with those of the original function at given points. Derive a remainder theorem.

8-8. Numerical Integration. Integration has been one of the most fruitful fields for the use of approximate methods. The most capable student of the calculus comes repeatedly upon functions which he cannot integrate. So seemingly innocent an integral as $\int \sqrt{2 - \sin^2 \theta} \, d\theta$ is beyond his powers until he has made a study of elliptic integrals. Other expressions, hardly more complicated than this, lead to functions which have not been studied at all. The mathematician is driven by necessity to some sort of approximate evaluation of integrals.

The integral between specified limits is the area under the graph of the function. A rough method of evaluating this area is to draw the graph on cross-section paper and count the squares under the curve, estimating the areas of fractional parts of squares which are only partly under the curve. Integrating machines, or planimeters, which evaluate an area when a pointer is moved around the boundary, may be used. However, much more precise results may be obtained by integrating interpolating functions, for the area may be found to any desired degree of accuracy.

Numerous useful formulas result from the use of Lagrange's formula. Suppose we wish to integrate $f(x)$ between limits α and β. We choose an interpolating polynomial $F(x)$ which takes on the same values as $f(x)$ at such points as to make $F(x)$ a good interpolating function throughout the interval $\alpha\beta$ and then integrate $F(x)$. We have from (8-3)

$$\int_\alpha^\beta F(x) \, dx = H_0 f_0 + H_1 f_1 + \cdots + H_m f_m \qquad (8\text{-}10)$$

where

$$H_i = \int_\alpha^\beta \frac{(x - x_0)(x - x_1) \cdots (x - x_{i-1})(x - x_{i+1}) \cdots (x - x_m)}{(x_i - x_0)(x_i - x_1) \cdots (x_i - x_{i-1})(x_i - x_{i+1}) \cdots (x_i - x_m)} \, dx \qquad (8\text{-}11)$$

We use (8-10) as an approximation to $\int_\alpha^\beta f(x) \, dx$. If $F(x)$ and $f(x)$ differ by less than ϵ in the interval $\alpha\beta$, then the error resulting from (8-10) is certainly less than $|\beta - \alpha|\epsilon$.

We observe that H_i is independent of the values of the function $f(x)$. It depends only on the values $x_0, x_1, \ldots, x_m$ and on the limits α and β. Hence, $H_0, \ldots, H_m$ can be computed once for all and can then be used for the integration of various functions between the limits α and β.

8-9. Simpson's Rule. As a first illustration we shall derive Simpson's rule, a formula of which we shall make much use in the following chapter. We shall get an expression for $\int_a^{a+2h} f(x)\, dx$ in terms of the values of the integrand at the limits of integration and at the mid-point of the interval over which the integration is made.

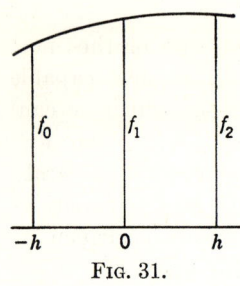

Fig. 31.

Let $f(x)$ take on the values f_0, f_1, f_2 at $x = a$, $a + h$, $a + 2h$, respectively. It will simplify the calculations somewhat if we translate the origin to the point $a + h$. We have then a function taking on the values f_0, f_1, f_2 at $-h, 0, h$, respectively (Fig. 31). The interpolating polynomial of the second degree at most taking on these three values is, from (8-3), setting $x_0 = -h$, $x_1 = 0$, $x_2 = h$,

$$\frac{(x-0)(x-h)}{(-h-0)(-h-h)} f_0 + \frac{(x+h)(x-h)}{(0+h)(0-h)} f_1 + \frac{(x+h)(x-0)}{(h+h)(h-0)} f_2$$

or

$$F(x) = \frac{1}{2h^2} [(x^2 - hx)f_0 - 2(x^2 - h^2)f_1 + (x^2 + hx)f_2] \quad (8\text{-}12)$$

On integrating, we have

$$\int_{-h}^{h} (x^2 - hx)\, dx = \frac{2h^3}{3} = \int_{-h}^{h} (x^2 + hx)\, dx$$

$$\int_{-h}^{h} (x^2 - h^2)\, dx = -\frac{4h^3}{3}$$

and

$$\int_{-h}^{h} F(x)\, dx = \frac{h}{3} (f_0 + 4f_1 + f_2)$$

Shifting the origin back to its original position, we have the approximate formula

$$\int_a^{a+2h} f(x)\, dx = \frac{h}{3} [f(a) + 4f(a+h) + f(a+2h)] \quad (8\text{-}13)$$

This is Simpson's rule.

In applying the rule to an extended interval of integration ab, we divide ab into an even number of equal intervals and use (8-13) for each pair.

We have approximately

$$\int_{a+2h}^{a+4h} f(x)\,dx = \frac{h}{3}[f(a+2h) + 4f(a+3h) + f(a+4h)]$$

and so on. Adding, we have the approximate formula

$$\int_a^{a+2rh} f(x)\,dx = \frac{h}{3}\{f(a) + 4f(a+h) + 2f(a+2h) + 4f(a+3h)$$
$$+ \cdots + 4f[a+(2r-1)h] + f(a+2rh)\} \quad (8\text{-}14)$$

On account of its simplicity and of its accuracy under suitable conditions, this is a favorite formula of numerical integration.

Simpson's rule gives an exact result if $f(x)$ is a polynomial of the second degree or less, for then $F(x)$ and $f(x)$ are identical. It is a curious fact that it is also exact if $f(x)$ is a polynomial of the *third* degree. Here $f'''(x) = k$, a constant, and the error in the interpolating function is, from (8-8),

$$R(x) = \frac{(x+h)x(x-h)}{3!}k = \frac{k}{6}(x^3 - h^2 x)$$

The error in the integral is

$$\int_{-h}^{h} R(x)\,dx = \frac{k}{6}\int_{-h}^{h}(x^3 - h^2 x)\,dx = 0$$

Hence, Simpson's rule is exact.

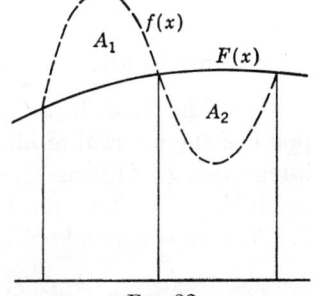

FIG. 32.

Now, a polynomial $f(x)$ of the third degree may be constructed which takes on the three prescribed values but which elsewhere differs widely from the polynomial $F(x)$ (see Fig. 32). But the difference, as we see from $R(x)$, is positive in one interval and negative in the other, and the excess A_1 of the integral in one interval is balanced by an equal defect A_2 in the other.

We shall now derive two additional formulas, based on the interpolating polynomial (8-12), for which we shall have use subsequently. Integrating (8-12) over the first interval, we find

$$\int_{-h}^{0} F(x)\,dx = \frac{h}{12}(5f_0 + 8f_1 - f_2)$$

On shifting the origin we may write the approximate formula

$$\int_a^{a+h} f(x)\,dx = \frac{h}{12}[5f(a) + 8f(a+h) - f(a+2h)] \quad (8\text{-}15)$$

Again, integrating from $-2h$ to $2h$, we find

$$\int_{-2h}^{2h} F(x)\,dx = \frac{4h}{3}(2f_0 - f_1 + 2f_2)$$

On shifting the origin, this gives us the approximate formula

$$\int_a^{a+4h} f(x)\, dx = \frac{4h}{3}[2f(a+h) - f(a+2h) + 2f(a+3h)] \quad (8\text{-}16)$$

Formula (8-15) is not exact for polynomials of the third degree, but formula (8-16) is. This we find on integrating the remainder term,

$$\int_{-2h}^{2h} R(x)\, dx = \frac{k}{6}\int_{-2h}^{2h}(x^3 - h^2 x)\, dx = 0$$

Example. Evaluate log 2 from $\int_1^2 \frac{dx}{x}$

Taking $h = 0.1$, and applying (8-14), we have

$$\log 2 = \int_1^2 \frac{dx}{x} = \frac{0.1}{3}\left(1 + \frac{4}{1.1} + \frac{2}{1.2} + \frac{4}{1.3} + \frac{2}{1.4} + \frac{4}{1.5} + \frac{2}{1.6} + \frac{4}{1.7}\right.$$
$$\left. + \frac{2}{1.8} + \frac{4}{1.9} + \frac{1}{2}\right) = 0.6931502$$

The correct value is 0.6931472.

8-10. The Error in Simpson's Rule. We take the origin at the midpoint of the interval of integration, then Simpson's rule results from the integration of $F(x)$ as given by (8-12). The rule results likewise from the integration of a cubic $G(x)$ taking on the prescribed values at $-h, 0, h$. The most general such cubic we have found to be of the form

$$G(x) = F(x) + K(x^3 - h^2 x)$$

We shall determine K so that $f(x)$ and $G(x)$ have the same slope at the origin; and we shall then find how good an approximation $G(x)$ is to the original function. We have

$$G'(x) = F'(x) + K(3x^2 - h^2)$$

and $G(x)$ will have the required slope if

$$F'(0) - h^2 K = f'(0)$$

from which K is determined.

We write

$$f(x) = G(x) + x^2(x^2 - h^2)P(x)$$

an equation determining $P(x)$ when x is different from 0 or $\pm h$.

We now apply Rolle's theorem to the function

$$\Phi(z) = f(z) - G(z) - z^2(z^2 - h^2)P(x)$$

where x is held fast. $\Phi(z)$ vanishes at the four points $z = -h, 0, h,$ and x. Hence the first derivative,

$$\Phi'(z) = f'(z) - G'(z) - (4z^3 - 2h^2 z)P(x)$$

INTERPOLATION AND NUMERICAL INTEGRATION 189

vanishes in each of the three intervals cut off by the four points. But $\Phi'(z)$ also vanishes at $z = 0$, owing to the equality of $G'(0)$ and $f'(0)$. Hence $\Phi'(z)$ vanishes at four distinct points at least.

Continuing, $\Phi''(z)$ vanishes at three points, $\Phi'''(z)$ vanishes at two points, and $\Phi^{(4)}(z)$ vanishes at one point, ξ, at least. Now the fourth derivative of $G(x)$ is zero, and

$$\Phi^{(4)}(z) = f^{(4)}(z) - 4!P(x)$$
$$f^{(4)}(\xi) - 4!P(x) = 0$$
$$P(x) = \frac{f^{(4)}(\xi)}{4!}$$

We have then

$$f(x) = G(x) + \frac{x^2(x^2 - h^2)}{4!} f^{(4)}(\xi)$$

where ξ lies between the greatest and least of the values $\pm h$ and x.

We now integrate,

$$\int_{-h}^{h} f(x)\, dx = \int_{-h}^{h} G(x)\, dx + \frac{1}{4!} \int_{-h}^{h} (x^4 - h^2 x^2) f^{(4)}(\xi)\, dx$$

The first term of the second member gives Simpson's formula. To the second term we apply the law of the mean for integrals.[1] Here $x^4 - h^2 x^2$ is negative or zero throughout, and we can remove $f^{(4)}(\xi)$ outside the integral sign, evaluating it for some $x = \zeta$ between $-h$ and h. Then the resulting ξ is some value, η, say, between $-h$ and h, and we have

$$\int_{-h}^{h} (x^4 - h^2 x^2) f^{(4)}(\xi)\, dx = f^{(4)}(\eta) \int_{-h}^{h} (x^4 - h^2 x^2)\, dx = -\tfrac{4}{15} h^5 f^{(4)}(\eta)$$

The result is

$$\int_{-h}^{h} f(x)\, dx = \frac{h}{3}(f_0 + 4f_1 + f_2) - \frac{h^5}{90} f^{(4)}(\eta)$$

By translating the origin we may write the following exact formula:

$$\int_{a}^{a+2h} f(x)\, dx = \frac{h}{3}[f(a) + 4f(a + h) + f(a + 2h)] - \frac{h^5}{90} f^{(4)}(\eta) \quad (8\text{-}17)$$

where η lies between a and $a + 2h$.

Problem. Get limits for the error in the computation of log 2 in Sec. 8-9. Here

$$f(x) = \frac{1}{x} \qquad f^{(4)}(x) = \frac{24}{x^5} \qquad h = 0.1$$

[1] The theorem used is the following: If $\varphi(x)$ does not change sign and if $\psi(x)$ is continuous in the interval ab, then

$$\int_{a}^{b} \varphi(x)\psi(x)\, dx = \psi(\zeta) \int_{a}^{b} \varphi(x)\, dx$$

where $a < \zeta < b$.

and the error in each application of Simpson's rule is

$$R = -\frac{(0.1)^5}{90} \cdot \frac{24}{\eta^5} = -\frac{0.00000267}{\eta^5}$$

where η lies in the interval integrated over. It thus appears, since η is positive, that the formula gives too great a value.

We applied Simpson's rule to five intervals. Taking upper and lower bounds for η in each interval, we find that the correction lies between

$$-0.00000267 \left(\frac{1}{1^5} + \frac{1}{1.2^5} + \frac{1}{1.4^5} + \frac{1}{1.6^5} + \frac{1}{1.8^5}\right) = -0.0000046$$

and

$$-0.00000267 \left(\frac{1}{1.2^5} + \frac{1}{1.4^5} + \frac{1}{1.6^5} + \frac{1}{1.8^5} + \frac{1}{2^5}\right) = -0.0000021$$

The actual correction is -0.0000030.

8-11. An Application of Legendre Polynomials. Formula (8-10) is based on the situation in which the interpolating polynomial and the given function have the same values for $x_0, x_1, \ldots, x_m$. The interpolating polynomial and hence, also, the formula of integration are exact when applied to polynomials of degree m or less.

By properly placing the points $x_0, \ldots, x_m$ it is possible to make the formula valid for polynomials of higher degree. Thus Simpson's rule, although based on a second-degree interpolating polynomial, is valid for third-degree polynomials. We shall now show how this accuracy may be greatly extended.

Let the interval of integration be $-1 \leqslant x \leqslant 1$, and let $x_0, x_1, \ldots, x_m$ be the roots of the Legendre polynomial $P_{m+1}(x)$. The resulting formula

$$\int_{-1}^{1} F(x) \, dx = H_0 f_0 + \cdots + H_m f_m$$

where $f_0, \ldots, f_m$ are the values of $F(x)$ at $x_0, \ldots, x_m$, respectively, is exact when $F(x)$ is a polynomial of degree m or less. We now show that it is exact when $F(x)$ is a polynomial of degree $2m + 1$ at most. On dividing $F(x)$ by $P_{m+1}(x)$, we may write

$$F(x) = P_{m+1}(x) Q(x) + R(x)$$

where $Q(x)$ and $R(x)$ are of degree m or less. From a basic property of Legendre polynomials [Eq. (7-21)] the integral of the first term in the second member is zero, and we have

$$\int_{-1}^{1} F(x) \, dx = \int_{-1}^{1} R(x) \, dx$$

The formula is exact when applied to $R(x)$. But at $x_0, \ldots, x_m$, $P_{m+1}(x) = 0$, and $F(x)$ and $R(x)$ have the same values, $f_0, \ldots, f_m$.

The insertion of these values of $f(x)$ into the formula gives an exact value of the integral of $R(x)$ and hence of $F(x)$.

Example. The roots of

$$P_3(x) = \tfrac{1}{2}(5x^3 - 3x)$$

are $0, \pm \sqrt{\tfrac{3}{5}}$. Evaluating H_0, H_1, H_2 by formula (8-11), we get

$$\int_{-1}^{1} F(x)\, dx = \tfrac{1}{9}[5F(-\sqrt{\tfrac{3}{5}}) + 8F(0) + 5F(\sqrt{\tfrac{3}{5}})]$$

Although derived by integrating a second-degree polynomial, this formula is exact for all polynomials of the fifth or lower degrees.

By a suitable change of variable this formula can be written

$$\int_{a-h}^{a+h} F(x)\, dx = \frac{h}{9}[5F(a - h\sqrt{\tfrac{3}{5}}) + 8F(a) + 5F(a + h\sqrt{\tfrac{3}{5}})]$$

EXERCISES

1. Apply the integration formula of the example to

$$\int_{-1}^{1} \frac{dx}{x+3}$$

and compare with the exact result.

2. Derive a formula of integration using $P_2(x)$, and apply to the preceding integral.

8-12. Finite Difference Integration Formulas. By integrating Newton's formula, we get the integral in terms of the differences appearing in a table. We have $dx = h\, dn$, and $n = 0, 1$ when $x = a, a + h$; whence, on integrating (8-6),

$$\int_a^{a+h} F(x)\, dx = h \int_0^1 \left[f(a) + n\Delta f(a) + \frac{n(n-1)}{2!}\Delta^2 f(a) \right.$$
$$\left. + \cdots + \frac{n(n-1)\cdots(n-m+1)}{m!}\Delta^m f(a) \right] dn$$
$$= h[f(a) + a_1\Delta f(a) + a_2\Delta^2 f(a) + \cdots + a_m\Delta^m f(a)]$$

where

$$a_1 = \int_0^1 n\, dn = \frac{1}{2} \qquad a_2 = \int_0^1 \frac{n(n-1)}{2}\, dn = -\frac{1}{12}$$

and so on. Calculations give

$$a_3 = \frac{1}{24} \qquad a_4 = -\frac{19}{720} \qquad a_5 = \frac{3}{160} \qquad a_6 = -\frac{863}{60{,}480}$$

The error may be found by integrating the remainder term in (8-9).

$$h \int_0^1 R(x)\, dn = h^{m+2} \int_0^1 \frac{n(n-1)\cdots(n-m)}{(m+1)!} f^{(m+1)}(\xi)\, dn$$

192   DIFFERENTIAL EQUATIONS

Here $n(n-1) \cdots (n-m)$ does not change sign in the interval 0, 1; hence we can apply the law of the mean for integrals. We have

$$h^{m+2}f^{(m+1)}(\eta) \int_0^1 \frac{n(n-1) \cdots (n-m)}{(m+1)!} dn = h^{m+2}a_{m+1}f^{(m+1)}(\eta)$$

where a_{m+1} has the value defined above and η lies between a and $a + mh$. Taking $m = 0, 1, 2, \ldots$, we have the following exact formulas:

$$\int_a^{a+h} f(x)\, dx = hf(a) + \tfrac{1}{2}h^2 f'(\eta)$$
$$= h[f(a) + \tfrac{1}{2}\Delta f(a)] - \tfrac{1}{12}h^3 f''(\eta)$$
$$= h[f(a) + \tfrac{1}{2}\Delta f(a) - \tfrac{1}{12}\Delta^2 f(a)] + \tfrac{1}{24}h^4 f'''(\eta)$$
$$= h[f(a) + \tfrac{1}{2}\Delta f(a) - \tfrac{1}{12}\Delta^2 f(a) + \tfrac{1}{24}\Delta^3 f(a)] - \tfrac{19}{720}h^5 f^{(4)}(\eta)$$
(8-18)

and so on. By applying one of these formulas to successive lines of a table we may integrate over any number of intervals.

Example. Evaluate $\int_{12}^{14} x^3\, dx$ from the table in Sec. 8-4. Here $h = 2$, $a = 12$, and we take the second row in the table. The fourth derivative is zero, so we have exactly

$$\int_{12}^{14} x^3\, dx = 2[1728 + \tfrac{1}{2}(1016) - \tfrac{1}{12}(336) + \tfrac{1}{24}(48)] = 4420$$

8-13. An Integration Formula Involving Derivatives. If the derivatives of the function to be integrated are easy to compute, we may find it convenient to use them in performing the integration. Let $f(x)$ take on the values f_0, f_1 at $x = 0$, $x = h$, and let the derivatives be f_0', f_1', respectively, at 0 and h.

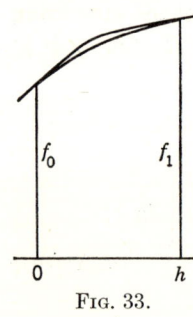

FIG. 33.

We shall set up as an interpolating function a polynomial taking on the same values as the function and having the same derivatives at the two points (Fig. 33). Since this imposes four conditions, a polynomial of the third degree will be required, in general.

A cubic which has the desired value and derivative at the origin is

$$F(x) = f_0 + f_0'x + Ax^2 + Bx^3$$

A and B can be so determined that $F(x)$ and its derivative,

$$F'(x) = f_0' + 2Ax + 3Bx^2$$

shall have the required values at h. Setting $x = h$, we have two equations,

(a) $\qquad f_0 + f_0'h + Ah^2 + Bh^3 = f_1$
(b) $\qquad f_0' + 2Ah + 3Bh^2 = f_1'$

INTERPOLATION AND NUMERICAL INTEGRATION 193

from which A and B can be found. The integration formula then is

$$\int_0^h F(x)\, dx = f_0 h + f_0' \frac{h^2}{2} + A \frac{h^3}{3} + B \frac{h^4}{4}$$

The particular combination appearing in this formula can be got without the trouble of solving for A and B. Multiplying (a) by $h/2$ and (b) by $-h^2/12$, and adding, we have

$$\frac{1}{2} h f_0 + \frac{5}{12} h^2 f_0' + A \frac{h^3}{3} + B \frac{h^4}{4} = \frac{1}{2} h f_1 - \frac{1}{12} h^2 f_1'$$

On solving this for $Ah^3/3 + Bh^4/4$ and substituting into the integral formula, we have

$$\int_0^h F(x)\, dx = \frac{h}{2}(f_0 + f_1) + \frac{h^2}{12}(f_0' - f_1')$$

On translating the origin to $x = a$, we get the approximate formula

$$\int_a^{a+h} f(x)\, dx = \frac{h}{2}[f(a) + f(a+h)] + \frac{h^2}{12}[f'(a) - f'(a+h)] \quad (8\text{-}19)$$

Let us apply this formula to successive intervals,

$$\int_{a+h}^{a+2h} f(x)\, dx = \frac{h}{2}[f(a+h) + f(a+2h)] + \frac{h^2}{12}[f'(a+h) - f'(a+2h)]$$

etc., and add. The derivatives cancel at intermediate points, and we have

$$\int_a^{a+mh} f(x)\, dx = h \left\{ \frac{1}{2} f(a) + f(a+h) + f(a+2h) + \cdots \right.$$
$$\left. + f[a + (m-1)h] + \frac{1}{2} f(a+mh) \right\} + \frac{h^2}{12}[f'(a) - f'(a+mh)] \quad (8\text{-}20)$$

We now get an expression for the error. We put

$$f(x) = F(x) + x^2(x-h)^2 P(x)$$

which defines $P(x)$; and we consider

$$\Phi(z) = f(z) - F(z) - z^2(z-h)^2 P(x)$$

$\Phi(z)$ vanishes for the three values 0, h, and x. Hence, $\Phi'(z)$ vanishes at two distinct points. But since $f(z)$ and $F(z)$ have the same derivative and the derivative of $z^2(z-h)^2$ is zero at $z = 0$ and $z = h$, then $\Phi'(z)$ vanishes at 0 and h also.

Then $\Phi'(z)$ vanishes at four distinct points; whence $\Phi''(z)$ vanishes at three points, $\Phi'''(z)$ vanishes at two points, and, finally, $\Phi^{(4)}(z)$ vanishes at one point, ξ. The fourth derivative of $F(z)$ is zero, and we have

$$\Phi^{(4)}(z) = f^{(4)}(z) - 4!P(x)$$
$$f^{(4)}(\xi) - 4!P(x) = 0$$
$$P(x) = \frac{f^{(4)}(\xi)}{4!}$$

The remainder term for the interpolating function $F(x)$ is then

$$R(x) = \frac{x^2(x-h)^2}{4!} f^{(4)}(\xi)$$

where ξ lies between the greatest and least of the values 0, h, and x.

By integrating $R(x)$ we have the correction to be added to the integration formula:

$$\int_0^h \frac{x^2(x-h)^2}{4!} f^{(4)}(\xi)\, dx$$

Since $x^2(x-h)^2$ does not change sign in the interval of integration, we can apply the law of the mean for integrals.

$$\int_0^h R(x)\, dx = \frac{f^{(4)}(\eta)}{4!} \int_0^h x^2(x-h)^2\, dx = \frac{h^5}{720} f^{(4)}(\eta) \qquad (8\text{-}21)$$

where η lies between the limits of integration.

The integration formula (8-20) is a special case of a formula known as the Euler-Maclaurin expansion, which expresses the integral in terms of the values of the function and of its first m derivatives (see No. 6, Sec. 8-20).

8-14. Numerical Differentiation. The device for numerical differentiation has probably already occurred to the reader. It consists in using the derivative of the interpolating polynomial in place of the derivative of the given function.

We first derive a formula for the derivative in terms of the entries of a row in a table of differences. Differentiating Newton's formula, (6-6), and noting that $dx = h\, dn$, we have

$$F'(x) = \frac{1}{h} \frac{d}{dn} \left[f(a) + n\Delta f(a) + \frac{n(n-1)}{2!} \Delta^2 f(a) + \cdots \right]$$

To fourth differences we have the approximate formula

$$f'(a + nh) = \frac{1}{h}\left[\Delta f(a) + \frac{2n-1}{2} \Delta^2 f(a) + \frac{3n^2 - 6n + 2}{6} \Delta^3 f(a) \right.$$
$$\left. + \frac{2n^3 - 9n^2 + 11n - 3}{12} \Delta^4 f(a) \right] \qquad (8\text{-}22)$$

INTERPOLATION AND NUMERICAL INTEGRATION 195

The terms in this formula become complicated for higher differences. They are particularly simple, however, for $n = 0$. The coefficient of $\Delta^r f(a)$ is

$$\left[\frac{d}{dn} \frac{n(n-1) \cdots (n-r+1)}{r!} \right]_{n=0}$$
$$= \lim_{n=0} \frac{1}{n} \frac{n(n-1) \cdots (n-r+1)}{r!} = \frac{(-1)^{r-1}}{r}$$

We thus have the simple approximate formula

$$f'(a) = \frac{1}{h} \left[\Delta f(a) - \frac{1}{2} \Delta^2 f(a) + \frac{1}{3} \Delta^3 f(a) - \frac{1}{4} \Delta^4 f(a) + \cdots \right] \quad (8\text{-}23)$$

We shall now consider the error in approximate differentiation. Let $F(x)$ be the interpolating polynomial taking on the same values as $f(x)$ at $x = x_0, x_1, \ldots, x_m$. We shall get a formula for the error in replacing $f'(x)$ by $F'(x)$ at any one of these $m + 1$ points. We define $P(x_i)$ by the equation

$$f'(x_i) = F'(x_i) + (x_i - x_0) \cdots (x_i - x_{i-1})(x_i - x_{i+1})$$
$$\cdots (x_i - x_m)P(x_i)$$

We now write

$$\Phi(z) = f(z) - F(z) - (z - x_0) \cdots (z - x_m)P(x_i)$$

$\Phi(z)$ vanishes at $x_0, \ldots x_m$; hence $\Phi'(z)$ vanishes at least once in each of the intervals cut off by $x_0, \ldots, x_m$. But $\Phi'(z)$ also vanishes at x_i, since

$$\Phi'(x_i) = f'(x_i) - F'(x_i) - \left\{ \frac{d}{dz} [(z - x_0) \cdots (z - x_m)] \right\}_{z=x_i} P(x_i)$$
$$= f'(x_i) - F'(x_i) - (x_i - x_0) \cdots (x_i - x_{i-1})(x_i - x_{i+1})$$
$$\cdots (x_i - x_m)P(x_i)$$

and this is zero because of the way $P(x_i)$ was defined. So $\Phi'(z)$ vanishes at $m + 1$ distinct points. Then $\Phi''(z)$ vanishes at m distinct points; and, continuing, $\Phi^{(m+1)}(z)$ vanishes at one point, ξ, at least.

Now

$$\Phi^{(m+1)}(z) = f^{(m+1)}(z) - (m+1)!P(x_i)$$

whence, since

$$\Phi^{(m+1)}(\xi) = 0$$
$$P(x_i) = \frac{f^{(m+1)}(\xi)}{(m+1)!}$$

We thus have the exact formula

$$f'(x_i) = F'(x_i)$$
$$+ \frac{(x_i - x_0) \cdots (x_i - x_{i-1})(x_i - x_{i+1}) \cdots (x_i - x_m)}{(m+1)!} f^{(m+1)}(\xi) \quad (8\text{-}24)$$

where ξ lies between the greatest and least of the values $x_0, \ldots, x_m$.
The remainder in (8-23) can now be written down. If mth differences be used, the interpolating parabola passes through $m + 1$ points on the original curve. We have

$$\frac{(x_0 - x_1) \cdots (x_0 - x_m)}{(m + 1)!} = \frac{(-1)^m m! h^m}{(m + 1)!} = (-1)^m \frac{h^m}{m + 1}$$

The exact formula is

$$f'(a) = \frac{1}{h}\left[\Delta f(a) - \frac{1}{2}\Delta^2 f(a) + \cdots + (-1)^{m-1}\frac{1}{m}\Delta^m f(a)\right]$$
$$+ (-1)^m \frac{h^m}{m + 1} f^{(m+1)}(\xi)$$

where ξ lies between a and $a + mh$.

Approximate formulas for the second and higher derivatives may be obtained by differentiating an interpolating function two or more times. Expressions for the remainders may be worked out. We shall not go further into these matters here.

EXERCISES

1. Evaluate π by integrating numerically

$$\frac{\pi}{4} = \int_0^1 \frac{dx}{1 + x^2}$$

2. Find the length of an ellipse of semiaxes 3 and 2:

$$x = 3 \cos t \qquad y = 2 \sin t$$

3. Show that a and c can be so chosen that the formula

$$\int_{-h}^{h} f(x)\, dx = c[f(a) + f(-a)]$$

is exact for polynomials of the third degree or less.

4. Which gives in general the better approximation to the area under a curve, the trapezoid whose parallel sides are the two end ordinates, or the rectangle on the interval whose height is the middle ordinate?

5. A curve has for $x = 0, 1, 2, 3, 4, 5, 6$ the ordinates $0, 1.17, 2.13, 2.68, 2.62, 1.77, -0.07$, respectively. Find its slope at each of the seven points and its length from $x = 0$ to $x = 6$.

SYMBOLIC METHODS

The present chapter would not be complete without some mention, however brief, of the calculus of operators. The methods to be described are of great importance for the reader who wishes to become expert in numerical work. They not only enable him to keep the fundamental formulas already derived in mind but supply a powerful instrument for working out new approximate formulas of all kinds.

8-15. Operators. An operator is a symbol placed before a function to indicate the application of some process to the function to produce a new function. The symbol Δ is an example. Our fundamental operators will be E, Δ, D, and k, any constant, to indicate the following processes:

$$Ef(x) = f(x + h)$$
$$\Delta f(x) = f(x + h) - f(x)$$
$$Df(x) = f'(x)$$
$$kf(x) = kf(x)$$

Here E means that the function is to be replaced by its value h units to the right, D indicates differentiation, and the constant operator merely multiplies the function by the given constant.

If an operator is applied to a function, a second operator is applied to the resulting function, and so on; the several operators are written as a product. Each new operator is written to the left of those preceding it, although, as will appear presently, the order of application is immaterial. Thus

$$\Delta Df(x) = \Delta f'(x) = f'(x + h) - f'(x)$$

If an operator is repeated n times, we indicate this by an exponent; thus E^3 means EEE. We thus define all positive integral powers of the operators, and we add the convention that an operator with power zero produces no change in the function, $D^0 f(x) = f(x)$. Products of powers of operators then combine according to the law of exponents as

$$D^2 D^3 f(x) = D^5 f(x)$$

Although we shall restrict the powers of D and Δ to the integral and, for the present, non-negative numbers, we shall admit all real powers of E. We define the general power as follows:

$$E^n f(x) = f(x + nh)$$

which agrees with the previous values when n is integral. These powers combine according to the law of exponents:

$$E^m E^n f(x) = E^m f(x + nh) = f(x + nh + mh) = E^{m+n} f(x)$$

The sum or difference of two operators applied to a function is defined to be the sum or difference of the functions resulting from the application of each operator, thus

$$(E^{-\frac{1}{2}} + D)f(x) = E^{-\frac{1}{2}} f(x) + Df(x) = f(x - \tfrac{1}{2}h) + f'(x)$$

8-16. The Algebra of Operators. The preceding conventions give a meaning to all operators formed from E^n, Δ, D, and k by addition, subtraction, and multiplication; that is, to all polynomials in E^n, Δ, and D. We now consider the question of divorcing the operators from the functions to which they apply, and working with them as independent quantities.

Two operators are called *equal* if, when applied to an arbitrary function, they produce the same results; for example, $\Delta = E - 1$.

These operators, which we shall represent by A, B, C, etc., can be combined by additions, subtractions, and multiplications according to the rules of algebra, provided they conform to the following five laws:
1. $A + B = B + A$.
2. $A + (B + C) = (A + B) + C$.
3. $AB = BA$.
4. $A(BC) = (AB)C$.
5. $A(B + C) = AB + AC$.

We shall not establish these laws in detail. Numbers 1 and 2 follow directly from the definition of addition of operators; 5 follows from the fact that our operators are distributive over the sum of two functions. We need prove 3 and 4 for the fundamental operators only. Thus, to prove $E^n D = D E^n$, we have

$$E^n D f(x) = E^n f'(x) = f'(x + nh) = Df(x + nh) = DE^n f(x)$$

Verification for the other operators is left as an exercise.

Since the laws of algebra are satisfied, we can combine operators according to the usual algebraic rules. Thus we can write such equations as the following:

$$(E^{\frac{1}{2}}\Delta - D)(E^{\frac{1}{2}}\Delta + D) = E\Delta^2 - D^2$$
$$2E^2 - 3E + 1 = (E - 1)(2E - 1) = \Delta(1 + 2\Delta) = \Delta + 2\Delta^2$$

8-17. The Fundamental Equations. The formula for parabolic interpolation, based on a table of differences, as well as numerous formulas for integration, differentiation, and summation can be derived quite readily from a pair of symbolic equations. They are the following:

$$E = 1 + \Delta = e^{hD} \tag{8-25}$$

The first of these has already been noted. The last operator is used to represent the series

$$e^{hD} = 1 + hD + \frac{h^2 D^2}{2!} + \frac{h^3 D^3}{3!} + \cdots$$

We have

$$Ef(x) = f(x + h) = f(x) + hf'(x) + \frac{h^2}{2!}f''(x) + \cdots$$
$$= \left(1 + hD + \frac{h^2}{2!}D^2 + \frac{h^3}{3!}D^3 + \cdots\right)f(x)$$

INTERPOLATION AND NUMERICAL INTEGRATION 199

provided the Taylor's expansion of $f(x + h)$ is valid. If $f(x)$ is a polynomial the series converges to the function; in this case the number of terms is finite.

8-18. Application to a Difference Table. We shall express our operators in powers of Δ. Consider a difference table made from the values $f(a), f(a + h), \ldots, f(a + mh)$. The first row contains the items, 1, Δ, $\Delta^2, \ldots, \Delta^m$ applied to $f(a)$ (see Sec. 8-4). To get general values of the function we apply E^n to $f(a)$. From (8-25), $E^n = (1 + \Delta)^n$.

THEOREM. *If the formula*

$$(1 + \Delta)^n = 1 + n\Delta + \frac{n(n-1)}{2!} \Delta^2 + \cdots$$

be applied to $f(a)$, the differences of order higher than m being discarded, there results the interpolating polynomial of lowest degree which is equal to the tabulated function at $x = a, a + h, \ldots, a + mh$.

For we observe that we have precisely Newton's formula, (8-6).
For the derivative, (8-25) gives

$$D = \frac{1}{h} \log (1 + \Delta)$$

Applying to $f(a)$, after expanding the second member in powers of Δ, we have

$$f'(a) = \frac{1}{h}\left(\Delta - \frac{\Delta^2}{2} + \frac{\Delta^3}{3} - \cdots\right) f(a)$$

which is formula (8-23). This is the derivative of the interpolating polynomial when differences beyond the mth are discarded.

The derivative of the interpolating polynomial at a general point is $1/h$ times the derivative of Newton's formula with respect to n (since $x = a + nh$, $dx = h\,dn$). We have

$$\frac{1}{h}\frac{d}{dn}(1 + \Delta)^n f(a)$$

Now it is immaterial whether we expand $(1 + \Delta)^n$ and then differentiate, or differentiate first and then expand. Following the latter course, we have

$$\frac{1}{h}(1 + \Delta)^n \log (1 + \Delta) f(a)$$

This formula, however, is simply $E^n D$, as put from (8-25) in terms of Δ. Likewise for higher derivatives of the interpolating polynomial, we have

$$E^n D^r = (1 + \Delta)^n \cdot \frac{1}{h^r} [\log (1 + \Delta)]^r$$

the second member to be expanded and applied to $f(a)$.

The preceding formulas are exact when applied to polynomials.

Example. Find from the table in Sec. 8-5 the derivative of sin x, x being in radians, for an angle of 36°. Here $h = 5° = 0.087266$ radian, and $n = \frac{6}{5}$.

$$E^{6/5}D = (1 + \Delta)^{6/5} \cdot \frac{1}{h} \log (1 + \Delta)$$

$$= \frac{1}{h}\left(1 + \frac{6}{5}\Delta + \frac{3}{25}\Delta^2 + \cdots\right)\left(\Delta - \frac{\Delta^2}{2} + \frac{\Delta^3}{3} - \cdots\right)$$

$$= \frac{1}{h}\left(\Delta + \frac{7}{10}\Delta^2 - \frac{11}{75}\Delta^3 + \cdots\right)$$

Applying to the table and neglecting the small fourth difference, we have

$$\frac{1}{0.087266}\left[0.07358 + \frac{7}{10}(-0.00437) - \frac{11}{75}(-0.00052)\right] = 0.8090$$

This agrees with the correct value, cos 36°, to four figures.

8-19. The Operators D^{-1} and Δ^{-1}. We now define negative integral powers of D and Δ. We define D^{-1}, or $1/D$, as the operator which when followed by D leaves the function unchanged; that is,

$$D^{-1}f(x) = \int f(x)\, dx + C$$

Similarly $1/\Delta$ applied to $f(x)$

$$u(x) = \frac{1}{\Delta} f(x)$$

is a function $u(x)$ such that

$$\Delta u(x) = f(x)$$

We have

$$u(x + h) = u(x) + \Delta u(x) = u(x) + f(x)$$
$$u(x + 2h) = u(x + h) + \Delta u(x + h) = u(x) + f(x) + f(x + h)$$
$$\cdots\cdots\cdots\cdots\cdots\cdots\cdots\cdots\cdots\cdots\cdots\cdots\cdots\cdots\cdots\cdots$$
$$u(x + nh) = u(x) + f(x) + f(x + h) + \cdots + f[x + (n-1)h]$$

Hence for integral n we have

$$E^n \Delta^{-1} f(x) = \sum_{r=0}^{n-1} f(x + rh) + \Delta^{-1} f(x)$$

Other negative integral powers of D and Δ are defined by repetitions of the preceding processes.

If we integrate the interpolating polynomial, derived from a table of differences, we have

$$h \int_p^q (1 + \Delta)^n f(a)\, dn = \left[h \frac{(1+\Delta)^n}{\log(1+\Delta)} f(a)\right]_p^q$$

$$= h \frac{(1+\Delta)^q - (1+\Delta)^p}{\log(1+\Delta)} f(a)$$

This may be written

$$\frac{E^q - E^p}{D} f(a)$$

where E and D are to be expressed in terms of Δ from (8-25), and the operator expanded in powers of Δ. If $f(x)$ is a polynomial the result is exact. If differences of order greater than m are discarded, we have the integral of the interpolating polynomial agreeing with the function at $a, \ldots, a + mh$.

Example. Evaluate

$$\int_{11}^{14} x^3 \, dx$$

from the table of Sec. 8-4.
Here $a = 10$, $h = 2$, and we have

$$\frac{E^2 - E^{1/2}}{D} = 2 \frac{(1 + \Delta)^2 - (1 + \Delta)^{1/2}}{\log (1 + \Delta)}$$

$$= 2 \frac{1 + 2\Delta + \Delta^2 - (1 + \tfrac{1}{2}\Delta - \tfrac{1}{8}\Delta^2 + \tfrac{1}{16}\Delta^3 - \tfrac{5}{128}\Delta^4)}{\Delta - \tfrac{1}{2}\Delta^2 + \tfrac{1}{3}\Delta^3 - \tfrac{1}{4}\Delta^4 + \cdots}$$

$$= \frac{3 + \tfrac{9}{4}\Delta - \tfrac{1}{8}\Delta^2 + \tfrac{5}{64}\Delta^3 + \cdots}{1 - \tfrac{1}{2}\Delta + \tfrac{1}{3}\Delta^2 - \tfrac{1}{4}\Delta^3 + \cdots}$$

$$= 3 + \tfrac{15}{4}\Delta + \tfrac{3}{4}\Delta^2 - \tfrac{3}{64}\Delta^3 + \cdots$$

Applying this to the table,

$$3(1{,}000) + \tfrac{15}{4}(728) + \tfrac{3}{4}(288) - \tfrac{3}{64}(48) = 5{,}943\tfrac{3}{4}$$

which is exact.

When the operator Δ^{-1} is applied to $f(a)$ in a difference table, the resulting values may be tabulated in a column preceding the values of the function and of which $f(a), f(a + h), \ldots$ are the differences. Taking $\Delta^{-1}f(a) = K$, any constant, we have

$$E^n \, \Delta^{-1}f(a) = \sum_{r=0}^{n-1} f(a + rh) + K$$

for n a positive integer. From this

$$\sum_{r=0}^{n-1} f(a + rh) = (E^n - 1) \, \Delta^{-1}f(a)$$

which may be expanded in powers of Δ and expressed in terms of the first row of the table. The result is exact for the summation of any number of terms if the function is a polynomial. The formula may also be derived as follows:

$$(1 + E + E^2 + \cdots + E^{n-1})f(a) = \frac{E^n - 1}{E - 1} f(a)$$

Example. Find the sum of the first n squares. Forming a difference table

x	x^2	Δ	Δ^2
1	1	3	2
2	4	5	
3	9		

we have

$$\frac{E^n - 1}{\Delta} = \frac{n\Delta + \frac{n(n-1)}{2}\Delta^2 + \frac{n(n-1)(n-2)}{6}\Delta^3 + \cdots}{\Delta}$$

Applying to the table, after the division by Δ, we get

$$n + \frac{n(n-1)}{2}(3) + \frac{n(n-1)(n-2)}{6}(2) = \frac{n(n+1)(2n+1)}{6}$$

8-20. Various Formulas. In this section we work out some examples by symbolic methods. These few cases suggest the wide applicability and power of the algebra of operators. All changes of operators are derived from Eqs. (8-25). The ratios of series in powers of Δ are found by long division. The various infinite series give exact results if the function to which they are applied is a polynomial.

1. For the integral $\int_a^{a+h} f(x)\, dx$, we have

$$\frac{E - 1}{D} f(a) = \frac{h\Delta}{\log(1 + \Delta)} f(a) = h \frac{\Delta}{\Delta - \frac{\Delta^2}{2} + \frac{\Delta^3}{3} - \cdots} f(a)$$

$$= h(1 + \tfrac{1}{2}\Delta - \tfrac{1}{12}\Delta^2 + \tfrac{1}{24}\Delta^3 - \tfrac{19}{720}\Delta^4 + \cdots) f(a)$$

This is the formula of Sec. 8-12.

2. To evaluate $\int_a^{a+nh} f(x)\, dx$ for any n, we have

$$\frac{E^n - 1}{D} f(a) = \frac{h[(1 + \Delta)^n - 1]}{\log(1 + \Delta)} f(a)$$

$$= nh \left[1 + \frac{n}{2}\Delta + \frac{n(2n-3)}{12}\Delta^2 + \frac{n(n-2)^2}{24}\Delta^3 \right.$$

$$\left. + \frac{n(6n^3 - 45n^2 + 110n - 90)}{720}\Delta^4 + \cdots \right] f(a)$$

3. When n is an integer, a common formula for the integral in 2 is

INTERPOLATION AND NUMERICAL INTEGRATION   203

$$\frac{E^n - 1}{D} f(a) = \frac{h(E^n - 1)}{\Delta - \frac{1}{2}\Delta^2 + \cdots} f(a)$$

$$= h(E^n - 1) \left(\frac{1}{\Delta} + \frac{1}{2} - \frac{1}{12}\Delta + \cdots \right) f(a)$$

$$= h \Big\{ \sum_{r=0}^{n-1} f(a + rh) + \tfrac{1}{2}[f(a + nh) - f(a)]$$

$$- \tfrac{1}{12}[\Delta f(a + nh) - \Delta f(a)] + \tfrac{1}{24}[\Delta^2 f(a + nh) - \Delta^2 f(a)]$$

$$- \tfrac{19}{720}[\Delta^3 f(a + nh) - \Delta^3 f(a)] + \cdots \Big\}$$

This expresses the integral in terms of the sum of n values of the function and of the entries in the lines $x = a$ and $x = a + nh$.

4. Simpson's Rule. For $n = 2$ the formula of 2 is

$$h(2 + 2\Delta + \tfrac{1}{3}\Delta^2 - \tfrac{1}{90}\Delta^4 + \cdots) f(a)$$

Dropping fourth and higher differences, we have a formula valid for polynomials of degree less than four:

$$h(2 + 2\Delta + \tfrac{1}{3}\Delta^2) f(a) = h[2 + 2(E - 1) + \tfrac{1}{3}(E - 1)^2] f(a)$$

$$= \frac{h}{3}(1 + 4E + E^2) f(a)$$

This is Simpson's rule.

5. The Three-eighths Rule. For $n = 3$ the formula of 2 is

$$3h(1 + \tfrac{3}{2}\Delta + \tfrac{3}{4}\Delta^2 + \tfrac{1}{8}\Delta^3 - \tfrac{1}{80}\Delta^4 + \cdots) f(a)$$

Dropping fourth and higher differences and putting in terms of E, we get

$$3h[1 + \tfrac{3}{2}(E - 1) + \tfrac{3}{4}(E - 1)^2 + \tfrac{1}{8}(E - 1)^3] = \frac{3h}{8}[1 + 3E + 3E^2 + E^3]$$

We have the three-eighths rule of Cotes,

$$\int_a^{a+3h} f(x)\, dx = \frac{3h}{8}[f(a) + 3f(a + h) + 3f(a + 2h) + f(a + 3h)]$$

6. The Euler-Maclaurin Series. In the preceding formulas we have expanded in powers of Δ; we now use (8-25) to expand in powers of D.

$$\frac{1}{\Delta} = \frac{1}{e^{hD} - 1} = \frac{1}{hD + \dfrac{h^2 D^2}{2!} + \dfrac{h^3 D^3}{3!} + \cdots}$$

$$= \frac{1}{hD} - \frac{1}{2} + \frac{1}{12} hD - \frac{1}{720} h^3 D^3 + \frac{1}{30{,}240} h^5 D^5 - \cdots$$

Multiplying by $E^m - 1$, where m is an integer, and applying to $f(a)$, we have the Euler-Maclaurin series:

$$\sum_{r=0}^{m-1} f(a + rh) = \frac{1}{h} \int_a^{a+mh} f(x)\, dx - \frac{1}{2}[f(a + mh) - f(a)]$$
$$+ \frac{h}{12}[f'(a + mh) - f'(a)] - \frac{h^3}{720}[f'''(a + mh) - f'''(a)]$$
$$+ \frac{h^5}{30{,}240}[f^{(5)}(a + mh) - f^{(5)}(a)] - \cdots$$

This may be used to get either the sum or the integral in terms of the other and of the derivatives.

7. Change of Interval. To change the interval of a table to th, we get the new difference δ from the formula

$$\delta = E^t - 1 = (1 + \Delta)^t - 1$$

and higher differences are powers of δ. If the new table is to begin at $a + ph$ the differences are multiplied by E^p. The new differences are useful for amplifying a table, for summing at nontabular intervals, etc.

Example. Using the table of Sec. 8-5 evaluate

$$\sin 35° + \sin 35°10' + \sin 35°20' + \cdots + \sin 35°50'$$

Here the 10' interval is $\frac{1}{30}$ of the interval of the table. We have

$$\frac{E^{6/5} - E}{\delta} = \frac{(1 + \Delta)^{6/5} - (1 + \Delta)}{(1 + \Delta)^{1/30} - 1}$$
$$= 6 + \tfrac{13}{2}\Delta + \tfrac{101}{360}\Delta^2 - \tfrac{59}{720}\Delta^3 + \cdots$$

Applying to the table,

$$6(0.50000) + \tfrac{13}{2}(0.07358) + \tfrac{101}{360}(-0.00437) - \tfrac{59}{720}(-0.00052) = 3.47708$$

Actual addition from a five-place table gives 3.47707.

EXERCISES

1. Find the sum of the first n cubes.

2. Sum to m terms the series

$$(3^2 + 8) + (5^2 + 11) + (7^2 + 14) + (9^2 + 17) + \cdots$$

3. Express Weddle's rule

$$\int_a^{a+6h} f(x)\, dx = \frac{3h}{10}(1 + 5E + E^2 + 6E^3 + E^4 + 5E^5 + E^6)f(a)$$

in terms of differences and check against the expansion of $(E^6 - 1)/D$.

4. Find the sum of the first n cubes using the entries in the table of cubes in Sec. 8-4.

5. Using the same table evaluate

$$\int_0^4 x^3\, dx \qquad \left[\frac{d(x^3)}{dx}\right]_{x=5} \qquad \left[\frac{d^2(x^3)}{dx^2}\right]_{x=13}$$

INTERPOLATION AND NUMERICAL INTEGRATION 205

6. Given the series
$$S = a(0) + a(1)x + a(2)x^2 + \cdots$$
show that the coefficient of x^n in the expansion of $(1 - x)^k S$ is, after a few terms, $\Delta^k a(n - k)$. If $a(n)$ is a polynomial in n, the series may be summed in finite form. Show that the radius of convergence in this case is 1.

7. Sum the series
(a) $(1 \cdot 2 + 3) + (2 \cdot 3 + 4)x + (3 \cdot 4 + 5)x^2 + \cdots$,
(b) $(1 \cdot 3 + 2^3) + (3 \cdot 5 + 3^3)x + (5 \cdot 7 + 4^3)x^2 + \cdots$.

8. What should be paid, at 6 per cent interest, for a perpetual annuity paying 1, 8, ..., n^3, ... dollars at the end of 1, 2, ..., n, ... years?

9. Derive the formula
$$\int_a^{a+4h} f(x) \, dx = \frac{2h}{45} (7 + 32E + 12E^2 + 32E^3 + 7E^4) f(a)$$

10. Using the data on page 181, find the number of hours the sun is above the horizon in December.

11. The square roots of 250, 260, 270, 280 are, respectively, 15.8114, 16.1245, 16.4317, 16.7332. Form a difference table and compute for 250 the first and second differences for the interval 1. Use these to fill in by addition the square roots of the integers between 250 and 260.

12. Derive Newton's formula for backward interpolation
$$f(a + nh) = f(a) + n\Delta f(a - h) + \frac{(n + 1)n}{2!} \Delta^2 f(a - 2h)$$
$$+ \frac{(n + 2)(n + 1)n}{3!} \Delta^3 f(a - 3h) + \cdots$$

13. Derive Gauss's formula of interpolation
$$f(a + nh) = f(a) + n\Delta f(a) + \frac{n(n - 1)}{2!} \Delta^2 f(a - h)$$
$$+ \frac{(n + 1)n(n - 1)}{3!} \Delta^3 f(a - h) + \cdots$$
the entries moving up one row at each even difference.

14. Derive Stirling's formula of interpolation
$$f(a + nh) = f(a) + n \frac{\Delta f(a) + \Delta f(a - h)}{2} + \frac{n^2}{2!} \Delta^2 f(a - h)$$
$$+ \frac{n(n^2 - 1)}{3!} \frac{\Delta^3 f(a - h) + \Delta^3 f(a - 2h)}{2} + \cdots$$

15. Derive the formula
$$\int_{-h}^{h} f(x) \, dx = 2h[f(a) + \tfrac{1}{6}\Delta^2 f(a - h) - \tfrac{1}{180}\Delta^4 f(a - 2h)$$
$$+ \tfrac{1}{1512}\Delta^6 f(a - 3h) + \cdots]$$

16. Derive the formula
$$\int_a^{a+h} f(x) \, dx = h \left[\frac{f(a) + f(a + h)}{2} - \frac{1}{12} \frac{\Delta^2 f(a - h) + \Delta^2 f(a)}{2} \right.$$
$$+ \frac{11}{720} \frac{\Delta^4 f(a - 2h) + \Delta^4 f(a - h)}{2}$$
$$\left. - \frac{191}{60,480} \frac{\Delta^6 f(a - 3h) + \Delta^6 f(a - 2h)}{2} + \cdots \right]$$

17. Derive Woolhouse's formula for summation at short intervals

$$\sum_{k=0}^{mr} f\left(a + \frac{kh}{m}\right) = m \sum_{s=0}^{r} f(a + hs) - \frac{m-1}{2}[f(a) + f(a + rh)]$$

$$- \frac{m^2 - 1}{12} \frac{h}{m} [f'(a + rh) - f'(a)] + \frac{m^4 - 1}{720} \frac{h^3}{m^3} [f'''(a + rh) - f'''(a)] + \cdots$$

18. Derive Lubbock's formula for the same sum, the terms involving derivatives in the preceding formula being replaced by

$$- \frac{m^2 - 1}{12m} \{\Delta f[a + (r - 1)h] - \Delta f(a)\} - \frac{m^2 - 1}{24m} \{\Delta^2 f[a + (r - 2)h]$$

$$+ \Delta^2 f(a)\} - \frac{(m^2 - 1)(19m^2 - 1)}{720 m^3} \{\Delta^3 f[a + (r - 3)h] - \Delta^3 f(a)\} + \cdots$$

CHAPTER 9

THE NUMERICAL SOLUTION OF DIFFERENTIAL EQUATIONS

The attention of the reader has been called repeatedly to the fact that the elementary methods of solution presented in the earlier chapters of this book are applicable to only a limited class of differential equations. The method of successive approximations, discussed in Chaps. 5 and 6, is not so limited. It is applicable when the conditions determining the solution are stated at a single point $x = a$. The method of procedure in solving an equation, such as

$$\frac{dy}{dx} = f(x,y) \qquad (9\text{-}1)$$

with the condition that $y = b$ when $x = a$, is direct and theoretically simple. However, in applying the process, difficulties in integration will arise unless $f(x,y)$ is a very simple function, such as a polynomial.

We circumvent these difficulties by the use of the numerical methods of the preceding chapter.[1] The method of successive approximations involves repeated integrations, and we can perform these quadratures to any desired degree of accuracy by the rules we have developed. Moreover, the methods to be here presented apply when the function $f(x,y)$ is given in graphical or tabular form, its analytical expression being quite unknown.

In the method to be used, we select an interval h and determine approximate values of the solution $y(x)$ for the equally spaced values a, $a + h$, $a + 2h$, and so on. Our principal tool will be Simpson's rule.

Questions of the convergence of the approximating process and of the faithfulness with which the final result represents the exact solution will

---

[1] The reader is referred to the following, which contain many references:
A. A. Bennett, W. E. Milne, and H. Bateman: "Numerical Integration of Differential Equations," Bulletin No. 92 of the National Research Council (1933).
H. Levy and E. A. Baggott: "Numerical Solutions of Differential Equations" (1934), Dover Publications.
W. E. Milne: "Numerical Solution of Differential Equations" (1953), John Wiley & Sons.

not be raised until later. At present, our concern is to make the reader familiar with the method.

9-1. The Start of the Solution. In solving

$$\frac{dy}{dx} = f(x,y) \tag{9-1}$$

where $y = b$ when $x = a$, we make an initial guess $y_0(x)$ and determine $y_1(x)$, $y_2(x)$, . . . from the recurrence formula

$$y_n(x) = b + \int_a^x f(x,y_{n-1})\, dx \tag{9-2}$$

The limit of $y_n(x)$ is the solution $y(x)$.

We choose an interval h. We know that $y(a) = b$. Our general method is not applicable until we have evaluated $y(a + h)$. The evaluation of $y(a + h)$ is called *starting the solution* and involves a somewhat different process from the subsequent steps.

In getting $y(a + h)$ we shall also get the intermediate value $y(a + h')$, where

$$h' = \tfrac{1}{2}h$$

This use of a smaller interval insures greater accuracy, which is desirable at the start.

For brevity we shall use the notation

$$f_n(x) = f[x,y_n(x)] \qquad f(x) = f[x,y(x)]$$

so that we write (9-2) in the form

$$y_n(x) = b + \int_a^x f_{n-1}(x)\, dx$$

We now apply the formulas (8-15) and (8-13) of the preceding chapter:

$$y_n(a + h') = b + \frac{h'}{12}[5f(a) + 8f_{n-1}(a + h') - f_{n-1}(a + h)] \tag{9-3}$$

$$y_n(a + h) = b + \frac{h'}{3}[f(a) + 4f_{n-1}(a + h') + f_{n-1}(a + h)] \tag{9-4}$$

These are the fundamental formulas of the starting process.

These formulas are used as follows. The value $f(a) = f(a,b)$ is known. We guess values of y at $a + h'$ and $a + h$, $y_0(a + h')$ and $y_0(a + h)$; and compute $f_0(a + h')$ and $f_0(a + h)$. We then use (9-3) and (9-4) to get $y_1(a + h')$ and $y_1(a + h)$.

We use these values to compute $f_1(a + h')$ and $f_1(a + h)$. Formulas (9-3) and (9-4) then give $y_2(a + h')$ and $y_2(a + h)$. We continue this process until $y_n(a + h')$ and $y_n(a + h)$ are sensibly equal to $y_{n-1}(a + h')$

THE NUMERICAL SOLUTION OF DIFFERENTIAL EQUATIONS

and $y_{n-1}(a + h)$, that is, until the values of y agree with those of the previous step to the desired number of decimal places.

We illustrate the method by solving

$$\frac{dy}{dx} = x - y$$

with the condition $y = 1$ when $x = 0$. This problem is used because it is easy to follow the numerical work and because, knowing the exact solution, we can check the accuracy of the results. Here $a = 0$, $b = 1$; and we shall take $h = \frac{1}{2}$, so $h' = \frac{1}{4}$. We have

$$f_n(x) = x - y_n(x)$$

The tabulated results of the starting process, which will be explained immediately, are as follows:

x	y_0	f_0	y_1	f_1	y_2	f_2	y_3	f_3	y_4
0	1.0	-1.0							
0.25	0.75	-0.50	0.81	-0.56	0.81	-0.56	0.807	-0.557	0.807
0.50	0.50	0	0.75	-0.25	0.71	-0.21	0.713	-0.213	0.713

The items 0.75 and 0.50 in the y_0 column are guesses. The slope at the beginning is -1, and these are the values y would have if the slope were constant. We could have made better guesses: for example, by observing that the curve is concave up at the beginning ($y'' = 1 - y' = 2$).

The entries in the f_0 column are computed from the values of x and y_0, $f_0 = x - y_0$.

The entries in the y_1 column are then computed by the use of formulas (9-3) and (9-4):

$$y_1(0.25) = 1 + \tfrac{1}{48}[5(-1) + 8(-0.50) - 0] = 0.81$$
$$y_1(0.50) = 1 + \tfrac{1}{12}[-1 + (-0.50)4 + 0] = 0.75$$

We compute the entries $f_1 = x - y_1$, and apply (9-3) and (9-4) again:

$$y_2(0.25) = 1 + \tfrac{1}{48}[5(-1) + 8(-0.56) + 0.25] = 0.81$$
$$y_2(0.50) = 1 + \tfrac{1}{12}[-1 + 4(-0.56) - 0.25] = 0.71$$

Repeating this process, we get the remaining entries in the table. Carrying the work to three decimals, we find that y_4 and y_3 are equal. We stop at this point with the values $y(0.25) = 0.807$, $y(0.50) = 0.713$.

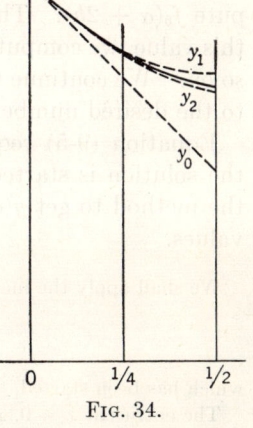

FIG. 34.

The approximations y_0, y_1, y_2 are shown drawn to scale in Fig. 34. The plotted points of each approximation are joined by a curve.

The exact solution of the equation is readily found to be

$$y = 2e^{-x} + x - 1$$

From this we find

$$y(0.25) = 0.8076 \qquad y(0.50) = 0.7131$$

EXERCISES

1. Carry through the process with the initial guess
$$y(0.25) = 0.9 \quad y(0.50) = 0.8$$

2. Start the solution with the interval $h = 1$.

9-2. The Subsequent Process. Having a starting value $y(a + h)$, we now proceed by the use of Simpson's rule alone. Suppose $y(a)$, $y(a + h)$, ..., $y[a + (m - 1)h]$ have been found, $m \geq 2$, and we wish the next entry. Applying the method of successive approximations from $x = a + (m - 2)h$, and putting, for brevity,
$$a + (m - 2)h = \alpha$$
we have the recurrence formula
$$y_n(\alpha + 2h) = y(\alpha) + \int_\alpha^{\alpha+2h} f(x, y_{n-1}) \, dx$$

Using Simpson's rule to evaluate this, we have the formula which forms the basis of the method:
$$y_n(\alpha + 2h) = y(\alpha) + \frac{h}{3}[f(\alpha) + 4f(\alpha + h) + f_{n-1}(\alpha + 2h)] \quad (9\text{-}5)$$

All the terms in the second member of (9-5) are known with the exception of $f_{n-1}(\alpha + 2h)$. We proceed as follows. Guessing $y_0(\alpha + 2h)$, we compute $f_0(\alpha + 2h)$. Then, applying (9-5), we have $y_1(\alpha + 2h)$. Having this value, we compute $f_1(\alpha + 2h)$ and apply (9-5) to get $y_2(\alpha + 2h)$, and so on. We continue the process until $y_n(\alpha + 2h)$ and $y_{n-1}(\alpha + 2h)$ agree to the desired number of decimal places.

Equation (9-5) requires that two previous entries be known. After the solution is started, so that $y(a)$ and $y(a + h)$ are known, we apply the method to get $y(a + 2h)$. We then advance step by step to further values.

We shall apply the method to the continuation of the solution of
$$\frac{dy}{dx} = x - y$$
which has been started.

The entries at $x = 0$, $x = 0.5$ have already been worked out. At $x = 1$, we have taking $\alpha = 0$ in Eq. (9-5),
$$y_n(1) = 1 + \tfrac{1}{6}[-1 + 4(-0.213) + f_{n-1}(1)]$$
$$= 0.6913 + \tfrac{1}{6}f_{n-1}(1)$$

With the guess $y_0 = 0.5$, we have $f_0 = 1 - y_0 = 0.5$ and
$$y_1(1) = 0.6913 + \tfrac{1}{6}(0.5) = 0.77$$

THE NUMERICAL SOLUTION OF DIFFERENTIAL EQUATIONS

x	y_0	f_0	y_1	f_1	y_2	f_2	y	f
0							1.000	−1.000
0.5							0.713	−0.213
1.0	0.5	0.5	0.77	0.23	0.730	0.270	0.736	0.264
1.5	1.0	0.5	0.94	0.56	0.947	0.553	0.946	0.554
2.0	1.279	0.721	1.269	0.731			1.271	0.729
2.5	1.668	0.832	1.663	0.837			1.664	0.836
3.0	2.103	0.897	2.099	0.901			2.100	0.900
3.5	2.561	0.939					2.560	0.940
4.0	3.039	0.961					3.037	0.963
4.5	3.521	0.979					3.522	0.978
5.0	4.015	0.985					4.014	0.986

Then $f_1 = 0.23$ and
$$y_2(1) = 0.6913 + \tfrac{1}{6}(0.23) = 0.730$$

We continue until the value of y_n is unaltered by the process. This value of y and the corresponding f are put in the last two columns.

At $x = 1.5$, we have from Eq. (9-5), taking $\alpha = 0.5$,

$$y_n(1.5) = 0.713 + \tfrac{1}{6}[-0.213 + 4(0.264) + f_{n-1}(1.5)]$$
$$= 0.8535 + \tfrac{1}{6}f_{n-1}(1.5)$$

We guess $y_0 = 1$, then $f_0 = 0.5$ and

$$y_1(1.5) = 0.8535 + \tfrac{1}{6}(0.5) = 0.94$$

and so on.

The remaining entries are computed in a similar way. The more accurate estimates used from $x = 2$ on in the y_0 column are based on a formula to be described in the next section.

Accurate values, computed from the exact solution, are as follows:

$x =$	1.0	1.5	2.0	2.5	3.0	3.5	4.0	4.5	5.0
$y =$	0.7358	0.9463	1.2707	1.6642	2.0996	2.5604	3.0366	3.5222	4.0135

EXERCISES

1. Change the interval in the problem of the text to $h = 1$ after $y(3)$ has been computed, finding $y(4)$ and $y(5)$.

2. Show that for the problem in the text the final approximate values may be got without the step-by-step process. Thus in

$$y_n(1) = 0.6913 + \tfrac{1}{6}[1 - y_{n-1}(1)]$$

we may put $y(1) = y_{n-1}(1) = y$ and solve for y.

3. Show that if Eq. (9-1) is linear, Eq. (9-5) may be used by solving a linear equation for $y(\alpha + 2h)$ at each step.

212   DIFFERENTIAL EQUATIONS

4. Show that if (9-1) is linear, the step-by-step method for the starting process may be replaced by the solution of a pair of linear equations.

5. Show that the starting values at $\frac{1}{4}$ and $\frac{1}{2}$ for the equation

$$dy/dx = -y^2$$

with $y(0) = 1$, could be found by solving the equations

$$z = 1 + \tfrac{1}{48}(-5 - 8z^2 + y^2)$$
$$y = 1 + \tfrac{1}{12}(-1 - 4z^2 - y^2)$$

6. Use successive approximations to get the solution near $x = 1$, $y = 2$ of the algebraic equations

$$x = 1 + 0.02x^2 + 0.01y^3$$
$$y = 2 + 0.03x - 0.01y^2$$

9-3. Aids to Good Guessing.

Each step in the preceding numerical process begins with an estimate of y_0. If h is small enough, any estimate which is not ridiculously wide of the mark will suffice. But the more accurately this estimate can be made, the greater will be the saving in the subsequent labor. This is particularly important in case $f(x,y)$ is a complicated function whose evaluation entails much computation. Good guessing will then effect much economy of time.

The Start. The properties of the solution at (a,b) assist us at the beginning. We know the slope here, $y'(a) = f(a,b)$. Assuming that the graph of the solution is a straight line with this slope, we have

$$y_0(x) = b + y'(a)(x - a)$$

From this we have the guesses

$$y_0(a + h') = b + y'(a)h'$$
$$y_0(a + h) = b + y'(a)h$$
(9-6)

These were the guesses used in the illustrative problem.

The second derivative at (a,b) may be obtained by differentiating $f(x,y)$,

$$\frac{d^2y}{dx^2} = \frac{d}{dx}f(x,y) = \frac{\partial f}{\partial x} + \frac{\partial f}{\partial y}\frac{dy}{dx} = \frac{\partial f}{\partial x} + f\frac{\partial f}{\partial y}$$

so

$$y''(a) = \left(\frac{\partial f}{\partial x} + f\frac{\partial f}{\partial y}\right)_{a,b}$$

The polynomial of the second degree with the same slope and the same second derivative is

$$y_0(x) = b + y'(a)(x - a) + \tfrac{1}{2}y''(a)(x - a)^2$$

Using this, we have the estimates

$$y_0(a + h') = b + y'(a)h' + \tfrac{1}{2}y''(a)h'^2$$
$$y_0(a + h) = b + y'(a)h + \tfrac{1}{2}y''(a)h^2$$
(9-7)

If we had used these in the illustrative problem, where $y'(0) = -1$, $y''(0) = 2$, we should have had

$$y_0(0.25) = 0.81 \qquad y_0(0.5) = 0.75$$

This would have saved one step in the computation.

The computer may have further information concerning the trend of the solution which can be used as an aid to good guessing. Or he may know something of the behavior of $f(x,y)$. He could begin the process by guessing values for f_0 and thence computing y_0.

The Subsequent Process. One of the most useful of estimating formulas, which cannot be used, however, until a few steps have been completed, is based on Eq. (8-16) of the preceding chapter. Applying that formula to

$$y(\alpha + 4h) = y(\alpha) + \int_\alpha^{\alpha+4h} f(x,y)\,dx$$

we have

$$y_0(\alpha + 4h) = y(\alpha) + \frac{4h}{3}[2f(\alpha + h) - f(\alpha + 2h) + 2f(\alpha + 3h)] \quad (9\text{-}8)$$

This formula does not involve $f(\alpha + 4h)$. It contains the three preceding values of f and the value of y preceding these. It is first applicable for an estimate of $y(a + 4h)$. From this point on, the entries in the second member of (9-8) are known.

In the illustrative example, we have

$$y_0(2) = 1 + \tfrac{2}{3}[2(-0.213) - 0.264 + 2(0.554)] = 1.279$$
$$y_0(2.5) = 0.713 + \tfrac{2}{3}[2(0.264) - 0.554 + 2(0.729)] = 1.668$$

For an estimate of $y_0(a + 2h)$ and $y_0(a + 3h)$, there are various methods. We may use Eq. (9-6) with a replaced by $a + h$ or $a + 2h$:

$$y_0(a + 2h) = y(a + h) + f(a + h)h$$

and

$$y_0(a + 3h) = y(a + 2h) + f(a + 2h)h$$

If $f(x,y)$ is easily differentiated we may employ (9-7),

$$y_0(a + 2h) = y(a + h) + f(a + h)h + \tfrac{1}{2}f'(a + h)h^2$$

with a similar expression for $y_0(a + 3h)$.

In the illustrative example we used a very rough guess. If we had employed the last formula we should have had, since $f' = 1 - y' = 1 - f$,

$$y_0(1) = 0.713 + \tfrac{1}{2}(-0.213) + \tfrac{1}{8}(1.213) = 0.76$$
$$y_0(1.5) = 0.736 + \tfrac{1}{2}(0.264) + \tfrac{1}{8}(0.736) = 0.96$$

which are much nearer the true values.

Various other estimating formulas may be worked out. The following may be found useful:

$$y_0(a + 2h) = b + \frac{2h}{3}[2f(a) - 4f(a + h') + 5f(a + h)] \quad (9\text{-}9)$$

$$y_0(a + 3h) = b + \frac{3h}{4}[f(a) + 3f(a + 2h)] \quad (9\text{-}10)$$

Their derivation is left to the reader. In the illustrative example they give

$$y_0(1) = 0.72 \qquad y_0(1.5) = 0.92$$

EXERCISES

1. Derive (9-10) by expansion of the expression $(E^3 - 1)/D$.
2. Derive (9-9) by symbolic methods.
3. Continue the solution in the text two additional steps.

9-4. Checking. To avoid overlooking actual mistakes in the work and to detect the accumulation of small errors it is desirable to check the work from time to time by the use of an integration formula which involves a considerable range of entries. Of numerous formulas that one could employ for this purpose we shall work out one which is useful. It is based on an interpolation formula involving six equally spaced entries.

Let the origin be placed at the mid-point of the range of entries and let $f_0, f_1, \ldots, f_5$ be the values of a function at the points $-5h'$, $-3h'$, $-h'$, h', $3h'$, $5h'$, respectively, where $h' = \frac{1}{2}h$. Lagrange's formula then gives for the interpolating polynomial taking on these values

$$F(x) = \frac{(x^2 - 9h'^2)(x^2 - h'^2)(x - 5h')}{-3840h'^5} f_0$$
$$+ \frac{(x^2 - 25h'^2)(x^2 - h'^2)(x - 3h')}{768h'^5} f_1 \quad (9\text{-}11)$$
$$+ \frac{(x^2 - 25h'^2)(x^2 - 9h'^2)(x - h')}{-384h'^5} f_2$$
$$+ \cdots$$

The last three terms have not been written down. It is easy to see, from symmetry, that when we integrate over an interval with the mid-point at the origin, the last three constants will duplicate the first three.

We have, on integrating (9-11),

$$\int_{-5h'}^{5h'} F(x) \, dx = \frac{5h}{288}(19f_0 + 75f_1 + 50f_2 + 50f_3 + 75f_4 + 19f_5)$$

This gives, on shifting the origin, the approximate formula

$$\int_a^{a+5h} f(x) \, dx = \frac{5h}{288}[19f(a) + 75f(a + h) + 50f(a + 2h)$$
$$+ 50f(a + 3h) + 75f(a + 4h) + 19f(a + 5h)] \quad (9\text{-}12)$$

THE NUMERICAL SOLUTION OF DIFFERENTIAL EQUATIONS 215

Using (9-12) for performing the integration over five intervals, we have

$$y(c + 5h) = y(c) + \frac{5h}{288}[19f(c) + 75f(c + h) + 50f(c + 2h)$$
$$+ 50f(c + 3h) + 75f(c + 4h) + 19f(c + 5h)] \quad (9\text{-}13)$$

If this be applied at every fifth entry in the table, it will serve as a check on the computed values.

In the illustrative problem we have

$$y(2.5) = 1 + \tfrac{5}{576}[19(-1) + 75(-0.213) + 50(0.264) + 50(0.554) + 75(0.729)$$
$$+ 19(0.836)] = 1.664$$

$$y(5) = 1.664 + \tfrac{5}{576}[19(0.836) + 75(0.900) + 50(0.940) + 50(0.963)$$
$$+ 75(0.978) + 19(0.986)] = 4.013$$

The work checks in a satisfactory manner.

EXERCISES

1. Derive the formula

$$\int_a^{a+3h} f(x)\, dx = \frac{h}{160}[-3f(a - h) + 69f(a) + 174f(a + h)$$
$$+ 174f(a + 2h) + 69f(a + 3h) - 3f(a + 4h)]$$

2. Use the preceding as a check on the illustrative example, getting $y(2)$ in terms of $y(0.5)$ and the first six entries in the f column.

3. Show that

$$\frac{E^5 - 1}{D} = 5h(1 + \tfrac{5}{2}\Delta + \tfrac{35}{12}\Delta^2 + \tfrac{15}{8}\Delta^3 + \tfrac{85}{144}\Delta^4 + \tfrac{19}{288}\Delta^5 + \cdots)$$

Use this to derive formula (9-12) of the text.

9-5. Practical Hints. The smaller the interval h, the more rapid is the convergence at each step. If $y_n(x)$ shows a tendency to oscillate instead of rapidly approaching a limit, a smaller interval should be used. With very short steps the convergence is rapid, but we get forward slowly; with very long steps, each may be laborious. The best stride strikes a balance between the two.

The smaller the interval h, the more faithfully will the approximate solution represent the true one. Quite apart from the question of convergence, h must be small if great accuracy is required.

The interval may be increased or decreased at any point if it appears desirable.

In the early stages of the computation, y_0, f_0, y_1, etc., should be obtained to only one or two significant figures. These early values are usually wide of the mark anyway, and nothing is gained by the extra labor involved in computing more figures. Only figures likely to be correct should be kept.

The numerical solution of a differential equation involves, at best. much arithmetical labor. If the computer has a calculating machine,

216   DIFFERENTIAL EQUATIONS

the drudgery will be materially lessened. Except for the calculation of $f(x,y)$, concerning which, of course, we can say nothing, all the processes can be performed with speed and accuracy on a machine.

EXERCISES

1. Taking $h = 10$, solve

$$\frac{dy}{dx} = \frac{y}{10} \qquad y(0) = 1$$

by approximate methods. Compare $y(10)$ and $y(20)$ with the correct values.

2. Find how much the solution of

$$\frac{dy}{dx} = x - y + \frac{1}{100(x+y)} \qquad y(0) = 1$$

differs from the solution of the problem in the text. Replace y by $y(x) + u$, where $y(x)$ is the solution of the text, and solve the resulting equation in u.

3. Horizontal cross sections of a vessel 5 ft high made 0, 1, 2, 3, 4, 5 ft below the level of the water have the areas 3.3, 3.0, 2.6, 2.1, 1.7, 1.0 sq ft, respectively. Find how long it takes the surface to fall 2 ft, due to an orifice of 1-sq in. effective area in the bottom.

9-6. Questions of Convergence. In Chap. 5 we proved the existence of a solution of

$$\frac{dy}{dx} = f(x,y)$$

where $y(a) = b$, in an interval I about a on the assumption that $f(x,y)$ satisfies a Lipschitz condition. In the following we refer to this treatment, particularly to Fig. 18.

Let a' ($> a$, say) be a point of I and divide aa' into intervals of length h. On applying the numerical process of the present chapter two questions arise:

I. Will the numerical process converge if h is small enough?

II. Can h be taken sufficiently small that the computed values of y in aa' differ from the true values by less than a preassigned amount?

We shall indicate, very briefly, the answers to these questions.

I. The steps of the convergence proof of Chap. 5 can be applied without much change here. Suppose the start has been made, the calculated points lying in R, and that we are calculating a new value. From the fundamental formula, (9-5), we have

$$\left| \frac{y_n(\alpha + 2h) - y(\alpha)}{2h} \right| = \frac{1}{6} |f(\alpha) + 4f(\alpha + h) + f_{n-1}(\alpha + 2h)|$$

$$< \frac{1}{6}(M + 4M + M) = M$$

if $y_{n-1}(\alpha + 2h)$ is in S. The slope of the line joining the point representing $y(\alpha)$ to that representing $y_n(\alpha + 2h)$ lies between M and $-M$, hence the latter point is in R. If $y_0(\alpha + 2h)$ is chosen in S the succeeding approximations lie in R.

THEOREM. *The continuing process based on formula (9-5) converges if $h < 3/A$, where A is the Lipschitz constant.*

Setting $n + 1$ for n in (9-5) and forming the difference, we have

$$|y_{n+1}(\alpha + 2h) - y_n(\alpha + 2h)| = \frac{h}{3} |f_n(\alpha + 2h) - f_{n-1}(\alpha + 2h)|$$

$$\leqslant \frac{hA}{3} |y_n(\alpha + 2h) - y_{n-1}(\alpha + 2h)|$$

The process will converge if $hA/3 < 1$, that is, $h < 3/A$.

In the problem in the text $A = 1$. The process will converge if $h < 3$. We used $h = 0.5$.

It can be shown in a similar way that the starting process based on Eqs. (9-3) and (9-4) will converge if

$$h < \frac{9 - 3\sqrt{5}}{A} = \frac{2.292}{A}$$

II. Omitting proofs, we shall merely state the answer to the second question. Let $y(x)$ represent the approximate solution and $Y(x)$ the exact solution.

THEOREM. *Given $\epsilon > 0$, then h can be taken sufficiently small that*

$$|y(x) - Y(x)| < \epsilon$$

for the entries in the interval aa'.

9-7. Equations of Higher Order and Systems of Equations. In Chap. 6 we have discussed the application of the method of successive approximations to systems of equations of the first order. It was also shown how single equations, or systems of equations, of higher order could be reduced to a system of the first order. In the numerical solution of such a system we perform the integrations by formulas of approximate quadrature.

We illustrate with two equations,

$$\frac{dy}{dx} = f(x,y,z)$$
$$\frac{dz}{dx} = g(x,y,z) \tag{9-14}$$

with the requirement that $y = b$, $z = c$ when $x = a$. The recurrence formulas for the method of successive approximations are [Eqs. (6-7)]

218   DIFFERENTIAL EQUATIONS

$$y_n(x) = b + \int_a^x f(x,y_{n-1},z_{n-1}) \, dx$$
$$z_n(x) = c + \int_a^x g(x,y_{n-1},z_{n-1}) \, dx \qquad (9\text{-}15)$$

Choosing an interval h, we estimate $y_0(a + h')$, $y_0(a + h)$, $z_0(a + h')$, $z_0(a + h)$, where $h' = \tfrac{1}{2}h$. We then compute f and g and get y_1 and z_1 by using Eqs. (9-3) and (9-4) for the approximate integrations. Writing

$$f_n(x) = f[x,y_n(x),z_n(x)]$$

with an analogous abbreviation for g, we make use of the recurrence formulas

$$y_n(a + h') = b + \frac{h'}{12}[5f(a) + 8f_{n-1}(a + h') - f_{n-1}(a + h)]$$
$$y_n(a + h) = b + \frac{h'}{3}[f(a) + 4f_{n-1}(a + h') + f_{n-1}(a + h)]$$
$$z_n(a + h') = c + \frac{h'}{12}[5g(a) + 8g_{n-1}(a + h') - g_{n-1}(a + h)] \qquad (9\text{-}16)$$
$$z_n(a + h) = c + \frac{h'}{3}[g(a) + 4g_{n-1}(a + h') + g_{n-1}(a + h)]$$

We alternately compute f and g and apply these formulas until y_n and z_n remain sensibly equal to y_{n-1} and z_{n-1}.

Having started the solution, we proceed by the use of Simpson's rule alone. We employ at each step two formulas analogous to (9-5):

$$y_n(\alpha + 2h) = y(\alpha) + \frac{h}{3}[f(\alpha) + 4f(\alpha + h) + f_{n-1}(\alpha + 2h)]$$
$$z_n(\alpha + 2h) = z(\alpha) + \frac{h}{3}[g(\alpha) + 4g(\alpha + h) + g_{n-1}(\alpha + 2h)] \qquad (9\text{-}17)$$

In practice, we should usually vary this process slightly. Having computed y_n from (9-16) or (9-17), we may put y_n instead of y_{n-1} into g in order to get z_n.

In estimating values and in checking, we may use the formulas suggested for the single equation of the first order.

Example. A particle starts from rest at one vertex of a unit square. It moves along one side due to an acceleration in that direction which is equal to its distance from the opposite vertex. Determine the motion.

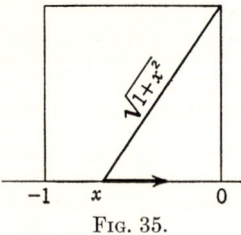

Fig. 35.

Taking the origin at the vertex toward which the particle moves (Fig. 35), the differential equation of the motion is

$$\frac{d^2x}{dt^2} = \sqrt{1 + x^2}$$

THE NUMERICAL SOLUTION OF DIFFERENTIAL EQUATIONS 219

where $x = -1$, $dx/dt = v = 0$ when $t = 0$. The equation may be written as the system

$$\frac{dv}{dt} = \sqrt{1 + x^2} = f(x)$$

$$\frac{dx}{dt} = v$$

Taking $h = \frac{1}{2}$, we start the solution.

t	x_0	f_0	v_1	x_1	f_1	v_2	x_2
0	−1.0	1.414	0				
0.25	−0.956	1.383	0.351	−0.956	1.383	0.351	−0.956
0.50	−0.825	1.296	0.687	−0.826	1.297	0.687	−0.826

We estimated x_0 by assuming that the acceleration is constant and equal to 1.4. Since v is not contained in $f(x)$, we do not estimate it. We compute $f_0 = \sqrt{1 + x_0^2}$ and get v_1 by formulas of the form (9-16); thus

$$v_1(0.25) = \tfrac{1}{48}[5(1.414) + 8(1.383) - 1.296] = 0.351$$

and so on. We get x_1 by an approximate integration of $v_1(x)$. We recompute $f(x)$ and repeat the process.
Having a start, we continue with Simpson's rule alone.

t	x_0	f_0	v_1	x_1	f	v	x
0	....	.....	.....	....	1.414	0	−1.0
0.5	....	.....	.....	....	1.297	0.687	−0.826
1.0	−0.320	1.050	1.275	−0.330	1.053	1.276	−0.329
1.5	0.441	1.093	1.787	0.437	1.091	1.787	0.437

The estimates of x_0 were made by using two terms of Taylor's series

$$x_0(t + h) = x(t) + h\frac{dx}{dt} + \frac{1}{2}h^2\frac{d^2x}{dt^2}$$

$$= x + hv + \tfrac{1}{2}h^2 f$$

$$x_0(1) = -0.826 + \tfrac{1}{2}(0.687) + \tfrac{1}{8}(1.297) = -0.320$$

The particle arrives at the origin at the time $t = 1.23$, approximately.

EXERCISES

1. How is the time of arrival at the origin found from the tabulated values of x?
2. Find the velocity at the origin.
3. Solve the problem in the text by exact methods, the final integral being left unevaluated.
4. Show that for a linear equation or a system of linear equations the step-by-step methods of the text may be replaced by the solution of a system of linear equations.

5. Find the quarter period of a pendulum of length $l = g$ which is held with the string horizontal and released, by solving numerically

$$\frac{d^2\theta}{dt^2} = -\sin\theta$$

The exact value is 1.8541 sec, whereas the approximate solution of Chap. 1 gives 1.5708 sec.

6. A body weighing 1 lb drops from rest through a resisting medium from a height of 10 ft. Measuring s downward, we have

$$\frac{d^2s}{dt^2} = 32 - f(v)$$

Suppose the acceleration due to resistance to have been determined experimentally for the following values of the velocity:

$$v = 0 \quad 5 \quad 10 \quad 15 \quad 20 \quad 25$$
$$f = 0 \quad 0.9 \quad 4.8 \quad 13.0 \quad 26.7 \quad 47.2$$

Find when the body strikes the ground and the velocity with which it strikes.

9-8. The Cauchy-Lipschitz Method. This is a method of great simplicity. It consists in the construction of a broken-line approximation

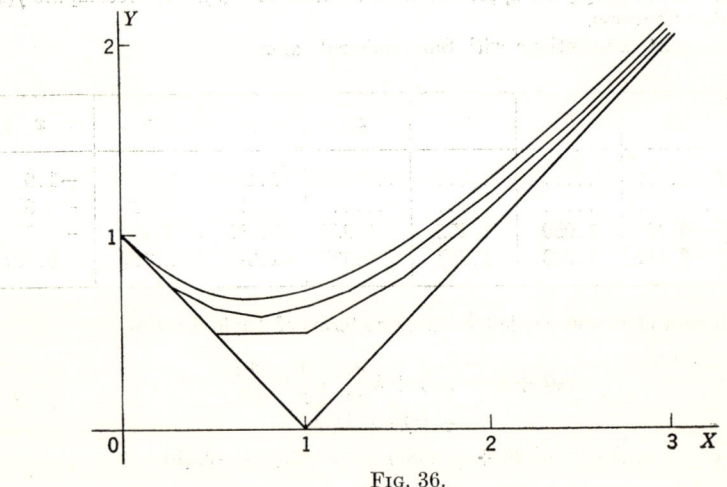

Fig. 36.

to the solution. Choosing an interval h, we construct a straight line through (a,b) with the requisite slope $f(a,b)$ and extend to meet the line $x = a + h$. This gives the approximation

$$y(a + h) = b + f(a,b)h \tag{9-18}$$

Through the point $[a + h, y(a + h)]$ we draw a line with the slope of the lineal element there, viz., $f[a + h, y(a + h)]$, and extend to the line $x = a + 2h$, and so on.

THE NUMERICAL SOLUTION OF DIFFERENTIAL EQUATIONS   221

In Fig. 36 are shown the results for the solution of $y' = x - y$ through the point (0,1) for $h = 1$, $h = \frac{1}{2}$, and $h = \frac{1}{4}$. The figure shows also the graph of the accurate solution.

For great accuracy a very small value of h would usually be necessary. As a practical method the process is useful only when very rough results are required.

As a theoretical instrument the Cauchy-Lipschitz process is important. It can be used to establish the existence of the solution in place of the method of Chap. 5.

9-9. The Runge-Kutta Formulas. The formula of the preceding section is the Taylor's expansion of $y(a + h)$, carried to two terms. Runge, Kutta, and others have proposed formulas agreeing with this expansion to various degrees.

In the series

$$y(a + h) = b + y'(a)h + \frac{y''(a)}{2!} h^2 + \frac{y'''(a)}{3!} h^3 + \frac{y^{iv}(a)}{4!} h^4 + \cdots$$

we can express the coefficients in terms of $f(x,y)$ and its partial derivatives. Using the familiar notation

$$p = f_x \quad q = f_y \quad r = f_{xx} \quad s = f_{xy} \quad t = f_{yy}$$

and letting f_0, p_0, etc., represent values at (a,b), we have

$$y' = f \quad y'(a) = f_0$$
$$y'' = f_x + f_y y' = p + qf \quad y''(a) = p_0 + f_0 q_0$$

etc. The preceding series is

$$y(a + h) = b + f_0 h + \frac{(p_0 + f_0 q_0)h^2}{2}$$
$$+ \frac{(p_0 q_0 + f_0 q_0{}^2 + r_0 + 2f_0 s_0 + f_0{}^2 t_0)h^3}{6} + \cdots \quad (9\text{-}19)$$

We propose to avoid the partial derivatives of the third and subsequent terms by setting up values of $f(x,y)$ in the neighborhood and getting combinations,

$$y(a + h) = b + h[\alpha_1 f(x_1, y_1) + \cdots + \alpha_n f(x_n, y_n)]$$

whose expansions reproduce these early terms.

Let us, for example, evaluate $f(x,y)$ at the mid-point of the first line segment of the preceding section, thus using the approximation

$$y(a + h) = b + hf(a + \tfrac{1}{2}h, b + \tfrac{1}{2}f_0 h)$$

or, to adopt the pattern of later formulas,

$$k_1 = hf_0$$
$$k_2 = hf(a + \tfrac{1}{2}h, b + \tfrac{1}{2}k_1) \qquad (9\text{-}20)$$
$$y(a + h) = b + k_2$$

We shall see how well this approximate solution agrees with the correct series (9-19). Making use of the Taylor's expansion

$$f(a + u, b + v) = f_0 + p_0 u + q_0 v + \frac{r_0 u^2 + 2s_0 uv + t_0 v^2}{2} + \cdots \qquad (9\text{-}21)$$

we find for (9-20)

$$y(a + h) = b + f_0 h + \frac{(p_0 + q_0 f_0)h^2}{2} + \frac{(r_0 + 2s_0 f_0 + t_0 f_0{}^2)h^3}{8} + \cdots \qquad (9\text{-}22)$$

This agrees with (9-19) through the term in h^2.

Example. Solve $y' = x - y$, with $a = 0$, $b = 1$, taking $h = \tfrac{1}{4}$. We have $f_0 = -1$, and (9-20) gives us

$$y(\tfrac{1}{4}) = 1 + \tfrac{1}{4}[\tfrac{1}{8} - (1 - \tfrac{1}{8})] = \tfrac{13}{16} = 0.8125$$

The correct value, as noted earlier, is 0.8076.

Among processes good to the next power of h, the following is frequently used:

$$k_1 = hf_0$$
$$k_2 = hf(a + \tfrac{1}{2}h, b + \tfrac{1}{2}k_1) \qquad (9\text{-}23)$$
$$k_3' = hf(a + h, b + 2k_2 - k_1)$$
$$y(a + h) = b + \frac{k_1 + 4k_2 + k_3'}{6}$$

We have k_2 from (9-22), whence

$$2k_2 - k_1 = f_0 h + (p_0 + q_0 f_0)h^2 + \frac{(r_0 + 2s_0 f_0 + t_0 f_0{}^2)h^3}{4} + \cdots$$

From this we find

$$k_3' = f_0 h + (p_0 + q_0 f_0)h^2 + \frac{(2p_0 q_0 + 2q_0{}^2 f_0 + r_0 + 2s_0 f_0 + t_0 f_0{}^2)h^3}{2} + \cdots$$

On substituting these series into the last equation of (9-23), we find (9-19) reproduced so far as the terms are written down.

In the preceding example, we have already $k_1 = -\tfrac{1}{4}$, $k_2 = -\tfrac{3}{16}$. Then

$$2k_2 - k_1 = -\tfrac{1}{8}$$

and

$$k_3' = \tfrac{1}{4}[\tfrac{1}{4} - (1 - \tfrac{1}{8})] = -\tfrac{5}{32}$$
$$y(\tfrac{1}{4}) = 1 + \tfrac{1}{6}[-\tfrac{1}{4} + 4(-\tfrac{3}{16}) - \tfrac{5}{32}] = 0.8073$$

THE NUMERICAL SOLUTION OF DIFFERENTIAL EQUATIONS   223

Probably the best-known process which is good through the term in h^4 is the following. We do not check it here because of the labor involved.

$$k_1 = hf_0$$
$$k_2 = hf(a + \tfrac{1}{2}h, b + \tfrac{1}{2}k_1)$$
$$k_3 = hf(a + \tfrac{1}{2}h, b + \tfrac{1}{2}k_2) \qquad (9\text{-}24)$$
$$k_4 = hf(a + h, b + k_3)$$
$$y(a + h) = b + \frac{k_1 + 2k_2 + 2k_3 + k_4}{6}$$

9-10. Finite Difference Methods. Let us suppose that by some of the methods already described we have started the solution, getting a set of values of y and of f for equally spaced values of x. We can then form a difference table for f and apply any one of a wide variety of formulas involving differences to get the next entry by evaluating

$$y(\alpha) = y(\alpha - h) + \int_{\alpha-h}^{\alpha} f(x,y)\, dx$$

We shall illustrate with the equation $y' = x - y$, $y(0) = 1$, and we suppose the values of y down to $x = 0.4$ in the following table have already been found. We compute the f's and form a table of differences.

x	y	f	∇	∇^2	∇^3	∇^4
0	1.	$-1.$				
0.1	0.90967	-0.80967	19033			
0.2	0.83746	-0.63746	17221	-1812		
0.3	0.78164	-0.48164	15582	-1639	173	
0.4	0.74064	-0.34064	14100	-1482	157	-16
		(-0.21305)	12759	-1341	141	$-16^*)$
0.5	0.71306	-0.21306	12758	-1342	140	-17

We have used backward differences, $\nabla f(\alpha) = f(\alpha) - f(\alpha - h)$, in forming the table down to the horizontal line. The formula to be used is

$$\int_{\alpha-h}^{\alpha} f\, dx = h[1 - \tfrac{1}{2}\nabla - \tfrac{1}{12}\nabla^2 - \tfrac{1}{24}\nabla^3 - \tfrac{19}{720}\nabla^4 - \tfrac{3}{160}\nabla^5 - \cdots]f(\alpha)$$
$$(9\text{-}25)$$

Taking $\alpha = 0.5$, we proceed as follows. Taking ∇^4 constant (the entry marked with an asterisk) and adding in, we have the trial line of entries for $x = 0.5$ shown between parentheses. Using (9-25), we have

$$y(0.5) = 0.74064 + 0.1[-0.21305 - \tfrac{1}{2}(0.12759) - \tfrac{1}{12}(-0.01341)$$
$$- \tfrac{1}{24}(0.00141)] = 0.71306$$

A recomputation gives $f = -0.21306$, so we would have the slightly more accurate corrected line appearing in the table.

Following the preceding pattern we could compute $y(0.6)$, $y(0.7)$, etc. The derivation of (9-25) is readily made by the manipulation of symbols. We have

$$\nabla = 1 - E^{-1} \qquad E = \frac{1}{1 - \nabla}$$

The integral is

$$\int_{\alpha-h}^{\alpha} f\,dx = \frac{1 - E^{-1}}{D} f(\alpha) = \frac{h\nabla}{\log E} f(\alpha)$$

We have

$$\frac{\nabla}{\log E} = \frac{\nabla}{-\log(1 - \nabla)} = \frac{\nabla}{\nabla + \nabla^2/2 + \nabla^3/3 + \nabla^4/4 + \cdots}$$

and the division of the numerator by the denominator gives the series in the second member of (9-25).

EXERCISES

Applying Runge-Kutta formulas to $y' = x - y$, $y(0) = 1$:
1. Use (9-20) to get $y(0.1) = 0.910$.
2. Use (9-24) to get $y(1) = 0.750$.
3. Use (9-24) to get $y(\tfrac{1}{2}) = 0.7135$.
4. Use (9-24) to get $y(\tfrac{1}{4}) = 0.8076$.
5. Get $y(0.1)$ and $y(0.2)$ as given in Sec. 9-10 by the use of Runge-Kutta formulas.
6. Apply the formulas

$$k_1 = hf_0$$
$$k_2' = hf(a + \tfrac{2}{3}h,\ b + \tfrac{2}{3}k_1)$$
$$k_3'' = hf(a + \tfrac{2}{3}h,\ b + \tfrac{2}{3}k_2')$$
$$y(a + h) = b + \frac{2k_1 + 3k_2' + 3k_3''}{8}$$

to get $y(\tfrac{1}{4}) = 0.8073$ for the problem in the text.

7. Prove that the preceding process is correct through terms in h^3.
8. Using the method of Sec. 9-10, extend the table to get $y(0.6) = 0.69762$.
9. Get $y(0.5)$ from the line of entries for $x = 0.4$, using

$$\int_{\alpha}^{\alpha+h} f\,dx = h[1 + \tfrac{1}{2}\nabla + \tfrac{5}{12}\nabla^2 + \tfrac{3}{8}\nabla^3 + \tfrac{251}{720}\nabla^4 + \tfrac{95}{288}\nabla^5 + \cdots]f(\alpha)$$

10. Derive the preceding formula by multiplying the series in (9-25) by E.
11. Derive the approximate formula

$$y''(a) = \frac{y(a - h) - 2y(a) + y(a + h)}{h^2}$$

Solve $y'' = y$, with the conditions $y(0) = y(1) = 1$ by estimating intermediate values and using this formula to readjust them. The following are correct:

$$y(\tfrac{1}{4}) = y(\tfrac{3}{4}) = 0.915 \qquad y(\tfrac{1}{2}) = 0.887$$

CHAPTER 10

PARTIAL DIFFERENTIAL EQUATIONS OF THE FIRST ORDER

10-1. Partial Differential Equations. In the differential equations appearing in the preceding chapters there has been in each case a single independent variable. In consequence the derivatives that appeared were ordinary derivatives. If there are two or more independent variables, the derivatives of the dependent variables with respect to these independent variables are partial derivatives. An equation involving such partial derivatives is called a *partial differential equation*. The order of a partial differential equation is that of the derivative of highest order appearing in the equation.

For example, the equations in three variables

(i) $$\frac{\partial z}{\partial x} + \frac{\partial z}{\partial y} = z$$

(ii) $$\frac{\partial^2 z}{\partial x^2} = 2y$$

are partial differential equations with one dependent variable z and two independent variables x and y. They are of the first and second orders, respectively.

A relation between the variables is a *solution*, if, on solving for the dependent variable and substituting into the differential equation, the equation is reduced to an identity in the independent variables. For example,

$$z = ae^x + be^y$$

is a solution of (i), for we have, on substituting, the identity

$$ae^x + be^y \equiv ae^x + be^y$$

Again

$$z = e^x(y - x)$$

is a solution, for, substituting,

$$e^x(y - x - 1) + e^x \equiv e^x(y - x)$$

225

Similarly, it can be verified that

$$z = y(x + \sin y)^2$$
$$z = x^2 y + e^y$$

are solutions of (ii).

A partial differential equation is much richer in solutions than an ordinary differential equation. Whereas the latter has solutions involving a few arbitrary constants, the former, as will appear, has solutions involving arbitrary functions. By particularizing the function we can insert any number of arbitrary constants.

A simple equation whose solution is immediate is

$$\frac{\partial z}{\partial x} = f(x,y)$$

The derivative of z with y held constant being given, z can be found by an integration in which y is held constant:

$$z = \int f(x,y) \, \partial x + C$$

Here C need be constant only so far as x is concerned, that is, C may be any function of y, $C = \varphi(y)$. The general solution is thus

$$z = \int f(x,y) \, \partial x + \varphi(y)$$

which contains an arbitrary function.

10-2. Derivation of the Equation of the First Order. We shall here restrict our study to the equation of the first order with one dependent variable z and two independent variables x and y. We shall employ the notation

$$p = \frac{\partial z}{\partial x} \qquad q = \frac{\partial z}{\partial y} \tag{10-1}$$

Given an equation in the variables and two arbitrary constants,

$$F(x,y,z,a,b) = 0 \tag{10-2}$$

we can set up a partial differential equation of the first order of which the given equation is a solution for any values of the constants. Differentiating partially with respect to x and y, we have

$$\begin{aligned}\frac{\partial F}{\partial x} + \frac{\partial F}{\partial z} p = 0 \\ \frac{\partial F}{\partial y} + \frac{\partial F}{\partial z} q = 0\end{aligned} \tag{10-3}$$

By the elimination of a and b from the three equations (10-2) and (10-3), we have a partial differential equation

$$f(x,y,z,p,q) = 0 \tag{10-4}$$

PARTIAL OF THE FIRST ORDER 227

That (10-2) is a solution of (10-4) follows from the same type of reasoning used in Chap. 1 for the ordinary differential equation. For given values of a and b, the value of z derived from (10-2) will have partial derivatives satisfying (10-3). Then (10-4), a combination of (10-2) and (10-3), will be satisfied. But (10-4) is free of the constants and is thus satisfied whatever the prescribed values of a and b.

In the preceding treatment we have made certain tacit assumptions, such as the appearance of z in (10-2) and the possibility of the elimination of a and b.

Example. Given

(iv) $$z = ax + by + ab$$

Differentiating, we have

$$p = a \qquad q = b$$

and eliminating a and b there results the partial differential equation

(iii) $$z = px + qy + pq$$

We shall now show that a partial differential equation may be set up, which has a given family of solutions involving an arbitrary function. Let u and v be given functions of the variables and consider

$$\varphi(u,v) = 0 \qquad (10\text{-}5)$$

where φ is an arbitrary function.

Differentiating partially with respect to x and y, we have

$$\frac{\partial \varphi}{\partial u}\left(\frac{\partial u}{\partial x} + \frac{\partial u}{\partial z}p\right) + \frac{\partial \varphi}{\partial v}\left(\frac{\partial v}{\partial x} + \frac{\partial v}{\partial z}p\right) = 0$$
$$\frac{\partial \varphi}{\partial u}\left(\frac{\partial u}{\partial y} + \frac{\partial u}{\partial z}q\right) + \frac{\partial \varphi}{\partial v}\left(\frac{\partial v}{\partial y} + \frac{\partial v}{\partial z}q\right) = 0 \qquad (10\text{-}6)$$

From these equations the arbitrary function φ may be eliminated. The result is an equation of the form

$$Pp + Qq = R \qquad (10\text{-}7)$$

where

$$P = \frac{\partial u}{\partial z}\frac{\partial v}{\partial y} - \frac{\partial u}{\partial y}\frac{\partial v}{\partial z}$$
$$Q = \frac{\partial u}{\partial x}\frac{\partial v}{\partial z} - \frac{\partial u}{\partial z}\frac{\partial v}{\partial x} \qquad (10\text{-}8)$$
$$R = \frac{\partial u}{\partial y}\frac{\partial v}{\partial x} - \frac{\partial u}{\partial x}\frac{\partial v}{\partial y}$$

The relation (10-5) is a solution of (10-7) whatever the function φ may be.

228   DIFFERENTIAL EQUATIONS

Example. Find a partial differential equation having as solution an arbitrary function of $z - x$ and $z - y$. Here

$$u = z - x \qquad v = z - y$$

and Eqs. (10-6) are

$$\frac{\partial \varphi}{\partial u}(-1 + p) + \frac{\partial \varphi}{\partial v} p = 0$$

$$\frac{\partial \varphi}{\partial u} q + \frac{\partial \varphi}{\partial v}(-1 + q) = 0$$

On performing the elimination we have

(v) $$p + q = 1$$

For example

$$(z - x)^2 - (z - y)^2 = 0$$

is a solution. Simplifying, we have

$$z = \tfrac{1}{2}(x + y)$$

EXERCISES

1. Show that the following values of z are solutions of (i):
   (a) $e^{\frac{1}{2}(x+y)}$ (b) $e^y(x - y)^3$
   (c) e^{2y-x} (d) $e^{y+f(x-y)}$
   (e) $e^x f(x - y)$

where f is an arbitrary function.

2. Show that the function in (e) can be particularized so as to give the solutions (a)–(d).

Find the differential equation of
3. The family $z = ax^2 + by^2$.
4. The family $z = \varphi(x + y)$.
5. The family $\varphi(x^2 z, y) = 0$.
6. All planes through the origin.
7. All spheres with centers on the x axis.
8. All right circular cones with x axis as axis.
9. All surfaces of revolution with x axis as axis. Compare with the preceding two problems.
10. The tangent planes to the unit sphere with center at the origin.
11. All surfaces cutting the family of cones

$$x^2 + y^2 - a^2 z^2 = 0$$

orthogonally.
12. All surfaces with the property that the radius vector to a point makes an angle of 45° with the tangent plane there.
13. Solve $q = x^2 y$.
14. Solve Eq. (ii) of the text.

10-3. Types of Solutions. Given an equation of the first order,

$$f(x,y,z,p,q) = 0 \qquad (10\text{-}4)$$

In the preceding sections we have dealt with two types of functions satisfying such an equation, each encompassing an infinitude of solutions, namely, the solution containing two arbitrary constants, and the solution containing an arbitrary function.

A solution containing two arbitrary constants is called a *complete solution* or *integral*.

A solution containing an arbitrary function is called a *general solution*.

Clearly a general solution gives a much broader body of solutions than a complete solution. We shall now show, however, how the complete solution may be used to derive a solution involving an arbitrary function.

Suppose that, in some manner, a complete integral has been found:

$$F(x,y,z,a,b) = 0 \tag{10-2}$$

The equation (10-4) results from the elimination of a and b from (10-2) and the derived equations

$$\frac{\partial F}{\partial x} + \frac{\partial F}{\partial z} p = 0$$
$$\frac{\partial F}{\partial y} + \frac{\partial F}{\partial z} q = 0 \tag{10-3}$$

Let a and b be no longer constants but functions of x, y, z connected by an arbitrarily chosen functional relation, say

$$b = \varphi(a) \tag{10-9}$$

We shall show that by a suitable choice of the function a the equation will be satisfied. Differentiating (10-2) partially, we have

$$\frac{\partial F}{\partial x} + \frac{\partial F}{\partial z} p + \left[\frac{\partial F}{\partial a} + \frac{\partial F}{\partial b} \varphi'(a)\right]\left(\frac{\partial a}{\partial x} + \frac{\partial a}{\partial z} p\right) = 0$$
$$\frac{\partial F}{\partial y} + \frac{\partial F}{\partial z} q + \left[\frac{\partial F}{\partial a} + \frac{\partial F}{\partial b} \varphi'(a)\right]\left(\frac{\partial a}{\partial y} + \frac{\partial a}{\partial z} q\right) = 0 \tag{10-10}$$

If now we select a so that

$$\frac{\partial F}{\partial a} + \frac{\partial F}{\partial b} \varphi'(a) = 0 \tag{10-11}$$

Eqs. (10-10) reduce to Eqs. (10-3). In the process of elimination it is immaterial whether a and b are constants or variables, so we are led again to Eq. (10-4).

Equation (10-11) involves x, y, z, and a [putting b in terms of a from (10-9)]. From this, a is found in terms of x, y, z and inserted into (10-2) to give the solution sought.

It will be noted that (10-11) is simply the partial derivative of F with respect to a, after b has been replaced by $\varphi(a)$.

The geometrical significance of this process will be brought out in the next section.

Example. For the equation

(iii) $$z = px + qy + pq$$

we had the complete solution

(iv) $$z = ax + by + ab$$

Put
$$b = \frac{1}{a}$$

(vi) $$z = ax + \frac{y}{a} + 1$$

Differentiating partially with respect to a, we have

$$0 = x - \frac{y}{a^2}$$

whence
$$a = \sqrt{\frac{y}{x}}$$

Substituting into (vi), we have the solution

$$z = 2\sqrt{xy} + 1$$

That this is a solution may be readily verified.

10-4. Geometric Interpretation. Letting x, y, z be the rectangular coordinates of a point in three-dimensional space, the locus of an equation in the variables

$$G(x,y,z) = 0$$

or
$$z = g(x,y) \tag{10-12}$$

is in general a surface. At a point $P(x,y,z)$ on the surface, let a lineal element with direction ratios dx, dy, dz be drawn. This line will be tangent to the surface, if and only if

$$dz = \frac{\partial z}{\partial x} dx + \frac{\partial z}{\partial y} dy = p\, dx + q\, dy \tag{10-13}$$

where p and q are found by differentiating the equation of the surface.

Two lines with direction ratios a_1, a_2, a_3 and b_1, b_2, b_3 are orthogonal if and only if

$$a_1 b_1 + a_2 b_2 + a_3 b_3 = 0$$

Hence (10-13) shows that the lineal elements at P which are tangent to the surface are all orthogonal to the line N through P with direction ratios p, q, -1. N is the normal to the surface at P.

If (10-12) is a solution of a partial differential equation

$$f(x,y,z,p,q) = 0 \tag{10-4}$$

then at each point of the surface the direction ratios of the normal p, q, -1 must be such that p and q satisfy (10-4). Conversely, any surface with this property is a solution.

We shall now generalize the notion of a direction field, presented in the first chapter. At a point $P(x,y,z)$, let p and q be a pair of values satisfying (10-4). We erect a line N at P with direction ratios p, q, -1. (We proceed from P a distance p in the x direction, thence a distance q in the y direction, thence a distance -1 in the z direction to a point Q. N is the line PQ.) At P we construct a small piece of plane perpendicular to N, a so-called *planar element*. We imagine a planar element constructed at P for each pair of values of p and q which satisfy (10-4) with the coordinates at P inserted for x, y, z. We imagine like planar elements constructed at all other points of space.

The resulting set of planar elements constitutes a planar field. A surface is an integral surface, and its equation is a solution of (10-4), if and only if at every point the surface touches a planar element belonging to the field.

Let us examine in more detail the situation at a particular point. At each point there is in general a one-parameter family of planar elements defined by the partial differential equation of the first order. Solve (10-4) for q, let us say,

$$q = Q(x,y,z,p)$$

At a given point P, the coordinates x, y, z are fixed. Then we may vary p arbitrarily, and for each p the value of q is determined.

As p varies, the planar element moves and in general envelops a surface. Since each planar element passes through P, the surface generated is a cone with vertex at P. We shall call this cone to which the planar elements at P are tangent the *conical element* at P. Then the necessary and sufficient condition that a surface be an integral surface of the partial differential equation is that it be tangent at each point to the conical element at the point.

It may happen that, as p varies, the planar element turns about a line (or there may be several such lines through which the planar elements pass). Here there is no true surface enveloped, but we may consider such a line as a degenerate cone. This case which, as we shall see, arises when the partial differential equation is linear will be considered in some detail later.

10-5. Envelopes. Before proceeding further with the geometric picture we shall now make a digression and recall the salient facts of the theory of envelopes.

Consider a one-parameter family of surfaces

$$\Psi(x,y,z,a) = 0 \qquad (10\text{-}14)$$

In the domain of values considered, let Ψ be single-valued and possess continuous first partial derivatives with respect to each of its variables.

We consider the surface (10-14) with a given value of a and the surface resulting from a slight change in the parameter

$$\Psi(x,y,z,a + da) = 0 \qquad (10\text{-}15)$$

The points whose coordinates satisfy both (10-14) and (10-15) make up in general the curve of intersection of the two surfaces. Now (10-15) may be written

$$\Psi(x,y,z,a) + \left[\frac{\partial \Psi(x,y,z,a)}{\partial a} + \epsilon\right] da = 0$$

where ϵ is an infinitesimal. Subtracting (10-14), we have along the curve

$$\frac{\partial \Psi}{\partial a} + \epsilon = 0$$

Letting da approach zero, we have for the ultimate position of this curve of intersection

$$\begin{aligned} \Psi(x,y,z,a) &= 0 \\ \frac{\partial \Psi(x,y,z,a)}{\partial a} &= 0 \end{aligned} \qquad (10\text{-}16)$$

The curve (10-16) is called a *characteristic curve* of the family. In geometrical language it is the curve of intersection of the surface (10-14) with an infinitely neighboring surface of the family.

Now let a be eliminated from Eqs. (10-16). There results a surface on which all the characteristic curves lie:

$$\theta(x,y,z) = 0 \qquad (10\text{-}17)$$

For (10-17) is satisfied by coordinates satisfying (10-16) with any value of a.

We shall now show that, barring certain singular points, the surfaces (10-14) and (10-17) are tangent all along the characteristic curve (10-16). It will suffice to show that the partial derivatives p and q are the same for both, for the tangent planes will thus coincide.

From (10-14) we have for p,

$$\frac{\partial \Psi}{\partial x} + \frac{\partial \Psi}{\partial z} p = 0$$

Now (10-17) is merely the first equation of (10-16) with a, no longer constant, inserted after solving from the second equation of (10-16). So for (10-17)

$$\frac{\partial \Psi}{\partial x} + \frac{\partial \Psi}{\partial z} p + \frac{\partial \Psi}{\partial a}\left[\frac{\partial a}{\partial x} + \frac{\partial a}{\partial z} p\right] = 0$$

But along the characteristic curve $\partial \Psi/\partial a = 0$ and the two equations for p are identical. Similarly q is the same for the two surfaces, and (10-17) is tangent to (10-14) along the characteristic curve lying on (10-14).

Because of this property, (10-17) is called the *envelope* of the family (10-14). Equations (10-16), with the understanding that a is to be eliminated, give us the envelope. Equations (10-16) with a retained give the curve along which the individual surface touches the envelope.

Certain extraneous loci may appear in this process. It is not difficult to show that singular points (if any occur), at which

$$\frac{\partial \Psi}{\partial x} = \frac{\partial \Psi}{\partial y} = \frac{\partial \Psi}{\partial z} = 0$$

lie on (10-16). Singular loci will thus appear in the result.

The preceding analysis, save for a few alterations, applies to families of curves in the plane

$$\Psi(x,y,a) = 0 \qquad (10\text{-}18)$$

The envelope results from the elimination of a from

$$\Psi = 0 \qquad \frac{\partial \Psi}{\partial a} = 0 \qquad (10\text{-}19)$$

Equations (10-19), with a fixed, determine the points of contact (*characteristic points*) of (10-18) with the envelope.

In the case of a two-parameter family of surfaces,

$$\Psi(x,y,z,a,b) = 0 \qquad (10\text{-}20)$$

we may proceed similarly to the derivation of the envelope. The envelope results from the elimination of a and b from the equations

$$\Psi = 0 \qquad \frac{\partial \Psi}{\partial a} = 0 \qquad \frac{\partial \Psi}{\partial b} = 0 \qquad (10\text{-}21)$$

Here (10-21), with a and b fixed, gives the points of contact of (10-20) with the envelope.

10-6. The Complete Integral. A knowledge of envelopes renders the correctness of our previous use of the complete integral intuitively obvious. The complete integral,

$$F(x,y,z,a,b) = 0 \qquad (10\text{-}2)$$

is a two-parameter family of surfaces, each of which is tangent at each of its points to the conical element of the point. Now it is clear that if we abstract from (10-2) any one-parameter family and form the envelope, the envelope will be an integral surface. For at each point the envelope is tangent to a surface of the family which in turn is tangent to the conical element.

We may form a one-parameter family by setting up a relation between b and a, say

$$b = \varphi(a)$$

Then the envelope is got by eliminating a between the equations

$$F = 0$$
$$\frac{\partial F}{\partial a} = \frac{\partial F}{\partial a} + \frac{\partial F}{\partial b} \varphi'(a) = 0$$

These are the formulas which we previously employed.

It may happen that all members of (10-2) touch an enveloping surface. This surface is found [see Eqs. (10-21)] by eliminating a and b from

$$F = 0 \quad \frac{\partial F}{\partial a} = 0 \quad \frac{\partial F}{\partial b} = 0 \quad (10\text{-}22)$$

This solution, which contains no arbitrary element, is called the *singular solution*.

Since, in determining envelopes, extraneous loci may appear, a check should be made to verify that the differential equation is satisfied.

Example. Find the singular solution of

$$z = px + qy + pq$$

Differentiating the complete integral

$$z = ax + by + ab$$

partially with respect to a and b, we have

$$0 = x + b \quad 0 = y + a$$

Eliminating a and b, we have

$$z = -xy$$

which satisfies the equation.

Through an aribitrary point $P_0(x_0, y_0, z_0)$ pass the members of a one-parameter family of the surfaces (10-2), namely, the family resulting from setting up the relation

$$F(x_0, y_0, z_0, a, b) = 0$$

between a and b. As a, say, varies, the tangent plane to the surface turns about P_0 and envelops the conical element of P_0. The envelope of the family is an integral surface with a conical point at P_0. The characteristic curves on the envelope are tangent to the rulings of the conical element (see Fig. 37).

This property of tangency of the characteristic curves to the rulings of the conical elements is a general one. Given any integral surface got as the envelope of a one-parameter family of integral surfaces. Let P be a point on the curve of intersection C of two nearby surfaces of the family. The two tangent planes at P, whose line of intersection is tangent to C, are tangent to the conical element at P. Now as the two surfaces approach coincidence, the limiting position of C is a characteristic curve. The tangent planes of the surfaces at a point of C are tangent to the conical element and approach coincidence. The limiting position of the line of intersection, assuming suitable continuity, is a ruling on the conical element.

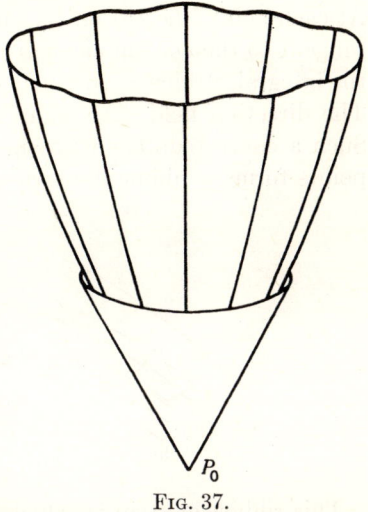

Fig. 37.

EXERCISES

In the complete solution (iv) of the text set up the following relations, and verify for each envelope that Eq. (iii) is satisfied.

1. $b = a$.   **2.** $b = 2a$.   **3.** $b = a^2$.

4. Find the integral surface of (iii) enveloped by all surfaces of (iv) through the point (0,0,1).

5. Show that the equation of Exercise 6, page 228, is the equation of all cones with the vertex at the origin.

6. Show how any surface of Exercise 9, page 228, may be got as the envelope of the family of Exercise 8; also of that of Exercise 7.

7. Find the equation of all cylinders whose generators have the direction ratios l, m, n by considering a suitable two-parameter family of planes.

8. Find the equation whose conical elements are right circular cones with vertical angles of 90° and axes parallel to the x axis.

From the geometric picture state certain families of planes, cones, and cylinders which are integral surfaces.

10-7. Characteristic Strips. Let us look more closely into the matters suggested by the last two paragraphs. Given the partial differential equation

$$f(x,y,z,p,q) = 0 \tag{10-4}$$

and some integral surface

$$z = g(x,y) \qquad (10\text{-}12)$$

At each point $P(x,y,z)$ on the surface, the tangent plane to the surface is tangent to the conical element along a ruling. These rulings tangent to the integral surface constitute a direction field on the surface (see Fig. 38). The direction field determines a family of curves lying on the surface. Such a curve together with small pieces of the tangent planes at all its points forms a ribbonlike structure lying on the integral surface.

FIG. 38.

This ribbon or strip is called a *characteristic strip*. It may be defined independently of the integral surface. We define a *strip* as a curve endowed with a planar element at each point, provided the planar element passes through the tangent to the curve and the orientation of the planar element varies continuously as we proceed along the curve. A strip is a characteristic strip of (10-4) if at each point the curve is tangent to a ruling of the conical element, and the planar element is tangent to the conical element along this ruling.[1]

10-8. Differential Equations of the Characteristic Strip. A point $(x + dx,\ y + dy,\ z + dz)$ in a tangent plane to the conical element at $P(x,y,z)$ satisfies the equation

$$dz = p\,dx + q\,dy \qquad (10\text{-}23)$$

where p and q satisfy

$$f(x,y,z,p,q) = 0 \qquad (10\text{-}4)$$

Here x, y, z are held fixed, and dx, dy, dz are the rectangular coordinates of a point in the plane.

The conical element is the envelope (see Sec. 10-5) of the family of planes. Considering p, say, as the parameter, we get the ruling of the conical element from (10-23) and the derivative of (10-23) with respect to p:

$$0 = dx + \frac{dq}{dp}\,dy \qquad (10\text{-}24)$$

---

[1] This definition is inadequate if the conical element degenerates into a line.

PARTIAL OF THE FIRST ORDER 237

where

$$\frac{\partial f}{\partial p} + \frac{\partial f}{\partial q}\frac{dq}{dp} = 0 \tag{10-25}$$

From the latter equations,

$$\frac{\partial f}{\partial q} dx - \frac{\partial f}{\partial p} dy = 0 \tag{10-26}$$

Putting (10-26) in the form

$$\frac{dx}{\partial f/\partial p} = \frac{dy}{\partial f/\partial q} = dt$$

where t is a parameter, we have for the coordinates of a point on the strip:

$$\begin{aligned} \frac{dx}{dt} &= \frac{\partial f}{\partial p} \\ \frac{dy}{dt} &= \frac{\partial f}{\partial q} \\ \frac{dz}{dt} &= p\frac{\partial f}{\partial p} + q\frac{\partial f}{\partial q} \end{aligned} \tag{10-27}$$

For a characteristic strip on the surface (10-12) p and q are known functions of x and y, hence (10-27) is a system of equations in x, y, z, t for the determination of x, y, z in terms of the parameter t.

Now, on the integral surface, (10-4) holds, and we have, on differentiating partially with respect to x and y,

$$\begin{aligned} \frac{\partial f}{\partial x} + \frac{\partial f}{\partial z}p + \frac{\partial f}{\partial p}\frac{\partial p}{\partial x} + \frac{\partial f}{\partial q}\frac{\partial q}{\partial x} &= 0 \\ \frac{\partial f}{\partial y} + \frac{\partial f}{\partial z}q + \frac{\partial f}{\partial p}\frac{\partial p}{\partial y} + \frac{\partial f}{\partial q}\frac{\partial q}{\partial y} &= 0 \end{aligned} \tag{10-28}$$

Along the strip p is a function of x and y and so of t, and we have

$$\frac{dp}{dt} = \frac{\partial p}{\partial x}\frac{dx}{dt} + \frac{\partial p}{\partial y}\frac{dy}{dt}$$

Recalling that $\partial p/\partial y = \partial q/\partial x$ and using the first two equations of (10-27), this equation may be written

$$\frac{dp}{dt} = \frac{\partial p}{\partial x}\frac{\partial f}{\partial p} + \frac{\partial q}{\partial x}\frac{\partial f}{\partial q}$$

The first equation of (10-28) then takes the form

$$\frac{dp}{dt} = -\left(\frac{\partial f}{\partial x} + \frac{\partial f}{\partial z}p\right)$$

Similarly the second equation reduces to

$$\frac{dq}{dt} = -\left(\frac{\partial f}{\partial y} + \frac{\partial f}{\partial z} q\right)$$

We have finally the following system of five ordinary equations for the characteristic strip:

$$\frac{dx}{dt} = \frac{\partial f}{\partial p}$$

$$\frac{dy}{dt} = \frac{\partial f}{\partial q}$$

$$\frac{dz}{dt} = p\frac{\partial f}{\partial p} + q\frac{\partial f}{\partial q} \qquad (10\text{-}29)$$

$$\frac{dp}{dt} = -\left(\frac{\partial f}{\partial x} + p\frac{\partial f}{\partial z}\right)$$

$$\frac{dq}{dt} = -\left(\frac{\partial f}{\partial y} + q\frac{\partial f}{\partial z}\right)$$

This set of equations constitutes a system of the type treated in Chap. 6. We shall assume that the functions that appear satisfy a Lipschitz condition. It follows that there is a unique solution of the system issuing from a prescribed set of values $(t_0, x_0, y_0, z_0, p_0, q_0)$ of the variables. If we select any planar element of the strip $(x_0, y_0, z_0, p_0, q_0)$ as initial element and assign to t the value t_0 there, Eqs. (10-29) determine the strip uniquely.

The system (10-29) is entirely independent of the integral surface (10-12) on which the characteristic strip lies. The knowledge of a single planar element tangent to an integral surface enables us to construct the characteristic strip issuing from that element and lying on the surface, without reference to the equation of the surface. This remark has the following important consequence:

Theorem. *If two integral surfaces of* (10-4) *are tangent at a point, they are tangent along the characteristic strip issuing from the common planar element at the point.*

For, from the preceding discussion, the characteristic strip lies on both integral surfaces. In other words, if two integral surfaces are tangent at an ordinary point (where a Lipschitz condition holds), they are tangent along a curve through the point.

Not all solutions of (10-29) are characteristic strips, for obviously we may take as initial planar element one not tangent to the conical element at the initial point. We have the following theorem:

Theorem. *Each solution of the system* (10-29) *is a strip along which* $f(x,y,z,p,q) \equiv k$, *a constant. A solution is a characteristic strip if and only if* $k = 0$.

PARTIAL OF THE FIRST ORDER 239

The condition that the planar element contain the tangent to the curve is that (10-23) be satisfied. Substituting for dx, dy, dz their values from (10-29) we see that this is identically satisfied:

$$\left(p \frac{\partial f}{\partial p} + q \frac{\partial f}{\partial q} \right) dt \equiv p \frac{\partial f}{\partial p} dt + q \frac{\partial f}{\partial q} dt$$

This result, together with the continuity of p and q, shows that the solution is a strip.

Along a solution,

$$\frac{df}{dt} = \frac{\partial f}{\partial x}\frac{dx}{dt} + \frac{\partial f}{\partial y}\frac{dy}{dt} + \frac{\partial f}{\partial z}\frac{dz}{dt} + \frac{\partial f}{\partial p}\frac{dp}{dt} + \frac{\partial f}{\partial q}\frac{dq}{dt}$$

On substituting for dx/dt, etc., their values from (10-29), we find

$$\frac{df}{dt} \equiv 0 \qquad f \equiv k, \text{ a constant}$$

If the initial planar element is tangent to the conical element, $f = 0$; hence $f \equiv 0$ along the strip, and each planar element is tangent to the conical element at the point.

Finally, we find similarly that (10-26) is identically satisfied, so that the curve of the strip is tangent at each point to a ruling of the conical element.

10-9. The Integral Surface through a Given Curve. The problem of finding an integral surface passing through a given curve, known as the *Cauchy problem*, can now be given a geometrical treatment.

Given a curve C with parametric equations,

$$x = x(\tau) \qquad y = y(\tau) \qquad z = z(\tau) \tag{10-30}$$

We assume that C is not a curve bearing a characteristic strip.

Through C let us pass a strip tangent at each point to the conical element of the point (Fig. 39). For a strip, p and q must satisfy the equation

$$\frac{dz}{d\tau} = p(\tau) \frac{dx}{d\tau} + q(\tau) \frac{dy}{d\tau}$$

Tangency to the conical element requires further

$$f[x(\tau), y(\tau), z(\tau), p(\tau), q(\tau)] = 0$$

Let these two equations be solved for $p(\tau)$ and $q(\tau)$, if possible, and we have the desired strip.

We next construct the characteristic strip issuing from each planar element of the strip borne by C. These strips form an integral surface, as illustrated in Fig. 39, which passes through C.

240   DIFFERENTIAL EQUATIONS

The preceding statement could be given a rigorous proof. It needs to be shown that the surface swept out by the curves bearing the characteristic strips is tangent at each point to the strip—so that the partial differential equation is satisfied. However, we shall take this fact for granted here.

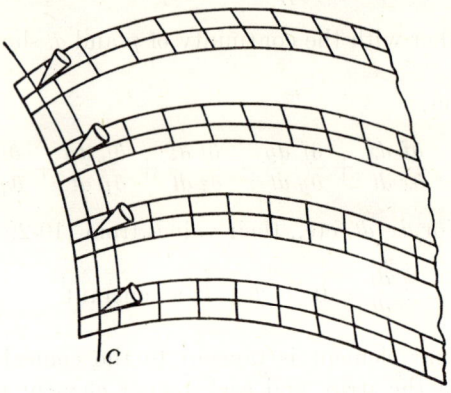

Fig. 39.

There may be several integral surfaces through C, one for each strip of the required character that we can construct on C. This number depends on the order of the conical element. Again, there may be no such surface, as Fig. 40 makes graphically evident.

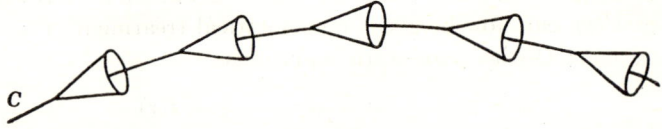

Fig. 40.

If C is a curve bearing a characteristic strip there is an infinite number of integral surfaces through C, all tangent along C. We may get such integral surfaces through suitable curves intersecting C.

Various other problems may be solved by means of the characteristic strips. For instance, let us find an integral surface which envelops a given surface,

$$z = h(x,y) \tag{10-31}$$

that is, which is tangent to the given surface along some curve.

The given surface touches the conical element of a point provided

$$f\left(x, y, z, \frac{\partial h}{\partial x}, \frac{\partial h}{\partial y}\right) = 0 \tag{10-32}$$

Equations (10-31) and (10-32) are surfaces which we shall assume to intersect in a curve C. (If there is no intersection our problem has no solution.) This curve, together with the tangent planes to (10-31) at each of its points, forms a strip touching the conical elements. If we construct the characteristic strip issuing from each planar element of the strip borne by C, there results an integral surface tangent to the surface (10-31) along C.

We shall now show how the Cauchy problem can be solved if a complete integral,

$$F(x,y,z,a,b) = 0 \qquad (10\text{-}2)$$

is known. At each point of the given curve (10-30), we shall find a surface of the family (10-2) tangent to the curve. This will give us, in general, a one-parameter family included in (10-2). It is then clear, geometrically, that the envelope of this family will be an integral surface passing through the given curve.

Let P be a point on the given curve defined by an assigned value of τ. The surface (10-2) will pass through P if

$$F[x(\tau),y(\tau),z(\tau),a,b] = 0 \qquad (10\text{-}33)$$

it will be tangent to the curve at P if

$$F_x[x(\tau),y(\tau),z(\tau),a,b]x'(\tau) + F_y[x(\tau),y(\tau),z(\tau),a,b]y'(\tau)$$
$$+ F_z[x(\tau),y(\tau),z(\tau),a,b]z'(\tau) = 0 \qquad (10\text{-}34)$$

These two equations define a and b so that the surface (10-2) shall have the desired behavior at P.

The elimination of τ between (10-33) and (10-34) gives a functional relation that subsists between a and b, say

$$b = \varphi(a)$$

as P moves along the curve. We then find the desired integral surface as the envelope of this one-parameter family of integral surfaces by the method previously explained.

If it happens that a and b are constants the problem is thereby solved, for the given curve lies on a surface of the family (10-2).

Example. Find an integral surface of

$$z = px + qy + pq$$

passing through the parabola

$$x = \tau \qquad y = \tau \qquad z = \tau^2$$

We know a complete integral,

$$z = ax + by + ab$$

Equations (10-33) and (10-34) become

$$\tau^2 = a\tau + b\tau + ab$$
$$a + b - 2\tau = 0$$

Eliminating τ, we have the functional relation

$$b = a(\pm 2\sqrt{2} - 3)$$

The insertion of this into the complete integral and differentiation with respect to a give

$$z = a[x + (\pm 2\sqrt{2} - 3)y] + a^2(\pm 2\sqrt{2} - 3)$$
$$0 = x + (\pm 2\sqrt{2} - 3)y + 2a(\pm 2\sqrt{2} - 3)$$

Eliminating a, we have

$$z = \frac{[x - (\pm 2\sqrt{2} + 3)y]^2}{4(3 \pm 2\sqrt{2})}$$

There are thus two surfaces resulting, according as the $+$ or $-$ sign is used.

The student may verify that each relation satisfies the differential equation and that the coordinates of a point on the given curve, for any τ, satisfy the equation of the surface.

The preceding treatment establishes, in a geometrical fashion, the adequacy of the complete integral as an instrument in setting up arbitrary solutions of the partial differential equation. Given an integral surface and a curve C, not bearing a characteristic strip, upon it. Using the complete integral to determine the integral surface through C, taking for granted the possibility of carrying out the process, we get the given integral surface. It thus appears that the complete integral suffices for setting up arbitrary solutions.

10-10. Complete Solutions. Charpit's Method. It is of the greatest importance, where possible, to find a complete solution. The various other solutions may then be got by setting up functional relations between the parameters and by elimination. In this section we shall discuss methods of arriving at a complete solution.

It is now clear, geometrically, that complete solutions exist in an infinite variety. We can get a two-parameter family of integral surfaces by passing integral surfaces through a suitably chosen two-parameter family of curves, for example, through all the lines in some plane.

Let Eqs. (10-29) be written in the form

$$\frac{dx}{f_p} = \frac{dy}{f_q} = \frac{dz}{pf_p + qf_q} = \frac{dp}{-f_x - pf_z} = \frac{dq}{-f_y - qf_z} \quad (10\text{-}35)$$

These will be called the *subsidiary equations* of (10-4).

Suppose we can form from (10-35) some equation which can be integrated and let

$$\varphi(x,y,z,p,q) = a \quad (10\text{-}36)$$

be the integral. Suppose the integral is such that we can solve (10-36) and the given equation

$$f(x,y,z,p,q) = 0 \quad (10\text{-}4)$$

PARTIAL OF THE FIRST ORDER 243

for p and q:
$$p = p(x,y,z,a) \qquad q = q(x,y,z,a)$$

Consider now the total differential equation
$$dz = p(x,y,z,a)\, dx + q(x,y,z,a)\, dy \qquad (10\text{-}37)$$

We shall show that this equation is integrable. The condition for this is that
$$\frac{\partial p}{\partial y} = \frac{\partial q}{\partial x} \qquad (10\text{-}38)$$

where
$$\frac{\partial p}{\partial y} = p_y + p_z q \qquad \frac{\partial q}{\partial x} = q_x + q_z p$$

We may compute these partial derivatives by differentiating (10-4) and (10-36) partially:

$$f_x + f_z p + f_p \frac{\partial p}{\partial x} + f_q \frac{\partial q}{\partial x} = 0$$

$$f_y + f_z q + f_p \frac{\partial p}{\partial y} + f_q \frac{\partial q}{\partial y} = 0$$

$$\varphi_x + \varphi_z p + \varphi_p \frac{\partial p}{\partial x} + \varphi_q \frac{\partial q}{\partial x} = 0$$

$$\varphi_y + \varphi_z q + \varphi_p \frac{\partial p}{\partial y} + \varphi_q \frac{\partial q}{\partial y} = 0$$

Multiplying by φ_p, φ_q, $-f_p$, $-f_q$, respectively, and adding, we have

$$(f_p \varphi_q - f_q \varphi_p)\left(\frac{\partial p}{\partial y} - \frac{\partial q}{\partial x}\right) = -(f_x + f_z p)\varphi_p$$
$$- (f_y + f_z q)\varphi_q + \varphi_x f_p + \varphi_y f_q + \varphi_z(pf_p + qf_q) \qquad (10\text{-}39)$$

Now we have from (10-36)
$$\varphi_x\, dx + \varphi_y\, dy + \varphi_z\, dz + \varphi_p\, dp + \varphi_q\, dq = 0$$

and thence from (10-35), of which (10-36) is an integral,
$$\varphi_x f_p + \varphi_y f_q + \varphi_z(pf_p + qf_q) + \varphi_p(-f_x - pf_z) + \varphi_q(-f_y - qf_z) = 0$$

The second member of (10-39) is thus zero. We have already tacitly assumed that $f_p \varphi_q - f_q \varphi_p \neq 0$, when we supposed that (10-4) and (10-36) could be solved for p and q. Equation (10-39) then reduces to (10-38), and (10-37) is integrable.

It is clear that the general solution of (10-37) is a solution of the partial differential equation. For the solution, $dz = z_x\, dx + z_y\, dy$, where $z_x = p(x,y,z,a)$, $z_y = q(x,y,z,a)$, and these values satisfy (10-4).

Furthermore, the integration of (10-37) introduces a second arbitrary constant, hence we have a complete solution.

The method just described is known as *Charpit's method*. We may summarize it briefly:

Find an integral of the system (10-35) from which, together with the given differential equation, p and q may be found. Substitute these values of p and q into

$$dz = p\,dx + q\,dy$$

and integrate.

Example. Solve:
$$p^2 + qy - z = 0$$

The subsidiary equations are

$$\frac{dx}{2p} = \frac{dy}{y} = \frac{dz}{2p^2 + qy} = \frac{dp}{p} = \frac{dq}{0}$$

One integral is
$$q = a$$

From the given equation
$$p = \sqrt{z - ay}$$

and we have to integrate
$$dz = \sqrt{z - ay}\,dx + a\,dy$$

Writing this
$$\frac{dz - a\,dy}{\sqrt{z - ay}} = dx$$

we have
$$2\sqrt{z - ay} = x + b$$

or
$$z = ay + \tfrac{1}{4}(x + b)^2$$

EXERCISES

Find a complete solution for the preceding example using:
  1. The first and fourth members of the subsidiary system.
  2. The second and fourth members.
  3. Show that an equation in the partial derivatives alone,
$$f(p,q) = 0$$
has the complete solution
$$z = ax + by + c$$
where $f(a,b) = 0$.
  4. Show that the characteristic strips in the preceding are lathlike in character. What sort of integral surface is formed by all the characteristic strips issuing from a point?
  5. Show that the equation $f(xp,yq) = 0$ is transformed into one in the derivatives alone by the change of variable $x = e^u$, $y = e^v$.
  6. Show how to find a complete solution if x and y are missing,
$$f(z,p,q) = 0$$
  7. Find a complete solution of
$$p^2 z + q^2 = 4$$

PARTIAL OF THE FIRST ORDER

8. Show how to find a complete solution of an equation of the form
$$\Phi(x,p) = \Psi(q,y)$$
9. Find a complete solution of
$$p^2 + q^2 = x - y$$
10. Find a complete solution of the analogue of Clairaut's equation
$$z = px + qy + f(p,q)$$
Show that the characteristic strip lies in a plane.

11. Find a complete solution and the singular solution of
$$z = px + qy + \sqrt{1 + p^2 + q^2}$$
Find an integral surface of
$$z = px + qy + pq$$
tangent to the surface:

12. $z = \tfrac{1}{2}(x^2 + y^2)$.   **13.** $z = xy + 2$.

Find complete solutions:
14. $p(q^2 + 4) + q(k - 2z) = 0$.   **15.** $(4p^2 + q^2)y = qz$.
16. $p^2 + q^2 - px - qy + \tfrac{1}{2}xy = 0$.

17. Find the most general function $f(x,y)$ which is equal to the sum of the squares of its first partial derivatives and such that $f(x,0) = x^2/8$.

18. What is the most general function $f(x,y)$ which is equal to the product of its first partial derivatives?

10-11. Linear Partial Differential Equations. A partial differential equation in three variables is called *linear* if it has the form

$$P(x,y,z)p + Q(x,y,z)q = R(x,y,z) \tag{10-40}$$

The first three of the five equations (10-29) for the characteristic strips are

$$\frac{dx}{dt} = P$$
$$\frac{dy}{dt} = Q \tag{10-41}$$
$$\frac{dz}{dt} = pP + qQ = R$$

These three equations are free of p and q and can be solved for the curves bearing the characteristic strips. We shall assume that P, Q, R are single valued (if multiple valued we shall concern ourselves with a single-valued branch) in some region in space and satisfy a Lipschitz condition. There is then one curve through each point of the region. We shall call these curves *characteristic curves*.

We have here the case, mentioned earlier, of a conical element which degenerates into a line. Any strip lying on a characteristic curve has coordinates (x,y,z,p,q) which satisfy the differential equation, for the tangent to the curve has, from (10-41), the direction ratios, P, Q, R. Since a planar element of the strip contains the tangent, we have

$$Pp + Qq + R(-1) = 0$$

that is, (10-40) is satisfied.

Among all such strips on a characteristic curve there is a one-parameter family of characteristic strips. These are defined by the last two equations of (10-29). It happens that we do not need these strips, so these equations have not been written down.

THEOREM. *The surface generated by a one-parameter family of characteristic curves is an integral surface.*

Through each point of the surface passes a characteristic curve lying on the surface. The tangent plane at the point contains the tangent to the curve; hence, from the remarks of the preceding paragraph, p and q satisfy (10-40).

THEOREM. *Through a given curve, not a characteristic curve, passes a unique integral surface.*

Through each point of the given curve C construct a characteristic curve. These curves generate an integral surface on which C lies. Furthermore, any integral surface through C must bear characteristic strips, and hence characteristic curves, issuing from C and so must coincide with the surface just constructed.

It should be borne in mind that the uniqueness of the integral surface through C is a consequence of our assumption that P, Q, R are single-valued functions. If this assumption be dropped there may be two or more characteristic curves through a point and more than one integral surface through a curve.

COROLLARY. *Any solution of* (10-40) *may be generated by a one-parameter family of characteristic curves.*

Given any integral surface and a curve C, not a characteristic curve, on the surface. The one-parameter family of characteristic curves through C generates the given surface.

THEOREM. *If two integral surfaces of* (10-40) *intersect at a point P, they intersect along the characteristic curve through P; if they are tangent at P, they are tangent along the characteristic curve through P.*

For, if P is on both surfaces the characteristic curve through P is on both surfaces. If the surfaces have a common planar element at P the characteristic strip issuing from this element lies on both surfaces.

We shall have achieved the general solution of (10-40), according to the preceding corollary, when we have the general solution of the system (10-41). Equations of this kind have been discussed at some length in Chap. 6.

To solve the linear partial differential equation (10-40), *we first solve*

PARTIAL OF THE FIRST ORDER   247

$$\frac{dx}{P} = \frac{dy}{Q} = \frac{dz}{R} \tag{10-42}$$

getting two independent integrals

$$u(x,y,z) = a \qquad v(x,y,z) = b \tag{10-43}$$

Then the general solution is

$$\varphi(u,v) = 0 \tag{10-44}$$

where φ is an arbitrary function.

The curves of intersection of these two one-parameter families of surfaces constitute the two-parameter family of characteristic curves.

We select a one-parameter family of characteristic curves by setting up a functional relation between the constants

$$\varphi(a,b) = 0$$

and the elimination of the constants gives the resulting integral surface. This gives as the general solution of (10-40) the formula (10-44).

The solution (10-44) may be written in the alternative forms, $u = g(v)$ or $v = h(u)$, where g and h are arbitrary functions.

Example. Solve:

$$z = qy - px + 2x$$

The equations of the characteristic curves are

$$\frac{dx}{-x} = \frac{dy}{y} = \frac{dz}{z - 2x}$$

From these we have

$$x\,dy + y\,dx = 0 \qquad (z - 2x)\,dx + x\,dz = 0$$
$$xy = a \qquad\qquad xz - x^2 = b$$

The general solution is then

$$xz - x^2 = f(xy)$$

or

$$z = x + \frac{1}{x} f(xy)$$

where f is an arbitrary function.

10-12. Integrating Factors. In Sec. 2-8 we found that the integrating factors μ of the differential equation

$$M(x,y)\,dx + N(x,y)\,dy = 0$$

are solutions of the partial differential equation

$$N \frac{\partial \mu}{\partial x} - M \frac{\partial \mu}{\partial y} = \left(\frac{\partial M}{\partial y} - \frac{\partial N}{\partial x} \right) \mu$$

This is linear. Any solution, other than $\mu \equiv 0$, supplies us with an integrating factor and enables us to solve the original equation by quadratures. It is not necessary to seek the general solution. The equations of the characteristic curves are

$$\frac{dx}{N} = \frac{dy}{-M} = \frac{d\mu}{\left(\frac{\partial M}{\partial y} - \frac{\partial N}{\partial x}\right)\mu}$$

As an example, consider the linear equation

$$\frac{dy}{dx} + P(x)y = Q(x)$$

or

$$(Py - Q)\,dx + dy = 0$$

We have the equations

$$\frac{dx}{1} = \frac{dy}{Q - Py} = \frac{d\mu}{P\mu}$$

From the first and third members we have the integral

$$\log \mu = \int P\,dx + a$$

Taking $a = 0$, we have the integrating factor

$$\mu = e^{\int P dx}$$

10-13. Equations in Three or More Independent Variables. We state here, for reference, the formulas for a partial differential equation of the first order in one dependent variable z and n independent variables $x_1, \ldots, x_n$. The analysis follows the general lines of the present chapter but is more complicated. The geometric pictures involve $n + 1$ dimensions and are less familiar.

Putting

$$\frac{\partial z}{\partial x_1} = p_1 \qquad \frac{\partial z}{\partial x_2} = p_2 \qquad \ldots \qquad \frac{\partial z}{\partial x_n} = p_n$$

the general equation of the first order has the form

$$f(x_1, \ldots, x_n, z, p_1, \ldots, p_n) = 0$$

The characteristic strips satisfy the $2n + 1$ differential equations

$$\frac{dx_i}{dt} = f_{p_i}$$

$$\frac{dz}{dt} = p_1 f_{p_1} + p_2 f_{p_2} + \cdots + p_n f_{p_n}$$

$$\frac{dp_i}{dt} = -f_{x_i} - p_i f_z$$

Any solution of this system

$$x_i = x_i(t) \qquad z = z(t) \qquad p_i = p_i(t)$$

issuing from a set of values (x_{i0}, z_0, p_{i0}) which satisfy the given equation is a characteristic strip. By the use of characteristic strips integral surfaces may be built up.

The linear equation

$$P_1 p_1 + P_2 p_2 + \cdots + P_n p_n = R$$

where $P_1, \ldots, P_n, R$ are functions of $x_1, \ldots, x_n, z$, may be solved from the system defining the characteristic curves:

$$\frac{dx_1}{P_1} = \frac{dx_2}{P_2} = \cdots = \frac{dx_n}{P_n} = \frac{dz}{R}$$

If

$$u_1 = a_1 \quad \ldots \quad u_n = a_n$$

are n independent integrals of this system, the general solution of the partial differential equation is

$$\varphi(u_1, \ldots, u_n) = 0$$

where φ is an arbitrary function.

Example. Solve:

$$y \frac{\partial u}{\partial x} - x \frac{\partial u}{\partial y} + yz \frac{\partial u}{\partial z} = x$$

From the system

$$\frac{dx}{y} = \frac{dy}{-x} = \frac{dz}{yz} = \frac{du}{x}$$

we have the combinations

$$x\,dx + y\,dy = 0$$
$$\frac{dz}{z} - dx = 0$$
$$du + dy = 0$$

which, on integration, give

$$x^2 + y^2 = a_1$$
$$ze^{-x} = a_2$$
$$u + y = a_3$$

The general solution then is

$$u = -y + \varphi(x^2 + y^2, ze^{-x})$$

where φ is an arbitrary function.

EXERCISES

Solve:
1. $ap + bq = c$.
2. $x^2 p + xyq = y^2$.
3. $(ny - mz)p + (lz - nx)q = mx - ly$.
4. $yp - xq = 1 + z^2$.

5. What function $f(x,y)$ is the sum of its two first partial derivatives?

6. Find all integrating factors of

$$2y\,dx + x\,dy = 0$$

By specializing the arbitrary function in the solution, get the integrating factors given on page 52.

7. Find all surfaces intersecting orthogonally the family of spheres with centers at the origin.

8. Find all surfaces intersecting orthogonally all right circular cylinders having the z axis as common axis.

9. Show how the characteristic curves (10-43) may be used to put an integral surface through a given curve.

10. Find an integral surface of

$$xp + yq = z$$

through the curve $y = x^2$, $z = 1$.

11. Find an integral surface of

$$yzp + xzq = xy$$

through the circle

$$x^2 + y^2 = 1 \qquad z = 0$$

12. What function $f(x,y,z)$ is the sum of its three first partial derivatives?

13. Solve:

$$(y + z + u)\frac{\partial u}{\partial x} + (z + u + x)\frac{\partial u}{\partial y} + (u + x + y)\frac{\partial u}{\partial z} = x + y + z$$

Find the most general function with the property:

14. $x\dfrac{\partial f}{\partial x} + y\dfrac{\partial f}{\partial y} + z\dfrac{\partial f}{\partial z} = nf.$ **15.** $x^2\dfrac{\partial f}{\partial x} + y^2\dfrac{\partial f}{\partial y} + z^2\dfrac{\partial f}{\partial z} = 0.$

CHAPTER 11

PARTIAL DIFFERENTIAL EQUATIONS OF THE SECOND ORDER

11-1. Certain Elementary Cases. It is not possible to give within the limits of an elementary book anything approaching a comprehensive treatment of the partial differential equation of the second order. We shall accordingly limit our study to equations of special kinds.

From the point of view of physical applications the partial differential equations of the second order are of the utmost significance. Such equations arise in the vibration of strings, membranes, and other bodies, in the problems of gravitational and electrical potential, in the flow of heat, in the behavior of moving liquids, in the movements of the air on which the design of our airplanes or the prediction of our weather depends, and in a host of other connections. Many of these problems are of the greatest difficulty. We shall, however, introduce a few simple applications.

We shall deal in the beginning with three variables. We shall employ the notation

$$\frac{\partial z}{\partial x} = p \quad \frac{\partial z}{\partial y} = q \quad \frac{\partial^2 z}{\partial x^2} = r \quad \frac{\partial^2 z}{\partial x\, \partial y} = s \quad \frac{\partial^2 z}{\partial y^2} = t \quad (11\text{-}1)$$

The most general equation of the second order is then of the form

$$f(x,y,z,p,q,r,s,t) = 0 \quad (11\text{-}2)$$

We begin with very simple cases. Consider

$$s = \frac{\partial^2 z}{\partial x\, \partial y} = 0 \quad (11\text{-}3)$$

Integrating with respect to x and introducing an arbitrary function of y for the constant of integration, we have

$$q = \frac{\partial z}{\partial y} = C(y)$$

Integrating with respect to y and replacing the integral of $C(y)$ by another arbitrary function, we get

$$z = A(y) + B(x) \quad (11\text{-}4)$$

252   DIFFERENTIAL EQUATIONS

The general solution thus contains two arbitrary functions.

More generally, the equation

$$s = f(x,y) \tag{11-5}$$

may be integrated similarly. The result of the two integrations is

$$z = \int[\int f(x,y)\,\partial x]\,\partial y + A(y) + B(x) \tag{11-6}$$

If the equation contains derivatives with respect to one variable only, we may integrate it as an ordinary equation of the second order, treating the other independent variable as a constant and introducing arbitrary functions of this variable in place of the constants of integration.

Thus

$$\frac{\partial^2 z}{\partial x^2} = y$$

gives on integration

$$z = \tfrac{1}{2}yx^2 + Ax + B$$

and the general solution is

$$z = \tfrac{1}{2}yx^2 + A(y)x + B(y)$$

The equation

$$\frac{\partial^2 z}{\partial x^2} - 2y\frac{\partial z}{\partial x} + 2y^2 z = 4y^3$$

with y held constant, is linear with constant coefficients. Its solution is readily found to be

$$z = e^{yx}[A(y)\cos yx + B(y)\sin yx] + 2y$$

EXERCISES

1. Find the general solution of

$$\frac{\partial^{m+n} z}{\partial x^m\,\partial y^n} = 0$$

Make a first integration by inspection and solve:

2. $\dfrac{\partial^2 z}{\partial x^2} - \dfrac{\partial^2 z}{\partial x\,\partial y} + \dfrac{\partial z}{\partial x} = 0.$ **3.** $\dfrac{\partial^2 z}{\partial x\,\partial y} + \dfrac{\partial z}{\partial y} = x.$ **4.** $y\dfrac{\partial^2 z}{\partial y^2} - \dfrac{\partial z}{\partial y} = xy^2.$

Eliminate the arbitrary functions and derive the partial differential equation of the second order:

5. $z = f(2x + 3y) + \varphi(x - y).$ **6.** $z = e^x f(y) + e^y \varphi(x).$

7. $z = e^x f(x + y) + e^{-x}\varphi(x + 2y).$ **8.** $\dfrac{\partial z}{\partial x} + \dfrac{\partial z}{\partial y} + z = f(x + y).$

9. Show that

$$\frac{\partial^2 z}{\partial x\,\partial y} + f\frac{\partial z}{\partial x} + g\frac{\partial z}{\partial y} + fgz = 0$$

where f and g are constants, can be reduced to the form (11-3) by putting $z = ue^{ax+by}$, and find the general solution.

PARTIAL OF THE SECOND ORDER 253

10. Show that if u and v satisfy the Cauchy-Riemann equations

$$\frac{\partial u}{\partial x} = \frac{\partial v}{\partial y} \qquad \frac{\partial u}{\partial y} = -\frac{\partial v}{\partial x}$$

then each is a solution of Laplace's equation

$$\frac{\partial^2 z}{\partial x^2} + \frac{\partial^2 z}{\partial y^2} = 0$$

Verify that the following functions are solutions of Laplace's equation

$$\frac{\partial^2 V}{\partial x^2} + \frac{\partial^2 V}{\partial y^2} + \frac{\partial^2 V}{\partial z^2} = 0$$

11. $A \log [(x-a)^2 + (y-b)^2]$.

12. $\dfrac{A}{r}$, where $r^2 = (x-a)^2 + (y-b)^2 + (z-c)^2$.

13. $A \log \dfrac{r+z-c}{r-z+c}$.

14. $\dfrac{A}{r} \log \dfrac{r+z-c}{r-z+c}$.

15. Show that if an equation of the second order has an intermediate integral containing an arbirary function φ,

$$v = \varphi(u)$$

where u and v are functions of x, y, z, p, q, the equation has the form

$$Rr + Ss + Tt + U(rt - s^2) = V$$

where $R, \ldots, V$ are functions of x, y, z, p, q.

11-2. Euler's Equation. The equation

$$a \frac{\partial^2 z}{\partial x^2} + b \frac{\partial^2 z}{\partial x\, \partial y} + c \frac{\partial^2 z}{\partial y^2} = 0 \qquad (11\text{-}7)$$

where a, b, c are constants, may be reduced to one of the simple cases of the preceding section by a change of variable. Suppose $a \neq 0$, and put

$$u = y + m_1 x$$
$$v = y + m_2 x \qquad (11\text{-}8)$$

Then

$$\frac{\partial z}{\partial x} = \frac{\partial z}{\partial u}\frac{\partial u}{\partial x} + \frac{\partial z}{\partial v}\frac{\partial v}{\partial x} = m_1 \frac{\partial z}{\partial u} + m_2 \frac{\partial z}{\partial v}$$

$$\frac{\partial^2 z}{\partial x^2} = m_1 \left(\frac{\partial^2 z}{\partial u^2}\frac{\partial u}{\partial x} + \frac{\partial^2 z}{\partial u\, \partial v}\frac{\partial v}{\partial x} \right) + m_2 \left(\frac{\partial^2 z}{\partial u\, \partial v}\frac{\partial u}{\partial x} + \frac{\partial^2 z}{\partial v^2}\frac{\partial v}{\partial x} \right)$$

$$= m_1{}^2 \frac{\partial^2 z}{\partial u^2} + 2 m_1 m_2 \frac{\partial^2 z}{\partial u\, \partial v} + m_2{}^2 \frac{\partial^2 z}{\partial v^2}$$

Similarly,

$$\frac{\partial^2 z}{\partial x\, \partial y} = m_1 \frac{\partial^2 z}{\partial u^2} + (m_1 + m_2) \frac{\partial^2 z}{\partial u\, \partial v} + m_2 \frac{\partial^2 z}{\partial v^2}$$

$$\frac{\partial^2 z}{\partial y^2} = \frac{\partial^2 z}{\partial u^2} + 2 \frac{\partial^2 z}{\partial u\, \partial v} + \frac{\partial^2 z}{\partial v^2}$$

Substituting into (11-7) we have

$$(am_1^2 + bm_1 + c)\frac{\partial^2 z}{\partial u^2} + (2am_1m_2 + b(m_1 + m_2) + 2c)\frac{\partial^2 z}{\partial u\, \partial v}$$
$$+ (am_2^2 + bm_2 + c)\frac{\partial^2 z}{\partial v^2} = 0 \quad (11\text{-}9)$$

If, now, we take for m_1 and m_2 the roots of the equation

$$am^2 + bm + c = 0 \quad (11\text{-}10)$$

which we shall suppose unequal, the first and third terms of the differential equation disappear, and we have

$$\frac{4ac - b^2}{a}\frac{\partial^2 z}{\partial u\, \partial v} = 0$$

We thus have, since $4ac - b^2 \neq 0$,

$$\frac{\partial^2 z}{\partial u\, \partial v} = 0$$

This is an equation of the type of (11-3). Its general solution is

$$z = A(u) + B(v)$$

The general solution of (11-7) is then

$$z = A(y + m_1 x) + B(y + m_2 x) \quad (11\text{-}11)$$

where A and B are arbitrary functions.

If (11-10) has equal roots, the use of the roots in (11-8) fails, since u and v are no longer independent variables. We may let m_1 be the root, $m_1 = -b/2a$, and choose $m_2(\neq m_1)$ arbitrarily. A calculation shows that the second coefficient in (11-9) vanishes, and the equation reduces to

$$\frac{\partial^2 z}{\partial v^2} = 0$$

Integrating, we have

$$z = A(u) + vB(u)$$

whence the general solution of (11-7) is

$$z = A(y + m_1 x) + (y + m_2 x)B(y + m_1 x) \quad (11\text{-}12)$$

We may thus solve (11-7) by solving the quadratic (11-10) and thence writing down the solution in one of the forms (11-11) or (11-12).

If $a = 0$, Eq. (11-10) is no longer a quadratic. In this case we may interchange the roles of x and y and proceed as before; or the solution may be effected without change of variable. An integration with respect

to y gives
$$b\frac{\partial z}{\partial x} + c\frac{\partial z}{\partial y} = \varphi(x)$$
an equation of the first order which is easily solved.

If Eq. (11-10) has complex roots, a, b, c being real, the roots have the form $\alpha \pm i\beta$; and the solution (11-11) is
$$z = A(y + \alpha x + i\beta x) + B(y + \alpha x - i\beta x)$$
For assumed functions this gives z ordinarily as a complex function of x and y. Both the real and imaginary parts must then be solutions.

We may write the result in an alternative form. Putting
$$A = \phi + i\psi \qquad B = \phi - i\psi$$
the solution takes the form
$$z = \phi(y + \alpha x + i\beta x) + \phi(y + \alpha x - i\beta x) \\ + i[\psi(y + \alpha x + i\beta x) - \psi(y + \alpha x - i\beta x)]$$
If ϕ and ψ are real when their arguments are real, the imaginary parts cancel from the combination written down leaving a real equation.

Examples. 1. Solve:
$$\frac{\partial^2 z}{\partial x^2} - 3\frac{\partial^2 z}{\partial x\, \partial y} + 2\frac{\partial^2 z}{\partial y^2} = 0$$
Here Eq. (11-10) is
$$m^2 - 3m + 2 = 0$$
The roots are 1, 2, and the general solution is
$$z = A(y + x) + B(y + 2x)$$
2. Solve:
$$\frac{\partial^2 z}{\partial x^2} - 2\frac{\partial^2 z}{\partial x\, \partial y} + 5\frac{\partial^2 z}{\partial y^2} = 0$$
Here
$$m^2 - 2m + 5 = 0 \qquad m = 1 \pm 2i$$
The solution may be written in either of the forms
$$z = A(y + x + 2ix) + B(y + x - 2ix)$$
or
$$z = \phi(y + x + 2ix) + \phi(y + x - 2ix) + i[\psi(y + x + 2ix) - \psi(y + x - 2ix)]$$
As a special case of the first form let us choose $A(t) = t^2$, $B(t) = t^3$; then
$$z = (y + x + 2ix)^2 + (y + x - 2ix)^3$$
From the real and imaginary parts of this we have two real solutions
$$z = (y + x)^2 - 4x^2 + (y + x)^3 - 12x^2(y + x) \\ z = 4x(y + x) - 6x(y + x)^2 + 8x^3$$

In the second form of solution put $\phi(t) = t^2$, $\psi(t) = 1/t$. The result is

$$z = (y + x + 2ix)^2 + (y + x - 2ix)^2$$
$$+ i\left(\frac{1}{y + x + 2ix} - \frac{1}{y + x - 2ix}\right)$$
$$= 2(y + x)^2 - 8x^2 + \frac{4x}{(y + x)^2 + 4x^2}$$

The change of variables (11-8) enables us to solve readily the equation of the form

$$a\frac{\partial^2 z}{\partial x^2} + b\frac{\partial^2 z}{\partial x\,\partial y} + c\frac{\partial^2 z}{\partial y^2} = f(x,y)$$

The second member is transformed into a function of u and v, and we have an equation of the form

$$\frac{\partial^2 z}{\partial u\,\partial v} = F(u,v)$$

or

$$\frac{\partial^2 z}{\partial v^2} = G(u,v)$$

according as the roots of (11-10) are unequal or equal. The equation is readily solved by direct integration [as Eq. (11-5) for the first case], after which u and v are replaced in terms of x and y.

Example. Solve:

$$\frac{\partial^2 z}{\partial x^2} - \frac{\partial^2 z}{\partial y^2} = xy + y^2$$

Here $m^2 - 1 = 0$, $m = \pm 1$; so

Then
$$u = y + x \qquad v = y - x$$
$$x = \tfrac{1}{2}(u - v) \qquad y = \tfrac{1}{2}(u + v)$$
$$\frac{4ac - b^2}{a} = -4$$

The transformed equation is then

$$-4\frac{\partial^2 z}{\partial u\,\partial v} = \frac{1}{4}(u^2 - v^2) + \frac{1}{4}(u + v)^2$$

From this we find at once, on integration,

$$z = -\frac{1}{16}\left(\frac{vu^3}{3} - \frac{uv^3}{3}\right) - \frac{1}{16}\frac{(u + v)^4}{12} + A(u) + B(v)$$

whence

$$z = A(y + x) + B(y - x) - \tfrac{1}{12}(xy^3 - x^3y + y^4)$$

EXERCISES

Solve:

1. $\dfrac{\partial^2 z}{\partial x^2} + \dfrac{\partial^2 z}{\partial y^2} = 0.$
2. $\dfrac{\partial^2 u}{\partial x^2} = a^2 \dfrac{\partial^2 u}{\partial t^2}.$

PARTIAL OF THE SECOND ORDER 257

3. Solve
$$a\frac{\partial^2 z}{\partial x^2} + b\frac{\partial^2 z}{\partial x\,\partial y} = 0$$
directly without change of variables.

4. Verify by substitution that the particular solutions of Example 2 found in the text satisfy the equation.

5. Show that for the case of equal roots
$$z = A(y + m_1 x) + xB(y + m_1 x)$$
is a solution.

6. Get the preceding by altering the functions in (11-12).

11-3. Homogeneous Linear Equations with Constant Coefficients.

The solution found in the previous section for Eq. (11-7) may be obtained in an entirely different manner. Let us try a solution of the form
$$z = A(y + mx)$$
Differentiating, we find
$$\frac{\partial^2 z}{\partial x^2} = m^2 A''(y + mx)$$
$$\frac{\partial^2 z}{\partial x\,\partial y} = m A''(y + mx)$$
$$\frac{\partial^2 z}{\partial y^2} = A''(y + mx)$$
Substituting in (11-7), we have
$$A''(y + mx)(am^2 + bm + c) = 0$$
This equation will be satisfied, whatever the function A, provided
$$am^2 + bm + c = 0$$
If this has distinct roots m_1, m_2, we have the two solutions
$$z = A(y + m_1 x) \qquad z = B(y + m_2 x)$$
We have then at once that
$$z = A(y + m_1 x) + B(y + m_2 x)$$
is a solution.

This method of procedure is applicable to the equation
$$a_0 \frac{\partial^n z}{\partial x^n} + a_1 \frac{\partial^n z}{\partial x^{n-1}\,\partial y} + \cdots + a_n \frac{\partial^n z}{\partial y^n} = 0 \qquad (11\text{-}13)$$
in which the a's are constants.

An equation is called *linear* if it is linear in the derivatives of highest order. It is called *completely linear* if z and all its derivatives appear linearly. Hence (11-13) is completely linear with constant coefficients.

Equation (11-13) is called *homogeneous* because of the fact that all terms contain derivatives of the same order.

Trying a solution of the form

$$z = A(y + mx)$$

we have

$$\frac{\partial^n z}{\partial x^n} = m^n A^{(n)}(y + mx)$$

$$\frac{\partial^n z}{\partial x^{n-1} \partial y} = m^{n-1} A^{(n)}(y + mx)$$

$$\cdots \cdots \cdots \cdots$$

$$\frac{\partial^n z}{\partial y^n} = A^{(n)}(y + mx)$$

Substitution into (11-13) gives

$$A^{(n)}(y + mx)(a_0 m^n + a_1 m^{n-1} + \cdots + a_n) = 0$$

This equation will be satisfied, whatever the function A, provided

$$f(m) = a_0 m^n + a_1 m^{n-1} + \cdots + a_n = 0 \qquad (11\text{-}14)$$

Suppose $a_0 \neq 0$ and let $m_1, m_2, \ldots, m_n$ be the roots of this equation. If these roots are distinct we have then a solution involving n arbitrary functions:

$$z = A_1(y + m_1 x) + A_2(y + m_2 x) + \cdots + A_n(y + m_n x) \qquad (11\text{-}15)$$

If (11-14) has repeated roots the functions in (11-15) are no longer all independent. If m_1 is a double root, say, $m_2 = m_1$, so that not only is $f(m_1) = 0$, but also $f'(m_1) = 0$, it can be shown that

$$z = x A_2(y + m_1 x)$$

is a solution. This can be put in place of the second function in (11-15). If m_1 is a triple root,

$$z = x^2 A_3(y + m_1 x)$$

is also a solution; and so on. The proof of these facts is left as exercises for the student.

If $a_0 = 0$, Eq. (11-14) has fewer than n roots. If the first r coefficients are zero, there are $n - r$ roots, and we have

$$z = A_1(y + m_1 x) + \cdots + A_{n-r}(y + m_{n-r} x)$$

Since, however, each term in Eq. (11-13) now contains the derivative of $\partial^r z / \partial y^r$, it is clear that we can add the terms

$$B_1(x) + B_2(x) y + \cdots + B_r(x) y^{r-1}$$

where the B's are arbitrary functions. This expression vanishes when differentiated r times with respect to y.

Examples. 1. Solve:
$$\frac{\partial^3 z}{\partial x^3} + \frac{\partial^3 z}{\partial x^2 \partial y} - \frac{\partial^3 z}{\partial x \partial y^2} - \frac{\partial^3 z}{\partial y^3} = 0$$

Equation (11-14) is
$$m^3 + m^2 - m - 1 = 0$$

which has the roots $1, -1, -1$. The solution then is
$$z = A_1(y + x) + A_2(y - x) + xA_3(y - x)$$

2. Solve:
$$\frac{\partial^3 z}{\partial x^2 \partial y} + a \frac{\partial^3 z}{\partial x \partial y^2} = 0 \qquad a \neq 0$$

We have
$$m^2 + am = 0 \qquad m = -a, 0$$
and
$$z = A_1(y - ax) + A_2(y) + B(x)$$

EXERCISES

1. Solve Exercises 1 and 2 of the preceding section by the present method.
2. Show that $xA(y + m_1 x)$ is a solution if m_1 is a double root.

Solve:

3. $2 \dfrac{\partial^3 z}{\partial x^3} - \dfrac{\partial^3 z}{\partial^2 x \, \partial y} - \dfrac{\partial^3 z}{\partial x \, \partial y^2} = 0.$ 4. $\dfrac{\partial^4 z}{\partial x^4} - \dfrac{\partial^4 z}{\partial y^4} = 0.$

5. Find the most general integral surface of
$$\frac{\partial^2 z}{\partial x \, \partial y} - 3 \frac{\partial^2 z}{\partial y^2} = 0$$
which passes through the parabola $y = x$, $z = x^2$.

11-4. The Completely Linear Equation. The most general completely linear equation of the second order in two independent variables is

$$A \frac{\partial^2 z}{\partial x^2} + B \frac{\partial^2 z}{\partial x \, \partial y} + C \frac{\partial^2 z}{\partial y^2} + D \frac{\partial z}{\partial x} + E \frac{\partial z}{\partial y} + Fz = G \quad (11\text{-}16)$$

where $A, B, \ldots, G$ are functions of x and y.

We shall limit the discussion to this equation, although the results are readily seen to be valid for the completely linear equation of any order and in any number of independent variables.

The equation resulting from setting the second member of (11-16) equal to zero will be called the *reduced equation:*

$$A \frac{\partial^2 z}{\partial x^2} + B \frac{\partial^2 z}{\partial x \, \partial y} + C \frac{\partial^2 z}{\partial y^2} + D \frac{\partial z}{\partial x} + E \frac{\partial z}{\partial y} + Fz = 0 \quad (11\text{-}17)$$

If we know a particular solution $\varphi(x,y)$ of (11-16), the substitution

$$z = u + \varphi(x,y)$$

leads, as is readily seen, to an equation of the form (11-17) in u. Hence, aside from the detail of getting a particular solution of (11-16), the important matters concern the solutions of (11-17).

If $f_1(x,y), \ldots, f_n(x,y)$ are solutions of the reduced equation, then

$$C_1 f_1 + \cdots + C_n f_n$$

where $C_1, \ldots, C_n$ are constants, is a solution. This is readily established by substitution in the equation. Also a series built from an infinite number of solutions

$$C_1 f_1 + C_2 f_2 + \cdots$$

is a solution in a region S, provided the series and the various derivative series required for substitution into the partial differential equation converge uniformly in S. Such series solutions are much used in the application to physical problems.

If we have a solution $f(x,y,a)$ of the reduced equation depending upon an arbitrary constant, then $\partial f/\partial a$ is also a solution. This may be verified directly, or it can be seen from the fact that the partial derivative is a limiting form of a linear combination of two solutions

$$\frac{\partial f}{\partial a} = \lim \frac{1}{h} [f(x, y, a + h) - f(x,y,a)]$$

Here a must not, of course, enter into the coefficients of the partial differential equation.

In certain cases we can get a solution of the reduced equation by differentiating a solution $f(x,y)$ partially with respect to one of the variables. If $A, \ldots, F$ do not contain x, we see that $f(x + h, y)$ is a solution. We have then the solution

$$\frac{\partial f}{\partial x} = \lim \frac{1}{h} [f(x + h, y) - f(x,y)]$$

Again from the solution $f(x,y,a)$ of the reduced equation we can get others by integration. It may be verified by differentiation under the sign of integration that

$$\int_\alpha^\beta \varphi(a) f(x,y,a) \, da$$

where $\varphi(a)$ is an arbitrary function and α, β are constants, is a solution. It may also be observed that the integral is the limiting form of a linear combination of solutions. This method may be extended to any number of arbitrary constants. Thus from the solution $f(x,y,a,b)$ we have the solution

$$\iint \varphi(a,b) f(x,y,a,b) \, da \, db$$

where the integral is extended over some region in the ab plane.

EXERCISES

1. Show that if a solution $f(x,y,a)$ of the reduced equation can be expanded in a suitably convergent series

$$f(x,y,a) = \varphi_0(x,y) + a\varphi_1(x,y) + a^2\varphi_2(x,y) + \cdots$$

then each of the functions $\varphi_n(x,y)$ is a solution.

2. Verify that

$$\frac{\partial^2 u}{\partial x^2} = \frac{\partial u}{\partial y}$$

has the solution e^{ax+a^2y}; and show that the equation has a sequence of polynomial solutions:

$$\varphi_n(x,y) = x^n + n(n-1)x^{n-2}y + \frac{n(n-1)(n-2)(n-3)}{2!} x^{n-4}y^2 + \cdots$$

3. Show that

$$V = \int_0^{2\pi} f(x \cos t + y \sin t + iz, t) \, dt$$

is a solution of Laplace's equation

$$\frac{\partial^2 V}{\partial x^2} + \frac{\partial^2 V}{\partial y^2} + \frac{\partial^2 V}{\partial z^2} = 0$$

4. Show that the gravitational potential due to the attraction of a number of particles

$$\mu \left(\frac{m_1}{r_1} + \frac{m_2}{r_2} + \cdots \right)$$

where m_i is the mass of the particle, (a_i, b_i, c_i) its coordinates, and

$$r_i^2 = (x - a_i)^2 + (y - b_i)^2 + (z - c_i)^2$$

satisfies Laplace's equation.

5. Show that the potential of a body at an exterior point (x,y,z),

$$\mu \iiint \frac{\rho(a,b,c) \, da \, db \, dc}{[(x-a)^2 + (y-b)^2 + (z-c)^2]^{\frac{1}{2}}}$$

where ρ is the density and the integral is extended over the body, satisfies Laplace's equation.

6. From the real part of the solution of Laplace's equation in two independent variables

$$f(x + iy) = \frac{ae^{i\psi} + x + iy}{ae^{i\psi} - (x + iy)}$$

show that Poisson's integral,

$$\frac{1}{2\pi}\int_0^{2\pi} U(\psi)\, \frac{a^2 - r^2}{a^2 - 2ar\cos(\theta - \psi) + r^2}\, d\psi$$

where $x = r\cos\theta$, $y = r\sin\theta$ and U is an arbitrary function, is a solution.

11-5. Linear with Constant Coefficients. In case the coefficients of z and its derivatives are constants it is convenient to use the symbols

$$D_1 = \frac{\partial}{\partial x} \qquad D_2 = \frac{\partial}{\partial y} \qquad (11\text{-}18)$$

The reduced equation has the form

$$F(D_1, D_2)z = 0 \qquad (11\text{-}19)$$

where F is a polynomial. For example,

$$\frac{\partial^2 z}{\partial x^2} - \frac{\partial^2 z}{\partial x\, \partial y} - 2\frac{\partial^2 z}{\partial y^2} + 3\frac{\partial z}{\partial x} + 2z = 0$$

is written

(i) $\qquad (D_1^2 - D_1 D_2 - 2D_2^2 + 3D_1 + 2)z = 0$

It can be shown that operators of this form conform to the laws of algebra (Sec. 8-16). Suppose F contains a linear factor, $D_1 + mD_2 + n$, say. By putting the remaining factor $G(D_1, D_2)$ on the left,

$$F(D_1, D_2)z = G(D_1, D_2)[(D_1 + mD_2 + n)z] = 0$$

we see that any solution of

$$(D_1 + mD_2 + n)z = 0$$

will satisfy the original equation. This last equation of the first order has the auxiliary system

$$\frac{dx}{1} = \frac{dy}{m} = \frac{dz}{-nz}$$

with the solutions

$$y - mx = a \qquad ze^{nx} = b$$

Hence we have the solution

$$z = e^{-nx}f(y - mx)$$

where f is an arbitrary function. Each linear factor contained in F gives a solution of this form.

Equation (i) may be factored as follows:

$$(D_1 + D_2 + 1)(D_1 - 2D_2 + 2)z = 0$$

Each linear factor gives a solution, and we add the two,

$$z = e^{-x}f(y - x) + e^{-2x}g(y + 2x)$$

In general it is not possible to factor $F(D_1, D_2)$ into linear factors, and the results are not so explicit. Suppose we try

$$z = e^{mx+ny}$$

We find

$$F(D_1, D_2)e^{mx+ny} = e^{mx+ny}F(m,n)$$

and we have a solution if

$$F(m,n) = 0$$

This defines n, say, as a function of m, $n = N(m)$, giving the solution

$$z = e^{mx+N(m)y}$$

By choosing various values for m we may form solutions in the form of arbitrary series

$$z = \sum_{k=1}^{\infty} A_k e^{m_k x + N(m_k) y}$$

For example,

(ii)
$$\frac{\partial^2 z}{\partial x^2} = \frac{\partial z}{\partial y}$$

leads to the equation

$$m^2 = n$$

and we have the solution

$$z = \sum A_k e^{m_k x + m_k^2 y}$$

Thus, taking $m_k = k$, we have

$$z = A_1 e^{x+y} + A_2 e^{2x+4y} + \cdots$$

Taking $m_k = ki$, and using the real part, we have

$$z = A_1 e^{-y} \cos x + A_2 e^{-4y} \cos 2x + \cdots$$

Many other solutions may be got by the methods of the preceding section.

Another method which is much used to get the individual terms in a series solution is to employ a product of functions each of which depends upon only one variable,

$$z = X(x)Y(y)$$

This works particularly well if the equation can be separated into a member involving D_1 and another in D_2,

$$G(D_1)z = H(D_2)z$$

On substituting, we have

$$\frac{1}{X}G(D_1)X = \frac{1}{Y}H(D_2)Y$$

These expressions, one in x alone the other in y alone, can be equal only if both are constant. Setting each equal to a constant we have two ordinary linear equations for the determination of X and Y.

In (ii) we have
$$\frac{1}{X}\frac{d^2X}{dx^2} = \frac{1}{Y}\frac{dY}{dy} = \lambda$$

Taking $\lambda = a^2$, we find
$$X = Ae^{ax} + Be^{-ax}$$
$$Y = Ce^{a^2 y}$$

taking $\lambda = -a^2$,
$$X = A \sin ax + B \cos ax$$
$$Y = Ce^{-a^2 y}$$

By altering a we get many solutions.

If the equation has a second member,
$$F(D_1, D_2)z = g(x,y) \tag{11-20}$$
we require a particular solution to add to any solution of the reduced equation. Without going into great detail we may say that the special methods for the ordinary equation (pages 91 to 96) may be suitably generalized.

1. For certain kinds of terms in g we may use the method of undetermined coefficients.

2. We may expand $1/F$ in powers of D_1 and D_2 and apply to g; this is useful when g is a polynomial.

3. If F can be reduced to linear factors we may solve a series of equations of the first order (generalizing methods 1 and 2 of Sec. 4-7).

4. If g has the form $e^{mx+ny}\varphi(x,y)$, we can use
$$\frac{1}{F(D_1, D_2)} e^{mx+ny}\varphi(x,y) = e^{mx+ny} \frac{1}{F(D_1 + m, D_2 + n)} \varphi(x,y)$$
and so on.

Example.
$$\frac{\partial^2 z}{\partial x^2} + 2\frac{\partial^2 z}{\partial y^2} + z = x^2 y + xye^{2x-y}$$

Using method 2 for the first term and method 4 for the second (expanding in series by division):
$$\frac{1}{D_1^2 + 2D_2^2 + 1} x^2 y = (1 - D_1^2 + \cdots)x^2 y = x^2 y - 2y$$
$$\frac{1}{D_1^2 + 2D_2^2 + 1} e^{2x-y}xy = e^{2x-y} \frac{1}{(D_1 + 2)^2 + 2(D_2 - 1)^2 + 1} xy$$
$$= e^{2x-y} \frac{1}{7 + 4D_1 - 4D_2 + \cdots} xy$$
$$= e^{2x-y}(\tfrac{1}{7} - \tfrac{4}{49}D_1 + \tfrac{4}{49}D_2 - \tfrac{32}{343}D_1 D_2 + \cdots)xy$$
$$= e^{2x-y}\left(\frac{xy}{7} - \frac{4y}{49} + \frac{4x}{49} - \frac{32}{343}\right)$$

The sum of these two expressions is a particular solution.

EXERCISES

1. Treat the homogeneous equation (11-13) of Sec. 11-3 by the method of operators.
2. Solve Exercise 9, Sec. 11-1, by present methods.
3. If $D_1 + mD_2 + n$ is a repeated factor of F, show that $xe^{-nx}f(y - mx)$ is a solution.

Solve:
4. $(D_1{}^2 - 2D_1D_2 + D_2 - 1)z = x^2 + y$.
5. $(D_1{}^2 - 4D_2{}^2 - 3D_1 + 6D_2)z = xy + e^{x+y}$. 6. $(D_1{}^2 + 2D_1 - D_2)z = 0$.
7. Get the solution of
$$[(D_1 + mD_2 + a)^2 + b^2]z = 0$$
in real form.
8. Show that the change of variable $x = e^u$, $y = e^v$ reduces a linear equation in which the coefficient of $\partial^{p+q}z/\partial x^p\,\partial y^q$ is $A_{pq}x^py^q$ to an equation with constant coefficients.
9. Solve:
$$x^2\frac{\partial^2 z}{\partial x^2} - 2xy\frac{\partial^2 z}{\partial x\,\partial y} + y^2\frac{\partial^2 z}{\partial y^2} = 0$$

Find all solutions of $(D_1{}^2 + D_2{}^2)z = 0$ which have the form:
10. $X(x)Y(y)$.   11. $X(x) + Y(y)$.   12. $U(x + y)V(x - y)$.

11-6. The Vibrating String. As an illustration both of setting up a partial differential equation and of the use of the series solution we consider the motion of a string tightly stretched between two points. If the string be displaced from a straight form and released, it will undergo a complicated system of movements.

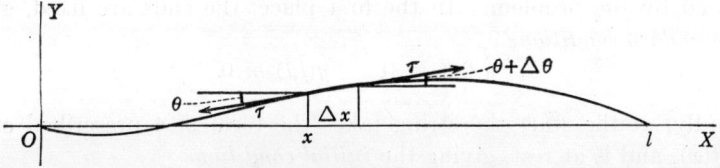

Fig. 41.

We treat the case in which the initial form of the string is a plane curve, whence the movement takes place in a plane. We assume that the deformation of any part of the string from its position of rest is small in both distance and direction. The tension τ is sensibly constant, and gravitation is neglected.

We let the position of rest be along the x axis, the ends of the string being the origin and the point $(l,0)$. Consider a portion of the string of length Δx approximately. A force τ acts along the tangent at each end. The components in the x direction are sensibly equal and opposite; so the resultant force is perpendicular to the x axis. It follows that each part vibrates along a perpendicular to the line of rest. The ordinate depends upon the abscissa and the time,

$$y = y(x,t)$$

The force in the y direction is

$$\tau \sin (\theta + \Delta\theta) - \tau \sin \theta$$

Here θ is small and we replace $\sin \theta$ by $\tan \theta$, so that, very nearly,

$$\sin \theta = \frac{\partial y(x,t)}{\partial x}$$

$$\sin (\theta + \Delta\theta) = \frac{\partial y(x + \Delta x, t)}{\partial x} = \frac{\partial y}{\partial x} + \frac{\partial^2 y}{\partial x^2} \Delta x$$

whence the force is

$$\tau \frac{\partial^2 y}{\partial x^2} \Delta x$$

We now set up the equation of motion (Sec. 1-17) for the piece. Its mass is $m \, \Delta x/l$, where m is the mass of the string; and we have

$$\frac{m \, \Delta x}{l} \frac{\partial^2 y}{\partial t^2} = \lambda \tau \frac{\partial^2 y}{\partial x^2} \Delta x$$

Putting $\lambda \tau l/m = a^2$, we have the equation

$$\frac{\partial^2 y}{\partial t^2} = a^2 \frac{\partial^2 y}{\partial x^2} \tag{11-21}$$

We seek a solution of this equation subject to certain conditions imposed by the problem. In the first place, the ends are fixed, giving the *boundary conditions*

$$y(0,t) \equiv 0 \qquad y(l,t) \equiv 0$$

Secondly, at the start the string takes the form of a prescribed curve, $y = f(x)$, and is at rest, giving the *initial conditions*

$$y(x,0) \equiv f(x) \qquad \left[\frac{\partial y(x,t)}{\partial t} \right]_{t=0} \equiv 0$$

As an example, if initially the string has the form at the top of Fig. 42, what will happen? Will the deformed part swing up and down where it stands, or will it move along the string in some fashion? We shall see.
Referring to (11-11), the general solution of (11-21) is

$$y = A(x + at) + B(x - at) \tag{11-22}$$

A and B being arbitrary functions. The first function gives a wave moving to the left, the second a wave moving to the right, each with velocity a.

The functions in (11-22) are determined by the given conditions. We have
$$y_t(x,t) = a[A'(x + at) - B'(x - at)]$$
The initial conditions then give
$$A(x) + B(x) = f(x)$$
$$A'(x) - B'(x) = 0$$
so
$$A(x) - B(x) = C$$
From these
$$A(x) = \tfrac{1}{2}[f(x) + C] \qquad B(x) = \tfrac{1}{2}[f(x) - C]$$
and (11-22) is
$$y = \tfrac{1}{2}[f(x + at) + f(x - at)] \qquad (11\text{-}23)$$

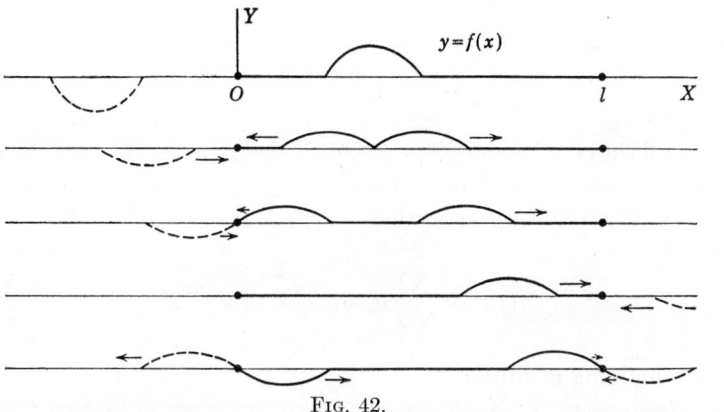

Fig. 42.

This formula requires values of $f(x)$ outside the interval $0 \leqslant x \leqslant l$ in which it is defined. However, the boundary conditions determine these values. At $x = 0$, we have
$$f(at) + f(-at) = 0$$
$$f(-at) = -f(at)$$
whence $f(x)$ is an *odd* function. Similarly
$$f(l - at) = -f(l + at)$$
or $f(x)$ is odd with respect to $x = l$. These requirements define $f(x)$ for all values of x, and the problem is solved.

In Fig. 42 are shown later positions of the string. The initial wave breaks into two waves, each of one-half the original magnitude, and they move in opposite directions. When a wave comes to the end of the string, it is reflected back with change of phase.

DIFFERENTIAL EQUATIONS

Fourier Series Solution. We shall now solve the preceding problem in series form, following the method mentioned in Sec. 11-5. We shall use terms of the form

$$y = X(x)T(t)$$

Substituting into (11-21), we have

$$\frac{1}{a^2 T}\frac{d^2 T}{dt^2} = \frac{1}{X}\frac{d^2 X}{dx^2} = -\alpha^2$$

whence

$$X = A\cos \alpha x + B \sin \alpha x$$
$$T = C \cos a\alpha t + D \sin a\alpha t$$

The product XT will satisfy both boundary conditions if $A = 0$, $\alpha = n\pi/l$, where n is an integer. It will satisfy the second initial condition,

$$XT'(0) = Xa\alpha(-C \sin 0 + D \cos 0) = 0$$

if $D = 0$. Hence

$$y = c_n \sin \frac{n\pi}{l} x \, \cos \frac{na\pi}{l} t$$

satisfies all the conditions except $y(x,0) \equiv f(x)$. From this, taking $n = 1, 2, \ldots$, we can form an infinite series which, assuming suitable convergence, satisfies the three conditions:

$$y(x,t) = \sum_{n=1}^{\infty} c_n \sin \frac{n\pi}{l} x \, \cos \frac{na\pi}{l} t \qquad (11\text{-}24)$$

The remaining condition,

$$y(x,0) = \sum c_n \sin \frac{n\pi}{l} x = f(x)$$

can now be satisfied. We shall take for granted that $f(x)$ can be expanded in a Fourier series of the type here given in the interval $0 \leqslant x \leqslant l$, c_n having the value

$$c_n = \frac{2}{l} \int_0^l f(x) \sin \frac{n\pi}{l} x \, dx$$

With these values of the constants, (11-24) is the required solution.

EXERCISES

1. Show that if the string is initially deformed into a sine curve,

$$y = A \sin \frac{n\pi}{l} x$$

PARTIAL OF THE SECOND ORDER 269

and released, each point of the string has simple harmonic motion. Find the period. Find the points which do not move (the nodes).

2. Taking $f(x) = h \sin x + k \sin 2x$, $l = \pi$, solve by both methods given in the text and compare.

3. A string six units long is plucked at $x = 3$ and released, the two parts of the string being initially straight lines. Plot the string for $t = 0, 1/a, \ldots, 6/a$.

4. If the string is initially in rectilinear pieces and at rest, does it continue to consist of rectilinear pieces?

5. Derive the solution (11-24) from (11-23), using the Fourier expansion of $f(x)$.

6. Solve the problem of the string plucked in the middle as in Exercise 3, but with $l = \pi$, using Fourier series:

$$y = \frac{8h}{\pi^2}\left(\sin x \cos at - \frac{\sin 3x \cos 3at}{3^2} + \frac{\sin 5x \cos 5at}{5^2} - \cdots\right)$$

7. If initially the string is straight but is moving according to the law

$$y_t(x,0) \equiv g(x)$$

show that

$$y = \frac{1}{2a}\int_{x-at}^{x+at} g(\tau)d\tau$$

where $g(x)$ is odd with respect to the ends.

8. Solve the preceding by Fourier series, given

$$g(x) = \sum d_n \sin \frac{n\pi x}{l}$$

9. Find the motion of a semi-infinite string, $x \geq 0$, lying along the x axis and in motion as follows:

$$y_t(x,0) = hxe^{-x^2}$$

10. Solve the problem of the string released from rest if the end at $x = 0$ is "flapping," that is, sliding without friction along the y axis, so that

$$y(l,t) \equiv 0 \qquad y_x(0,t) \equiv 0$$

Apply the preceding to the case $f(x) = h \cos \frac{\pi x}{2l}$.

11-7. The Vibrating Membrane. If a membrane bounded by a fixed contour in the xy plane be slightly deformed and released, it will vibrate. By considering the forces acting on an element of area, it can be shown that the distance z from the xy plane satisfies an equation of the form

$$\frac{\partial^2 z}{\partial t^2} = c^2\left(\frac{\partial^2 z}{\partial x^2} + \frac{\partial^2 z}{\partial y^2}\right) \qquad (11\text{-}25)$$

Rectangular Membrane. If the bounding edge of the membrane is the rectangle of the figure, we have the boundary conditions on the solution $z(x,y,t)$,

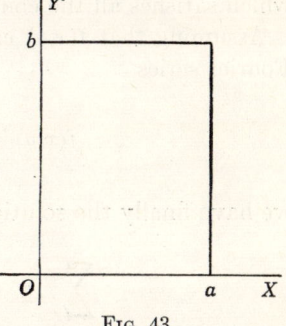

Fig. 43.

$$z(0,y,t) \equiv 0 \qquad z(a,y,t) \equiv 0$$
$$z(x,0,t) \equiv 0 \qquad z(x,b,t) \equiv 0$$

and the initial conditions

$$z(x,y,0) \equiv f(x,y) \qquad \left[\frac{\partial z}{\partial t}\right]_{t=0} \equiv 0$$

where $z = f(x,y)$ is the equation of the surface at the moment of release.

Setting
$$z = X(x)Y(y)T(t)$$

in (11-25), we have

$$\frac{1}{c^2 T}\frac{d^2 T}{dt^2} = \frac{1}{X}\frac{d^2 X}{dx^2} + \frac{1}{Y}\frac{d^2 Y}{dy^2}$$

This equation is possible only if the terms are suitable constants:

$$\frac{1}{X}\frac{d^2 X}{dx^2} = -\alpha^2 \qquad \frac{1}{Y}\frac{d^2 Y}{dy^2} = -\beta^2$$

$$\frac{1}{c^2 T}\frac{d^2 T}{dt^2} = -(\alpha^2 + \beta^2)$$

whence
$$X = A \cos \alpha x + B \sin \alpha x$$
$$Y = C \cos \beta y + D \sin \beta y$$
$$T = E \cos c\sqrt{\alpha^2 + \beta^2}\, t + F \sin c\sqrt{\alpha^2 + \beta^2}\, t$$

The boundary conditions require that

$$A = 0 \qquad \alpha = \frac{m\pi}{a}$$
$$C = 0 \qquad \beta = \frac{n\pi}{b}$$

where m and n are integers. The second initial condition gives $F = 0$. We have the solution

$$z = c_{mn} \sin \frac{m\pi}{a} x \, \sin \frac{n\pi}{b} y \, \cos c\pi \sqrt{\frac{m^2}{a^2} + \frac{n^2}{b^2}}\, t$$

which satisfies all the conditions but one.

Assuming that $f(x,y)$ can be expanded in the rectangle in the double Fourier series

$$f(x,y) = \sum_{m,n=1}^{\infty} c_{mn} \sin \frac{m\pi}{a} x \, \sin \frac{n\pi}{b} y$$

we have finally the solution

$$z = \sum_{m,n=1}^{\infty} c_{mn} \sin \frac{m\pi}{a} x \, \sin \frac{n\pi}{b} y \, \cos c\pi \sqrt{\frac{m^2}{a^2} + \frac{n^2}{b^2}}\, t$$

PARTIAL OF THE SECOND ORDER   271

The Drum. For a circular membrane of radius a and center at the origin the use of polar coordinates

$$x = r \cos \theta \qquad y = r \sin \theta$$

is convenient. It can be shown that (11-25) is transformed into

$$\frac{\partial^2 z}{\partial t^2} = c^2 \left(\frac{\partial^2 z}{\partial r^2} + \frac{1}{r}\frac{\partial z}{\partial r} + \frac{1}{r^2}\frac{\partial^2 z}{\partial \theta^2} \right) \qquad (11\text{-}26)$$

We shall treat the case in which the initial displacement is a function of r alone. We see from symmetry that z will be independent of θ. We have the boundary condition

$$z(a,t) \equiv 0$$

and the initial conditions

$$z(r,0) = f(r) \qquad \left(\frac{\partial z}{\partial t}\right)_{t=0} \equiv 0$$

where $f(r)$ is the prescribed deformation function.

Substituting

$$z = T(t)R(r)$$

we have

$$\frac{1}{c^2 T}\frac{d^2 T}{dt^2} = \frac{1}{R}\left(\frac{d^2 R}{dr^2} + \frac{1}{r}\frac{dR}{dr}\right) = -\mu^2$$

This gives

$$T = A \cos \mu c t + B \sin \mu c t$$

and the second initial condition requires that $B = 0$.

To determine R, set

$$u = \mu r$$

then

$$u \frac{d^2 R}{du^2} + \frac{dR}{du} + uR = 0$$

This is Bessel's equation (Sec. 7-9) with $n = 0$. We have the solution

$$R = CJ_0(u) = CJ_0(\mu r)$$

The second solution of Bessel's equation has a singularity at the origin and is thereby ruled out. The boundary condition requires that

$$J_0(\mu a) = 0 \qquad \mu = \frac{\alpha_n}{a}$$

where α_n is a root of J_0.

We have in

$$z = c_n \cos \frac{\alpha_n c}{a} t \; J_0\left(\frac{\alpha_n}{a} r\right)$$

a solution satisfying two of the three conditions. Letting $\alpha_1, \alpha_2, \ldots$ be the successive roots of J_0, we set up the series

$$z(r,t) = \sum_{n=1}^{\infty} c_n \cos \frac{\alpha_n c}{a} t \; J_0\left(\frac{\alpha_n}{a} r\right)$$

We wish to determine the constants so that

$$z(r,0) = \sum c_n J_0\left(\frac{\alpha_n}{a} r\right) = f(r)$$

Putting $r = av$ we require in the interval $0 \leqslant v \leqslant 1$ an expansion of the form

$$f(av) = \Sigma c_n J_0(\alpha_n v)$$

This is the problem treated in Sec. 7-11, the constants having the values found there.

EXERCISES

1. Solve the problem of the rectangular membrane with a prescribed initial velocity function as well as a deformation.

2. Find the period and the nodal lines when the membrane is put in the form

$$z = A \sin \frac{m\pi}{a} x \; \sin \frac{n\pi}{b} y$$

and released.

3. Solve the like problem for the drum with the initial deformation

$$z = A J_0 \left(\frac{\alpha_n}{a} r\right)$$

4. Show that for a square membrane ($b = a$) all points vibrate with the same period if the initial surface is

$$z = A \sin \frac{\pi x}{a} \sin \frac{2\pi y}{a} + B \sin \frac{2\pi x}{a} \sin \frac{\pi y}{a}$$

Discuss the nodal lines if $B = A$, $B = -A$, and in general.

11-8. The Conduction of Heat. The preceding examples illustrate the use of various series in setting up solutions satisfying prescribed conditions. In Sec. 11-4 the use of definite integrals was mentioned. We now give an example.

Given an infinite uniform bar of small cross section σ, so insulated that there is no transfer of heat at the surface. We take the x axis along the bar and give the temperature $u(x,t)$ an arbitrary initial distribution

$$u(x,0) = f(x)$$

It is required to find $u(x,t)$.

The rate at which heat flows along the bar is known to be proportional to the rate of change of temperature $\partial u/\partial x$, the heat moving in the direction of temperature drop. Now

$$\left(\frac{\partial u}{\partial x}\right)_{x+\Delta x} = \frac{\partial u}{\partial x} + \frac{\partial^2 u}{\partial x^2}\Delta x$$

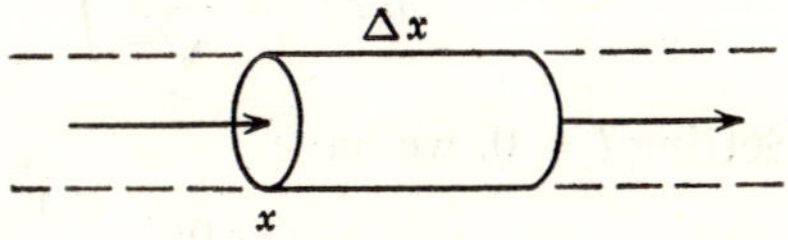

approximately; so the rate of increase of heat in the element is proportional to the final term in this equation. The increase of temperature is proportional to the increase of heat divided by the volume $\sigma\,\Delta x$. This gives

FIG. 44.

$$\frac{\partial u}{\partial t} = a^2 \frac{\partial^2 u}{\partial x^2} \tag{11-27}$$

where a is a suitable constant.

Setting $u = X(x)T(t)$ we have

$$\frac{1}{X}\frac{d^2 X}{dx^2} = \frac{1}{a^2 T}\frac{dT}{dt} = -\alpha^2, \text{ say}$$

whence we have the solution

$$u_1 = e^{-\alpha^2 a^2 t} \cos \alpha(x - \lambda)$$

We now form a solution by integration with respect to α,

$$u_2 = \int_0^\infty e^{-\alpha^2 a^2 t} \cos \alpha(x - \lambda)\, d\alpha$$

In evaluating this, we shall use the formula

$$\int_0^\infty e^{-y^2} \cos 2by\, dy = \tfrac{1}{2}\sqrt{\pi}\, e^{-b^2} \tag{11-28}$$

which will be found in most tables of integrals. Putting $\alpha a \sqrt{t}$ for y and $(x - \lambda)/2a \sqrt{t}$ for b, we find that

$$u_2 = \frac{\sqrt{\pi}}{2a \sqrt{t}} e^{-\frac{(x-\lambda)^2}{4a^2 t}}$$

We multiply this by $\dfrac{f(\lambda)}{\pi}$, where $f(x)$ is the initial temperature, and integrate again,

$$u(x,t) = \frac{1}{2a \sqrt{t\pi}} \int_{-\infty}^\infty f(\lambda) e^{-\frac{(x-\lambda)^2}{4a^2 t}}\, d\lambda$$

This final result (assuming the convergence of the integral) is the solution sought. To show that it satisfies the initial condition, put

$$\lambda = x + 2a\sqrt{t}\,y,$$

$$u(x,t) = \frac{1}{\sqrt{\pi}} \int_{-\infty}^{\infty} f(x + 2a\sqrt{t}\,y) e^{-y^2}\, dy$$

Setting $t = 0$, we have

$$u(x,0) = \frac{1}{\sqrt{\pi}} f(x) \int_{-\infty}^{\infty} e^{-y^2}\, dy$$

The integral here is twice the integral (11-28) with $b = 0$; that is, $\sqrt{\pi}$; and we have

$$u(x,0) = f(x)$$

EXERCISES

1. Let the bar be initially at 0° except for positive temperatures along a short section. Show that immediately after the start the temperature is positive throughout the whole bar. We observe from this that heat does not flow along the bar in a wavelike motion.

2. The ends of a section of the bar are held at constant but unequal temperatures. Find the ultimate distribution of temperature in the section.

3. Carry through the method of the text for the infinite lamina without passage of heat across the surface. The equation is

$$\frac{\partial u}{\partial t} = a^2 \left(\frac{\partial^2 u}{\partial x^2} + \frac{\partial^2 u}{\partial y^2} \right)$$

4. From the general heat equation

$$\frac{\partial u}{\partial t} = a^2 \left(\frac{\partial^2 u}{\partial x^2} + \frac{\partial^2 u}{\partial y^2} + \frac{\partial^2 u}{\partial z^2} \right)$$

and the initial distribution $u = f(x,y,z)$ in infinite space, derive the solution

$$u = \frac{1}{8a^3(\pi t)^{\frac{3}{2}}} \int_{-\infty}^{\infty} \int_{-\infty}^{\infty} \int_{-\infty}^{\infty} f(\lambda,\mu,\nu) G\, d\lambda\, d\mu\, d\nu$$

where

$$G = e^{-\frac{(x-\lambda)^2 + (y-\mu)^2 + (z-\nu)^2}{4a^2 t}}$$

5. The rate at which heat is lost, owing to slight surface radiation into the surrounding air at constant temperature u_0 is proportional to the difference of temperature $u - u_0$ and to the surface area of the element. Show that this leads to an equation of the form

$$\frac{\partial u}{\partial t} = a^2 \frac{\partial^2 u}{\partial x^2} - h(u - u_0)$$

6. Show that the change of variable

$$u = u_0 + e^{-ht} v$$

reduces the preceding equation to the form given in the text. Find u for the infinite bar with the initial distribution $u = f(x)$.

7. Find the ultimate distribution of temperatures in the section of the bar of Exercise 2 when there is radiation.

11-9. Laplace's Equation. If there is a fixed distribution of temperature on the surface of a uniform body, the temperature inside ultimately acquires a steady state, $\partial u/\partial t = 0$, and Laplace's equation is satisfied. We shall consider the two-dimensional case,

$$\frac{\partial^2 u}{\partial x^2} + \frac{\partial^2 u}{\partial y^2} = 0 \qquad (11\text{-}29)$$

and get the solution of a particular problem by means of series.

Let the faces of a long bar of rectangular cross section (use Fig. 43) be held at prescribed temperatures which are constant along any line parallel to an edge. We then have a two-dimensional problem. We shall prescribe $u = f(x)$ along the face $y = 0$, and $u = 0$ along the faces $x = 0$, $x = a$, and $y = b$. The case in which more general temperatures are given for the latter faces may be handled by assigning nonzero temperatures to one face at a time and adding the resulting four temperatures.

We substitute
$$u = X(x)Y(y)$$
into (11-29), whence
$$Y\frac{d^2 X}{dx^2} + X\frac{d^2 Y}{dy^2} = 0$$
$$\frac{1}{X}\frac{d^2 X}{dx^2} = -\frac{1}{Y}\frac{d^2 Y}{dy^2} = -\alpha^2$$

the negative constant being chosen so that the boundary conditions can be satisfied. From these we have the solution

$$u = XY = (A\cos\alpha x + B\sin\alpha x)(Ce^{\alpha y} + De^{-\alpha y})$$

We can satisfy the conditions on three faces. Taking $A = 0$, we have $u = 0$ along $x = 0$. Taking $\alpha = n\pi/a$, we have $u = 0$ along $x = a$. Taking $Y = \sinh\alpha(b - y)$, which we get by a suitable choice of C and D, we have $u = 0$ along $y = b$. The resulting solution

$$u_n = k_n \sin\frac{n\pi x}{a} \sinh\frac{n\pi(b-y)}{a}$$

for $n = 1, 2, \ldots$, satisfies the boundary conditions except on the face $y = 0$.

We solve the problem by forming the series

$$u = \sum k_n \sin\frac{n\pi x}{a} \sinh\frac{n\pi(b-y)}{a} \qquad (11\text{-}30)$$

and so choosing the constants k_n that, when $y = 0$,

$$\sum k_n \sin \frac{n\pi x}{a} \sinh \frac{n\pi b}{a} = f(x)$$

This we do by expanding $f(x)$ in a sine series,

$$f(x) = \sum c_n \sin \frac{n\pi x}{a}$$

and taking

$$k_n = \frac{c_n}{\sinh (n\pi b/a)}$$

11-10. Approximate Solutions. Laplace's equation arises in many connections. Besides being the equation of temperature in the steady state, it is also the equation of electric, magnetic, and gravitational potential, as well as that of the velocity potential of a moving fluid. It is desirable to have a method for the approximate solution of the difficult problems that arise. Fortunately there is a process of great simplicity.

In the equation

$$\frac{\partial^2 u}{\partial x^2} + \frac{\partial^2 u}{\partial y^2} = 0 \qquad (11\text{-}29)$$

we shall use the approximations

$$\frac{\partial^2 u(x,y)}{\partial x^2} = \frac{u(x+h, y) - 2u(x,y) + u(x-h, y)}{h^2}$$

$$\frac{\partial^2 u(x,y)}{\partial y^2} = \frac{u(x, y+h) - 2u(x,y) + u(x, y-h)}{h^2}$$

We justify these by Taylor's series. Thus

$$u(x+h, y) = u + u_x h + \frac{1}{2} u_{xx} h^2 + \frac{1}{3!} u_{xxx} h^3 + \frac{1}{4!} u_{xxxx} h^4 + \cdots$$

u and its derivatives in the second member being evaluated at (x,y). Replacing h by $-h$ and evaluating the first approximation above, we have

$$\frac{1}{h^2}[u(x+h, y) - 2u(x,y) + u(x-h, y)] = u_{xx} + \frac{1}{12} u_{xxxx} h^2 + \cdots$$

which differs from u_{xx} by terms in h^2 and higher powers.

On substituting these approximate values into (11-29) and rearranging, we have our major result:

$$u(x,y) = \tfrac{1}{4}[u(x+h, y) + u(x-h, y) + u(x, y+h) + u(x, y-h)] \qquad (11\text{-}31)$$

This formula is easily remembered, for it states that at any point u is approximately equal to the average of its values at four points at equal distances to the right, the left, above, and below.

Given a region with values of u prescribed along the boundary, to find the values of u inside (the Dirichlet problem). We rule the region into squares by horizontal and vertical lines, and we seek approximate values at the vertices. Where the lines meet the boundary the values of u are known. We suppose that these occur at vertices; otherwise an estimate of u will be made at a vertex falling close to the boundary.

We now estimate, or guess, values of u at vertices lying along *alternate* diagonal lines. This first set of vertices constitutes about half of those lying in the region. We next use (11-31) to compute values at the remaining vertices. We use these latter values, with (11-31), to redetermine the first set of values, and so on. The process will ultimately converge, although sometimes the convergence will seem distressingly slow.

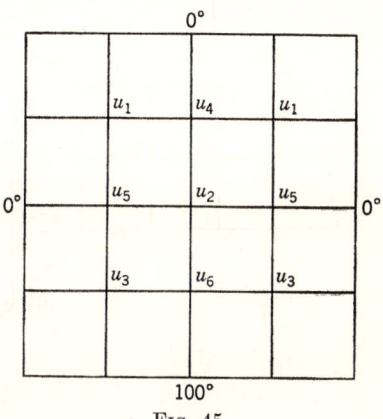

Fig. 45.

In the accompanying table we give the details of the process for the square post of Fig. 45, where one face is held at 100° and the others at 0°. Certain values are equal from symmetry. The initial estimates are marked by asterisks. We took $u_2 = 25$ as the average of the temperatures of the four faces, and we chose u_1 and u_3 as the averages of the values at the vertices of the quarter squares in which the points lie. Then $u_4 = \frac{1}{4}(6 + 6 + 25 + 0) = 9.2$, $u_5 = \frac{1}{4}(6 + 25 + 44 + 0) = 18.7$, etc.

u_1	u_2	u_3	u_4	u_5	u_6
6*	25*	44*	9.2	18.7	53.2
7.0	25.0	42.9	9.75	18.72	52.70
7.12	24.97	42.85	9.80	18.73	52.67
7.13	24.98	42.85	9.81	18.74	52.67
7.14	24.99	42.85	9.82	18.74	52.67
A: 6.63	25.00	43.37	9.37	17.86	54.92
B: 6.88	25.01	43.13	9.61	18.39	53.64
C: 6.80	25.00	43.20	9.54	18.20	54.06

The three lines following the table contain more accurate results. A is got by inserting points in the middles of the squares of the figure and

278   DIFFERENTIAL EQUATIONS

making our network with 45° lines. The results in the table are used as starting values. *B* is got by giving *h* a value half of that in the figure, so that there are 49 points in the network instead of 9. *C* is the set of exact values, computed from the series in Exercise 3 below.

EXERCISES

1. Find exactly the temperature of the square post with $f(x) = 100 \sin \dfrac{\pi x}{a}$.
2. Solve the preceding approximately, using the network of Fig. 45. The results should be $u_1 = 5.83$, $u_2 = 21.33$, $u_3 = 33.18$, $u_4 = 8.25$, $u_5 = 15.08$, $u_6 = 46.92$.
3. Solve exactly the problem treated in Sec. 11-10, getting the series

$$u = \frac{400}{\pi}\left[\frac{\sin\dfrac{\pi x}{a}\sinh\dfrac{\pi(a-y)}{a}}{\sinh \pi} + \frac{\sin\dfrac{3\pi x}{a}\sinh\dfrac{3\pi(a-y)}{a}}{3\sinh 3\pi} + \cdots\right]$$

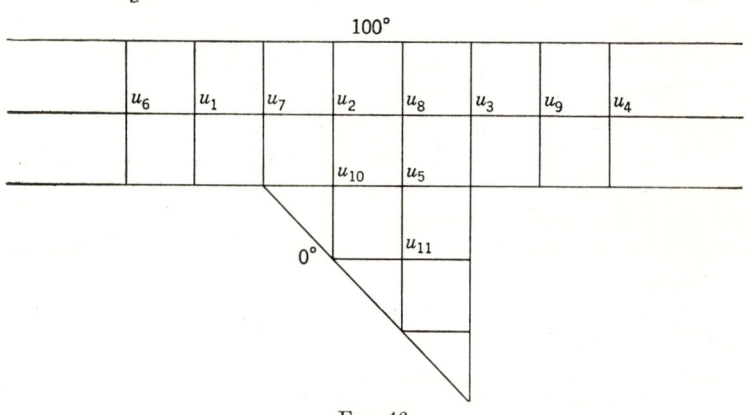

Fig. 46.

4. Show that at the center of a small square oriented in any way *u* is approximately equal to the average of its values at the vertices.
5. Apply the preceding to get line *A* of the table in the text.
6. Show that for the exact solution or any of the approximate solutions of the problem in Sec. 11-10 we should have

$$u_2 = 25 \qquad u_1 + u_3 = 50 \qquad u_4 + 2u_5 + u_6 = 100$$

7. Show that for the three-dimensional case $u(x,y,z)$ is approximately the average of its values at the six points $(x \pm h, y, z)$, $(x, y \pm h, z)$, $(x, y, z \pm h)$.
8. Solve approximately the Dirichlet problem for the infinite region of Fig. 46; the results being $u_1 = 50.5$, $u_2 = 57.2$, $u_3 = 52.0$, $u_4 = 50.1$, $u_5 = 20.5$, $u_6 = 50.1$, $u_7 = 51.9$, $u_8 = 57.4$, $u_9 = 50.5$, $u_{10} = 19.4$, $u_{11} = 5.1$.
9. Given a distribution of temperature, not necessarily in the steady state, at time *t*, and make the approximation

$$u_t = \frac{u(x,y,t_1) - u(x,y,t)}{t_1 - t}$$

Show that formula (11-31) gives an approximation to the temperature at time $h^2/4a^2$ later.

ANSWERS

Chapter 1

Pages 7-9
1. $s = 6e^{-5t}$.
2. $s = 3(2 - 5t)^2/2$.
3. $s = 6(1 - 10t)^{\frac{1}{2}}$.
4. $s = 10e^{-3t} - 4$.
5. $s = 6(1 - 2t)/(3t + 1)$.
6. $s = 8[3 - 4 \tan (12t/5)]/[4 + 3 \tan (12t/5)]$.
7. $1/(C - x)$ and 0, $1/(1 - x)$.
8. $\pm (2ax + C)^{\frac{1}{2}}$.
9. $\sin (x + C)$.
10. $\pm (2x/3 + C)^{\frac{3}{2}}$ and 0.
11. $(1 + y)(1 + x) = C$.
18. (a) $x^2 + y^2 = c^2$. (b) $x^2 + \log Cy = 0$. (c) $y = Cx$. (d) $y^2 = C \sin 2x$.
22. $y = C(x - 1)e^x$.
23. $e^{x^2} - 2(y + 1)e^{-y} = C$.
24. $x(y + 1) = Ce^{y-x}$.
25. $2x/(y - x) + \log |x| = C$.
26. $\log (x^2 + y^2) - 2 \arctan (y/x) = C$.
27. $x/y + \log |x| = C$.
28. $y^3(3xy + 2) = Cx^3$.
30. $y = Ce^{2x} - x/2 - \frac{1}{4}$.
31. $y = Ce^{-3x} + e^x/4$.
32. $y = (C + x)e^{-x}$.

Pages 11, 12
1, 2. $y = x - 1$.
4. $x^2 - xy = y$.
5. $(x - x_0)(lx + my + q) = y - y_0$.
12. $x^2 + y^2 = Cx^{\frac{4}{3}}$.
13. $r = C \cos^2 \theta$.

Pages 13, 14
1. $y = x - 2x^3/3 + \cdots$.
2. $y = x^2/2 + x^3/6 + x^4/24 + \cdots$.
3. $y = 2x + 2x^3/3 + \cdots$.
4. $y = x^3/3 + x^7/63 + \cdots$.
5. $y = x + x^2/2 - x^4/24 + \cdots$.
6. $y = x$.
7. $y = x^2/2 - x^3/3 + x^4/8 + \cdots$.
8. $y = 1 + x + x^2 + 4x^3/3 + 7x^4/6 + \cdots$.
9. $y = 1 + 2(x - 1) + 3(x - 1)^2 + 11(x - 1)^3/3 + 29(x - 1)^4/6 + \cdots$.
10. $y = x + x^3/3 + 2x^5/15 + 17x^7/315 + \cdots$.
11. $y = 1 + x + x^2/2! + \cdots + x^n/n! + \cdots = e^x$.
12. $y = C - Cx + (C + 1)x^2/2! + \cdots + (-1)^n(C + 1)x^n/n! + \cdots$
    $= (C + 1)e^{-x} + x - 1$.
13. $y = x + 3x^2/2! + 7x^3/3! + \cdots + (2^n - 1)x^n/n! + \cdots = e^{2x} - e^x$.
14. $y = x - x^3/3! + \cdots + (-1)^n x^{2n+1}/(2n + 1)! + \cdots = \sin x$.
15. $y = 1 - x^2/2! + \cdots + (-1)^n x^{2n}/(2n)! + \cdots = \cos x$.
16. $y = a + bx - ax^2/2! - bx^3/3! + \cdots = a \cos x + b \sin x$.

Pages 15, 16
1. (a) $2 + 3dy/dx = 0$. (b) $3 - 2dy/dx = 0$.
2. $(x - 1)y' = y - 1$.
3. $y' = y/2x$.
4. $y' = -y^2/x^2$.
5. $y' = (3y - x + 3)/(3x - y + 3)$.
6. $y' = 2xy/(x^2 - y^2 - 1)$.
7. $y \sin x + y' \cos x = 1$.
8. $y' = 1$.
9. $(y - xy')^2 = 1 + y'^2$.
10. $(3y - 4x)^2(1 + y'^2) = (3 + 4y')^2$, $3y = 4x \pm 5$.
11. $xyy'^2 + (x^2 - y^2 + b^2 - a^2)y' - xy = 0$.
12. $y = y'x + 1/4y'^2 + \frac{1}{2}$.

279

Page 18
1. $y = Cx + \cos C$. 2. $y^2 + 4x = 0$. 3. $xy - y^2 = \frac{1}{4}$.
5. $e^y + Ce^{Cx} = 0, y = -1 - \log x$.
6. With axes along fences, $y = -Kx + (200K)^{\frac{1}{2}}, xy = 50$.
7. $y = -Kx + 40K/(1 + K^2)^{\frac{1}{2}}, x^{\frac{2}{3}} + y^{\frac{2}{3}} = 40^{\frac{2}{3}}$.

Page 20
1. $y'' = 0$. 2. $(x^2 + y^2)y'' + 2(y - xy')(1 + y'^2) = 0$.
3. $yy'' + y'^2 + 1 = 0$. 4. $(1 + y'^2)y''' - 3y'y''^2 = 0$.
5. $xyy'' - y'(y - xy') = 0$. 6. $y'''(y - xy') + 3xy''^2 = 0$.
7. $y'' - y' - 2y = 0$. 8. $y'' + y = 0$. 9. $y'' + 4y = 0$.
10. $y'' - 2y' + 2y = 0$. 11. $y'' + y' \tan x = 0$. 12. $y''^2 + y'^2 = 1$.
13. $x^2y'' - (4x + x^2)y' + (6 + 2x)y + 2x^4e^{2x} = 0$. 14. $xy' = y \log y$.
15. $(y - 1)y'' = 2y'^2$. 17. $y''' = 0$. 18. $x^2y'' + 2(y - xy') = 0$.
19. $(x^2 - 1)y' = 2xy$. 20. $(1 - x^2)y' + 2xy = x^2 + 1$.
21. $xy' - 2y + x = 0$.

Page 21
1. $y = Ce^{x/a}$. 2. $y = x + C$. 3. $(x + c)^2 + y^2 = a^2, y = \pm a$.
4. $y^2 + 2x^2 = C$. 5. $y = Cx, y = C/x$.

Pages 23, 24
1. $3y - 2x = K$. 2. $x^2 + 2y^2 = K$. 3. $xy = K$. 4. $x^2 + 4y^2 = K$.
5. $2x^2 + 3y^2 = K$. 6. $y = Kx^{a^2/b^2}$. 7. $x^2 + y^2 - 2 \log y = K$.
8. $16y^3 = (3x - K)^2$. 9. $x = Ke^{y^2}$. 10. $\sin y = Ke^{-x}$.
16. $x^2 - y^2 = K, z = c$. 17. Radial straight lines.
18. $e^{-y} \cos x = K, z = c$. 19. $x^2 - y^2 = K, z = xy$.
20. $y = Kx^4$ on $x^2 + 4(y^2 + z^2) = 64$. 21. $x^2 - y^2 = K$.

Page 26
1. $y = 2a \tan (ax + b)$; or $\log |(y - a)/(y + a)| = ax + b$; or $y = 2/(C - x)$;
and $y = C$. 2. $y = Cx^2 + K$. 3. $y = 1/(Cx + K)$.
4. $y = Ce^{ax} + Ke^{-ax}$. 5. $(x + a)(y + b) = 4$.

Pages 30–32
1. $s = gt^2/2$. 2. $s = v_0(1 - e^{-ht})/h, s = v_0/h$, never.
3. $hs = \log (1 + v_0ht), s = \infty$, never.
4. $s = v_0t - hv_0^{\frac{1}{2}}t^2/2 + h^2t^3/12, s = 2v_0^{\frac{3}{2}}/3h, t = 2v_0^{\frac{1}{2}}/h$.
5. 5.89 mi/sec. 6. 6.96 mi/sec.
7. 4.92 mi/sec. After 1 hr 25 min. 8. $s = -v/h - (g/h^2) \log (1 - hv/g)$.
9. $hs = \log (e^{at} + e^{-at})/2, a^2 = gh$. 10. $m = m_0e^{-kt}$.
11. $x = x_0e^{at}$. 12. $x = x_0e^{at-bt^2}$. 13. 13.5%.
14. $x = Tx_0/[x_0 + (T - x_0)e^{-kt}$. 15. $r = r_0 - kt$. 16. 6.18%.
18. (b) 15,500 yr. (c) 6,500 yr. 19. 2 hr 19 min.
21. 3 min 24 sec, 1 min. 22. $x = 1.085$ ft.
23. $t = \pi r^2 h^{\frac{3}{2}}/20a$. 26. 55 min, 118 min.
27. $x' = k(A - ax)(B - bx), \log (B - bx)/(A - ax) = (aB - bA)kt + C, aB \neq bA$.
29. Go 3 miles toward spot, then follow spiral $r = e^{\theta/\sqrt{8}}$.

Chapter 2

Pages 37, 38
1. $y = Ce^x - 2$. 2. $y = Ce^{-x} + e^x$. 3. $y = Cx - 1$.
4. $y = Ce^x - 1$. 5. $y = Ce^{-x} + x$. 7. $y = Ce^{-x} + x^2 - 2x + 4$.
8. $y = Cx^2 + x^5/3$. 9. $y = (C + x)e^{3x} - e^{-3x}/6$.

10. $y = (Ce^{2x} - e^x)^2$.
11. $y = (C + x)e^{2x} - \frac{1}{2}$.
12. $x^2 + y^2 = Cy + 1$.
13. $y = e^{3x} + C/x^2$.
14. $y = Cx^{\frac{1}{2}} - x + x^3/5$.
15. $y = 2ax/3 - 1/x + C/x^{\frac{1}{2}}$.
16. $y = 1/(Ce^{-x} + x^2 - x + 2)$.
17. $y = C \cos x - 2 \cos^2 x$.
18. Minimum on $y = 2e^x$.
19. Minimum on $y = x + 1$.
20. Minimum $x > 0$, maximum $x < 0$, on $y = x^2 + 2$.
21. $x = Ce^{-y} + y - 2$.
22. $i = i_0 e^{-Rt/L}$.
23. $i = (i_0 - E_0/R)e^{-Rt/L} + E_0/R$.
24. $i = [i_0 - E_0/(R - kL)]e^{-Rt/L} + E_0 e^{-kt}/(R - kL)$.
25. $i = i_0 e^{-Rt/L} + E_0[R \sin \omega t + L\omega(e^{-Rt/L} - \cos \omega t)]/(L^2\omega^2 + R^2)$.
27. 52 min.   28. 131 min.   29. $y = (Cx + x^3)^2$.
30. $y^2 = (e^x + Ce^{-x})/2x$.
31. $\sin y = be^{-ax^2/2}(\int e^{ax^2/2}x^2 \, dx + C)$.
32. $y + 3 = (Cx^2 - 2x^3)^{-\frac{1}{2}}$.

Page 40
1. $y = x + 5x/(C - x^5)$.
2. $(y - x)/(y + x) = Ke^{2x^5/5}$.
3. $(y + x - 2)/(y + x) = Ke^{2x}$.
4. $(y - a)/(y - b) = Ke^{(a-b)x}$.
5. $(y - x^2)/(y - x) = C(x + 1)$.
7. $y = e^x \tan (x + C)$.

Page 43
1. $(x^2 + y^2)^3 = Cx^2$.
2. $l_1 x^2 - 2l_2 xy - m_2 y^2 = C$.
3. $(y + ax)^2 = Cx$.
4. $y = xe^{Cx}$.
5. $y = x/(Cx^2 - 1)$.
6. $y = x \tan Cx^n$.
7. $x^4 y^4(x^2 + xy + 5y^2) = C$.
8. $(x^2 - y^2)^2 = Cxy$.
9. $r = Ke^{\theta/\tan A}$.
10. $x^2 + y^2 = Cy$.
11. $x^2 + y^2 = C(x - y \tan \alpha)$.

Page 49
1. $2x^2 + 3y^2 = C$.
2. $x^2 - y^2 = C$.
3. $y = Cx^3$.
4. $y = x + Cx^2$.
5. $r = Ce^\theta$.
7. $(2x + y + 1)^{\frac{1}{2}} - \log [1 + (2x + y + 1)^{\frac{1}{2}}] = x + C$.
8. $y = x \log Cx$.
9. $x + y + 1 = \tan (x + C)$.
10. $x^2 - 4xy - y^2 + 12x - 4y = C$.
11. $(x - y - 1)(x + y + 3)^2 = C$.
12. $(x - y)^2 + 8x + 2 \log (x - y + 2) = C$.

Page 52
1. $y^3 + 3xy - x^3 = C$.
2. $lx^2 + 2mxy + ny^2 + 2ax + 2by = C$.
3. $x^2/2 + x/y + 2 \log y = C$.
4. $x^2 - 2x^2 y - 3y^2 - 10x + 4y = C$.
5. $2x^3 - 9x^2 y - 15xy^2 + y^3 = C$.
6. $xe^{y^2} + \csc y \cot x = C$.
7. $\sin 3t \cos 2\theta = C$.
8. $x^2 y^4 + x \sin y = C$.
10. $x^2 y^2(4y^2 - x^2) = C$.

Pages 57, 58
2. $\mu = e^{\int f(y)dy}$.
3. $(M_y - N_x)/(M - N) = f(x + y)$.
4. $(M_y - N_x)/(My - Nx) = f(x^2 + y^2)$.
5. $(y + ax)^2 = Cx$.
6. $x + y = Cx^2 y$.
7. $\log x - xy + y^2/2 = C$.
9. $7x + 3y^2 = Cx^7 y^9$.
10. $x(x^2 y^3 + 5)^2 = Cy$.

Page 58
1. $y = x(\log x + 2x^{\frac{1}{2}} + C)$.
2. $y = Cx - 1/C$, $y^2 + 4x = 0$.
3. $y^2 = 2x^2 \log Cx$.
4. $x^2 + y^2 = Ce^{2x}$.
5. $(x - 1)^2 + y^2 = Ce^{-2x}$.
6. $y = x \tan (C - x)$.
7. $x^4 + y^4 = (2xy + C)^2$.
8. $xy^2 + \sin xy = C$.
9. $x + y + 1 = \tan (y + C)$.
10. $x^2 + 2xy^3 - 2xy - y^3 = C$.
11. $2xy = Cx^2 + 1/C$.
12. $(x + y)^2(2x + y + 2)^3 = C$.
13. $\sin x + \sin y + \sin (x + y) = C$.
14. $y = 2ax/3 - 1/x + C/x^{\frac{1}{2}}$.
15. $y = e^x + 1/(Ce^{-x} - 2x + 2)$.
16. $y = Ce^{2x} - e^x$.
17. $x^2 - 2xy^3 + xy = C$.
18. $y + 1 = x \log Cx$.

19. $x^4 + 8x^3y + 2x^2y^2 = C$.
20. $xy(y^2 - 1) = C$.
21. $\log(1 + y) = Ce^{-x}$.
22. $x^2 + 2x - 2xy = C$.
23. $y = C(x \pm 1)$.
24. $xy + 1/x + 1/y = C$.
25. $y = e^x(1 + \tan x) + C \sec x$.
26. $x^2 - 2xy - 10x - 3y^2 + 4y = C$.
27. $(x + y - 2)^3 = C(x - y + 6)$.
28. $y = x^3/5 - x + Cx^{\frac{1}{2}}$.
29. $2x^3 - 9x^2y - 15xy^2 + y^3 = C$.
30. $x + 2y + 3\log(2x - y) = C$.
31. $y^2 = 2Cx + C^2$.
32. $y = (Cx + 1)/(C - x)$.
33. $y = kx^4/(4 - n) + Cx^n$.
34. $y^2 = Cx^2(x + 3y)$.
35. $y = x + e^{-x^2}/(C + \int e^{-x^2} dx)$.
36. $r = ae^{-\theta}/\sqrt{2}$, distance = side a.

Chapter 3

Page 62
1. $y = A + Be^{3x} - 2x$.
2. $y = Ae^{2x} + Be^{-2x} - (\sin x)/5$.
3. $y = A \sin 2x + B \cos 2x + e^x/5 + \frac{1}{4}$.
4. $y = A + Bx^2 + x^5/15$.
5. $y = Ax + Be^x$.
6. $x^2y'' - 2xy' + 2y = 0$.
7. $x^2y'' + xy' - y = 0$.
8. $y'' - 3y' + 2y = 0$.
9. $y'' - y = 3e^{2x}$.

Page 65
1. $e^x + e^{-x} - 2\cosh x = 0$.
2. $\sin^2 x + \cos^2 x - \frac{100}{100} = 0$.
4. $(n - m)x^{m+n-1}$.
5. $(n - m)e^{(m+n)x}$.
6. 0.
7. $-ae^{2mx}$.
10. $y = e^{-x^2/2}(Ax + B)$.

Pages 68, 69
1. $y = Ae^{-3x} + Be^{-5x}$.
2. $y = e^{-2x}(A \sin x + B \cos x)$.
3. $y = (A + Bx)e^{3x}$.
4. $y = A + Bx$.
5. $y = Ae^x + Be^{2x}$.
6. $y = Ae^x + Be^{-3x/2}$.
7. $y = A + Be^{-3x/2}$.
8. $y = A \sin \sqrt{\frac{3}{2}} x + B \cos \sqrt{\frac{3}{2}} x$.
9. $y = e^{-x/8}\left(A \sin \dfrac{\sqrt{15}\,x}{8} + B \cos \dfrac{\sqrt{15}\,x}{8}\right)$.
10. $y = e^{-3x/4}\left(A \sin \dfrac{\sqrt{23}\,x}{4} + B \cos \dfrac{\sqrt{23}\,x}{4}\right) - \dfrac{5}{4}$.
11. $y = 2e^x + e^{5x}$.
12. $y = (1 + 2x)e^{-2x} - 1$.

Page 70
4. Move (a) to left, (b) to right.

Pages 75, 76
1. $y = A \sin x + B \cos x + x^3 - 5x$.
2. $y = A + Be^{-x} + x^3/3$.
3. $y = A \sin \sqrt{2}\,x + B \cos \sqrt{2}\,x + e^{2x}/6 + x/2$.
4. $y = Ae^{-x} + Be^{-4x} + (8x^2 - 20x + 21)/32 - (\cos 2x)/10$.
5. $y = A \sin x + (B - x/2) \cos x$.
6. $y = (A + Bx)e^{2x} + e^x + (x + 1)/4$.
7. $y = A \sin 2x + B \cos 2x + (2x^2 - 1)/8 + (\cos x)/3$.
8. $y = (A + Bx + x^3/3)e^x$.
9. $y = A \sin x + B \cos x + (3 \sin 3x - \cos 3x)e^{2x}/40$.
10. $y = (A + x)e^x + Be^{-x}$.
11. $y = (A + x/2) \sin 2x + (B + \log \cos 2x)(\cos 2x)/4$.
12. $y = A \sin x + [B + \log(\sec x + \tan x)] \cos x$.
13. $y = A \sin x + [B + \log(\csc x + \cot x)] \cos x - 2$.
14. $y = (A + Bx - \log x)e^{-x}$.
15. $(x - 1)^2 y'' - 2(x - 1)y' + 2y = 2$, $y = A(x - 1) + B(x^2 - 1) + 1$.
21. $y = (Ax^2 + B)e^x$.
22. $y = Ax + B(1 - x^2)^{\frac{1}{2}}$.
24. $y = e^{-bx}(A + B\int e^{-(x+a-2b)^2/2} dx)$.
25. $y = Ae^{2x} + B \sin x$.

ANSWERS 283

26. $y = A/x + B(x - 2)$.
28. $y = Ax + Be^x + x^2 + 1$.
30. $y = (A + Be^x)e^{-e^x} + e^{2x} - 4e^x + 6$.
27. $y = (A + Bx^{1-b})/(x - 1)$.
29. $y = (A + Be^x + e^{2x})x^2$.
31. $y = (Ae^{4x} + Be^{-4x} - x/16)e^{x^2}$.

Page 77
1. $y = Ax^{\frac{1}{2}} + Bx^{\frac{3}{2}}$.
3. $y = A/x + Bx^{\frac{1}{2}}$.
5. $y = A/x + Bx^2 + x^3/4 - x/2$.
7. $y = A(x + 1)^2 + B(x + 1)^3$.
2. $y = (A + B \log x)/x$.
4. $y = A + Bx^3 + x^5/10$.
6. $y = (A + B \log x)x + x^2$.
8. $y = Ax^2 + Bx^3 + x + \frac{5}{6}$.

Page 79
1. $y = (A + Bx - \sin x)/(x^2 + 2)$.
3. $y = (Ae^{-x} + Be^{-2x})/x^2$.
2. $y = (Ae^{-2x} + B + x)/(x + 3)$.

Pages 82, 83
3. $p^2 = n^2 - 2k^2$.
5. $s = -e^{-4t}(\sin 4t + \cos 4t)$.
7. $y = (21 - \cos 8\sqrt{2}\,t)/4$, 0.56 sec.
9. 2.24 sec.
4. $s = -\cos 4\sqrt{2}\,t$, 1.11 sec.
6. $s = e^{-8t} - 2e^{-4t}$.
8. $y = (1 - \cos 8\sqrt{2}\,t)/4$, 0.56 sec.

Chapter 4

Page 85
3. $y = A \cos x^2 + B \sin x^2$.

Page 89
1, 2. 0.

Page 91
1. $y = Ae^{ax} + Be^{-ax} + C \sin ax + D \cos ax$.
2. $y = A + Be^x + Ce^{2x}$.
3. $y = (A + B \sin x + C \cos x)e^x$.
4. $y = (A + Bx + Cx^2 + Dx^3)e^{-x}$.
5. $y = A \sin x + B \cos x + C \sin 2x + D \cos 2x$.
6. $y = (A + Bx)e^{-x} + C \sin x + D \cos x$.
7. $y = Ae^{2x} + (B + Cx + Dx^2)e^{-x}$.
8. $y = e^{ax/\sqrt{2}}(A \sin ax/\sqrt{2} + B \cos ax/\sqrt{2})$
    $+ e^{-ax/\sqrt{2}}(C \sin ax/\sqrt{2} + D \cos ax/\sqrt{2})$.
9. $y = Ae^x + Be^{-x} + C \sin x + D \cos x + e^{x/\sqrt{2}}(E \sin x/\sqrt{2} + F \cos x/\sqrt{2})$
    $+ e^{-x/\sqrt{2}}(G \sin x/\sqrt{2} + H \cos x/\sqrt{2})$.

Page 96
1, 2, 3. $y = e^{2x}(9x^2 - 24x + 26)/27$.
5. $y = x^3 + 7x^2 + 30x + 55$.
6. $y = 8x^2 - 2x^3 + x^4/2$.
7. $y = -x^2/8$.
8. $y = 1 - x^2$.
9. $y = x^2/2 - 2x$.
10. $y = (\log x)/2$.
11. $y = x^2/2a^2$.
12. $y = (bx^3 - 3x^2)/6b^2$.
13. $y = \sin x + \cos x - 1$.
14. $y = e^{2x}/16$.
16. $y = e^{2x}(\sin 3x - 5 \cos 3x)/234$.
17. $y = (\cos x - \sin x)/2$, $y = (\cos x + \sin x)/2$.

Page 99
1. $y = -x^3/3 - 2x$.
2. $y = Ax + Bx^2 + C/x$.
3. $y = Ax + Bx^2 + x^{-\frac{3}{2}}[C \sin (\sqrt{15} \log x)/2 + D \cos (\sqrt{15} \log x)/2]$.
4. $y = Ax + Bx^2 + Cx^3$.
5. $y = -(6 \log x + 11)/36$.
6. $y = A/(2x + 3) + (2x + 3)[B \sin \log (2x + 3) + C \cos \log (2x + 3)]$.
7. $y = (A + B \sin x + C \cos x)x^2$.
8. $y = A(x^4 - 12x^2 + 24) + B \sin x + C \cos x$.

Pages 107, 108

1. $y = 2$.
2. $y = t^2 - 2t + 3 - e^{-2t}$.
3. $y = 4e^{-2t} + t - 1$.
4. $y = -t \cos t$.
5. $y = t^3 - 3t^2 + 1 - e^{-t}$.
6. $y = e^{-2t}(2 \cos t + 7 \sin t)$.
7. $y = e^t(\sin 2t - 2 \cos 2t) + 2$.
8. $y = e^t(t^2 - 2t + 3)$.
9. $y = e^t$.
10. $y = e^t + t + 1$.
11. $y = 3 - 2e^{t/2} \cos \sqrt{3}\, t/2 - e^{-t}$.
19. $y = (k' \sin nt)/n$.

Chapter 5

Page 111

1. $y_4 = 1 - x + x^2 - x^3/3 + x^4/12 - x^5/60$.
2. $y_4 = e^{-x} + x^2/2 - x^3/6 + x^4/24 - x^5/120$.
3. $y_4 = \cos x - x + 3x^2/2 - x^3/3 + x^4/24 - x^5/120$.
4. $y_4 = -1 + 2e^{-x} + x + x^4/24 - x^5/120$.
5. $y_4 = 1 - x + x^2 - x^3/3 + x^4/12$.
6. $y_4 = 1 - x + x^2 - x^3/3 + x^4/8 - x^5/60$.
7. $y_3 = 2 + 2x + x^2 + x^3/3$.
8. $y_3 = x + x^3/3 + 2x^5/15 + x^7/63$.
9. $y_3 = 1 + x + x^2 + x^3 + 2x^4/3 + x^5/3 + x^6/9 + x^7/63$.
10. $y_3 = (10 + 28x + 168x^2 + 112x^3 - 896x^4 + 1344x^5 - 896x^6 + 256x^7)/63$.
11. $y = -x$.
12. $y = 1 - x^2/2 + x^3/6 + x^4/24 - \cdots$.

Page 119

1. To left, same side if $f_y < 0$.
3. $|y_n - Y| \leqslant NA^n|x - a|^n/n![1 - |x - a|/(n + 1)]$.

Chapter 6

Pages 123, 124

6. $y = ae^{x/3} + be^{-x/3} - 9x - 3$, $z = ae^{x/3} + 2be^{-x/3} - 12x + 9$.
7. $y_3 = x^2/6 + x^3/6 + x^4/216$, $z_3 = x - x^2/2 + 13x^3/54$.
8. $y = x + 2x^2 + 13x^3/6 + \cdots$, $z = 1 + 3x + 7x^2/2 + 10x^3/3 + \cdots$, $u = 1 + x^2 + x^3/3 + \cdots$.

Page 127

1. $dx/x = dy/y = dz/2z$.
2. $dx/(y - 3z) = dy/(2z - x) = dz/(3x - 2y)$.
3. $dx/(x - a) = dy/(y - b) = dz/(z - c)$.
4. $dx/2xz = dy/2yz = dz/(z^2 - x^2 - y^2)$.
5. $dx = dy/(6y - 5z) = dz/(2y - z)$.   6. $dx/x = dy/y = dz/z$.
7. $dx/2xy = dy/2(z - x^2) = dz/(-y)$.   8. $dx/(1 - y^2 - z^2)^{\frac{1}{2}} = dy/y = dz/z$.
10. $dx/2xy = dy/(y^2 - x^2 + 1) = dz/0$.

Page 130

1. $y = ax$, $z = bx$.   2. $x = a \sin (z/k + b)$, $y = a \cos (z/k + b)$.
3. $z - y = a(x - y)$, $(x + y + z)(x - y)^2 = b$.
4. $y = ax$, $z^2 = 4xy + b$.   5. $y^2 + z^2 = a^2$, $2a/x + \log[(a + y)/(a - y)] = b$.
6. $x^2 + 2xy - y^2 = a$, $z^2 - 2xy = b$.   7. $x + y = a$, $2z - x^2 \sin (x + y) = b$.
8. $x + y - 2z = a$, $x^2 - 2ax + 4xy + y^2 = b$.
9. $1/(x + y) = 1/(x - y) + C$.   10. $x^2 + y^2 = a^2$, $z = b$.
11. $x + az = A$, $y + bz = B$.   12. $x^2 + y^2 + 2z^2 = a^2$, $y = bx$.
13. $x^2 + y^2 + z^2 = ax$, $y = bx$.   14. $x^2 - y^2 = a$, $z = b(x + y)$.
15. $ae^x + be^{-x}$, $ae^x - be^{-x}$.   16. $(2ax + b)^{\frac{1}{2}}$, $(2ax + b)^{\frac{1}{2}}/a$.

ANSWERS

17. $x^2 + y^2 = z^2$.   **18.** $z^2 = 8x^2 - 4y^2 + 1$.

Page 132
1. $y = kx^2/2H$.
2. $y = (k^2x + 2H^2)(k^2x^2 + 4H^2x)^{\frac{1}{2}}/4kH^2 - k^{-2}\log[k^2x + 2H^2 + k(k^2x^2 + 4H^2x)^{\frac{1}{2}}]$.
3. Of form $y = a(e^{mx} + e^{-mx})$.

Page 133
1. $x = s/k$, $y = r/k$.   2. $rX^2 + sY^2 = C$, period $= 2\pi/\sqrt{rs}$.
3. $k(x + y) - s\log x - r\log y = C$.   4. $y = -x + (s\log x)/k + C$.
5. Both die out.
6. $k(\theta x + y) - s\log x - r\log y = C$, $0 < \theta < 1$.
7. $y = Ke^{-st}$, $x = Ce^{rt+ky/s}$.   8. $y = a - x$, $x = a/(1 + be^{akt})$.

Page 141
3. $f(x)$, $Ce^x - f(x)$.   5. $z = x + y/x + C$.
6. $z - kxy = Cy$.   7. $(x + z)y = C(y + z)$.   8. $x^2 - yz^2 + \log yz^3 = C$.
10. $y + 2 = K(x - 3)$.   11. $z = C$, $x^2 = 1$.   12. $r = Ce^\theta$.
13. $y = mx - n$, $z = nx$.   14. $x^2 + y^2 + z^2 = a^2$, $xy + z = b$.
15. $nx - lz = C(ny - mz)$.
16. $x = C$, $y^2 + z^2 = a^2$; $y = K$, $x^2 + (z - 1)^2 = b^2$.
17. $x^2 + 3y^2 + 2z^2 = C$.   18. $f(x)$, $e^{-x}[C + 2\int e^x f(x)\,dx] - f(x)$.

Pages 143, 144
2. $y = a\sin x + b\cos x - x^2 + 4$,
   $z = a\sin x + b\cos x + cx + d + x^4/12 + x^3/6 - x^2$, $w = x^2 + x + 1$.
3. (a) $x = x_0 e^{-at}$, $y = ax_0(e^{-at} - e^{-bt})/(b - a)$, $z = x_0 - x - y$.   (b) $x = x_0 e^{-at}$,
   $(b - a)y = [by_0 - a(x_0 + y_0)]e^{-bt} + ax_0 e^{-at}$, $z = x_0 + y_0 + z_0 - x - y$.
4. $x_k = C_1 e^{-a_1 t} + C_2 e^{-a_2 t} + \cdots + C_k e^{-a_k t}$, if $a_i \neq a_j$.

Pages 148, 149
1. (a) 36,000,000 miles.   (b) 248 yr.   5. $\sqrt{2}$.
7. $x = v_0 t \cos\alpha$, $y = v_0 t \sin\alpha - gt^2/2$.
8. $x = (v_0 \cos\alpha)(1 - e^{-ht})/h$, $y = (v_0 h \sin\alpha + g)(1 - e^{-ht})/h^2 - gt/h$.
9. $x = A\sin at + B\cos at$, $y = C\sin at + D\cos at$;
   or $(Cx - Ay)^2 + (Dx - By)^2 = (AD - BC)^2$; period $= 2\pi/a$.
10. $x = Ae^{at} + Be^{-at}$, $y = Ce^{at} + De^{-at}$; or $(Cx - Ay)(By - Dx) = (AD - BC)^2$,
    a hyperbola.   11. A spiral, $(Cx - Ay)^2 + (Dx - By)^2 = e^{-ht}(AD - BC)^2$.

Chapter 7

Pages 154, 155
1. (a) $y = 1 + x^3/3! + 1 \cdot 4x^6/6! + 1 \cdot 4 \cdot 7x^9/9! + \cdots$ (b) $y = x + 2x^4/4!$
   $+ 2 \cdot 5x^7/7! + 2 \cdot 5 \cdot 8x^{10}/10! + \cdots$.   2. $y = 1 - 10x^2 + 35x^4/3$.
4. Radius at least: (a) 1, (b) 2, (c) $\sqrt{13}$, (d) $\sqrt{13}$, (e) ∞.
5. (a) $y = 1 + kx^2/2! + k(k + 2)x^4/4! + k(k + 2)(k + 4)x^6/6! + \cdots$
   (b) $y = x + (k + 1)x^3/3! + (k + 1)(k + 3)x^5/5! + \cdots$
6. $y = x - x^2/2 + x^3/3(2!)^2 - x^4/4(3!)^2 + \cdots$ 7. $y = 1 - x + x^2/6$
   $- \cdots + (-1)^n 2^n x^n/(2n)! + \cdots = \cos\sqrt{2x}$, $y = \sqrt{x}[1 - x/3 + x^2/30$
   $- \cdots + (-1)^n 2^n x^n/(2n + 1)! + \cdots] = (\sin\sqrt{2x})/\sqrt{2}$.
8. $y = 1 + x/2 + x^2/8 + \cdots + x^n/2^n n! + \cdots = e^{x/2}$, $y = \sqrt{x}$.
9. $y = x^2(1 - x^2/3! + x^4/5! - x^6/7! + \cdots) = x\sin x$,
   $y = x(1 - x^2/2! + x^4/4! - x^6/6! + \cdots) = x\cos x$.

10. $y = x^2[1 + x^4/3! + x^8/5! + \cdots + x^{4n}/(2n+1)! + \cdots] = (e^{x^2} - e^{-x^2})/2$,
$y = 1 + x^4/2! + x^8/4! + \cdots + x^{4n}/(2n)! + \cdots = (e^{x^2} + e^{-x^2})/2$.
12. 0, λ at 0; 0, μ at 1; $(1 - \lambda - \mu - \nu)/2$, $(1 - \lambda - \mu + \nu)/2$ at ∞.

Page 158
5. $y = F(1,1,-\tfrac{1}{2},x)$.   **6.** $y = F(2,1,3,x)$.
7. $y = F[\tfrac{2}{3}, \tfrac{1}{3}, -\tfrac{1}{2}, (x+1)/2]$.   **8.** $y = F[1, 1, \tfrac{1}{2}, (x-1)/2]$.
9. $y = (A + Bx^{1-\beta})/(1-x)$.

Pages 164, 165
1. $P_5(x) = (63x^5 - 70x^3 + 15x)/8$.   **3.** $x^4 = 8P_4(x)/35 + 4P_2(x)/7 + P_0(x)/5$.
5. $P_{2n}(x) = [(-1)^n 1 \cdot 3 \cdots (2n-1)/2 \cdot 4 \cdots 2n]F(-n, n + \tfrac{1}{2}, \tfrac{1}{2}, x^2)$;
$P_{2n+1}(x) = [(-1)^n 3 \cdot 5 \cdots (2n+1)/2 \cdot 4 \cdots 2n]xF(-n, n + \tfrac{3}{2}, \tfrac{3}{2}, x^2)$.
9. $F(x) = 15[4/\pi^2 - \tfrac{1}{5} - (12/\pi^2 - 1)x^2]/\pi = 0.9802 - 1.0307x^2$.
13. $x = P_0(x)/2 + 5P_2(x)/8 - 3P_4(x)/16 + \cdots$
$+ [(-1)^{n-1} 1 \cdot 3 \cdots (2n-3)/2 \cdot 4 \cdots (2n+2)](4n+1)P_{2n}(x) + \cdots$.

Pages 172, 173
4. $y = \sqrt{x}\, J_1(2\sqrt{x})$.   **5.** $y = x^{-n}[AJ_n(x) + BJ_{-n}(x)]$.
6. $y = AJ_{\sqrt{2}}(ix) + BJ_{-\sqrt{2}}(ix)$.   **7.** $y = \sqrt{x}\,[AJ_{\frac{1}{3}}(2kx^{\frac{3}{2}}) + BJ_{-\frac{1}{3}}(2kx^{\frac{3}{2}})]$.
8. $y = x^{-1}[AJ_0(bx) + BK_0(bx)]$.   **9.** $y = e^x[AJ_0(x) + BK_0(x)]$.

Chapter 8

Page 178
1. $(m^h - 1)^n m^x$.
2. $\Delta^r x^{(n)} = h^r n(n-1) \cdots (n-r+1)x^{(n-r)}$,
$\Delta^r x^{(-n)} = (-1)^r h^r n(n+1) \cdots (n+r-1)x^{(-n-r)}$.
3. $\Delta[u(x)v(x)] = u(x+h)\,\Delta v(x) + v(x)\,\Delta u(x)$,
$\Delta[u(x)/v(x)] = [v(x)\,\Delta u(x) - u(x)\,\Delta v(x)]/v(x)v(x+h)$.
5. (a) $y = A3^x + B5^x$. (b) $y = A2^x + B$.

Pages 181, 182
2. (a) 6:55, 4:56. (b) Jan. 10, 7:04. (c) Dec. 3, 4:54. (d) Dec. 23, 9 hr 59 min.
3. 8,210 ft.   **5.** Taking $a = 23°$, the errors are: 795, 17, 1, 0.
6. Errors: 195 28930 8, 24697 4, 53 6, 2, 0.

Page 185
1. $h \leqslant 0.0424$.   **2.** $h \leqslant 5.27°$.

Page 191
2. $\int_{-1}^{1} F(x)\,dx = F(-1/\sqrt{3}) + F(1/\sqrt{3})$.

Page 196
2. Length = 15.9.   **3.** $c = h$, $a = h/\sqrt{3}$.
4. The rectangle.   **5.** Length = 8.53.

Pages 204, 205
1. $n^2(n+1)^2/4$.   **2.** $m(8m^2 + 33m + 61)/6$.
3. Difference = $h(\Delta^6/140 + \cdots)$.
7. (a) $(5 - 2x + 2x^2)/(1-x)^3$. (b) $(11 - 2x - 3x^2)/(1-x)^4$.
8. $520,480.
10. 311 hr 43 min.
11. $\delta = 0.031576$, $\delta^2 = -0.000059$.

Chapter 9

Page 212
6. $x = 1.104$, $y = 1.997$.

ANSWERS

Page 214
3. $y(5.5) = 4.508$, $y(6) = 5.005$.

Page 216
1. $y(5) = 1.643$, $y(10) = 2.714$, $y(20) = 7.428$. Correct: 1.649, 2.718, 7.389.
2. $u(0) = 0$, $u(0.25) = 0.002$, $u(0.5) = 0.004$, $u(1.0) = 0.005$, $u(1.5) = 0.005$, $u(2.0) = 0.004$, $u(2.5) = 0.004$, $u(3.0) = 0.003$. 3. 54 min.

Pages 219, 220
2. $v = 1.52$.
3. $t = \int_{-1}^{x} \left(x(1 + x^2)^{\frac{1}{2}} + \sqrt{2} + \log\{[x + (1 + x^2)^{\frac{1}{2}}]/(\sqrt{2} - 1)\}\right)^{-\frac{1}{2}} dx$.
6. 0.87 sec, 19.3 feet per second.

Chapter 10

Page 228
3. $2z = px + qy$. 4. $p = q$. 5. $xp + 2z = 0$.
6. $z = px + qy$. 7, 8, 9. $y + zq = 0$.
10. $z = px + qy \pm (1 + p^2 + q^2)^{\frac{1}{2}}$. 11. $z(px + qy) + x^2 + y^2 = 0$.
12. $2(px + qy - z)^2 = (p^2 + q^2 + 1)(x^2 + y^2 + z^2)$.
13. $z = x^2y^2/2 + \phi(x)$. 14. $z = x^2y + A(y)x + B(y)$.

Page 235
1. $4z + (x + y)^2 = 0$. 2. $8z + (x + 2y)^2 = 0$.
3. $27z = 2y^3 - 9xy \pm 2(y^2 - 3x)^{\frac{3}{2}}$. 4. $(z - 1)^2 = 4xy$.
7. $lp + mq - n = 0$. 8. $p^2 - q^2 = 1$.

Pages 244, 245
1. $z = (x/2 + a)^2 + by$. 2. $z = y(ax - a^2y + b)$. 4. A cone.
6. Solve for ϕ from $f(z,\phi,a\phi) = 0$, then $x + ay + b = \int dz/\phi$.
7. $z = (3x + ay + b)^{\frac{2}{3}} - a^2/9$.
8. Solve for σ and τ from $\Phi(x,\sigma) = a$, $\Psi(\tau,y) = a$, then $z = \int \sigma\, dx + \int \tau\, dy + b$.
9. $3z/2 = (x - a)^{\frac{3}{2}} - (a - y)^{\frac{3}{2}} + b$. 10. $z = ax + by + f(a,b)$.
11. $z = ax + by + (1 + a^2 + b^2)^{\frac{1}{2}}$, $x^2 + y^2 + z^2 = 1$.
12. $z = (x - y)^2/4$; also $z = -xy$. 13. $z = 2\sqrt{xy} + 1$.
14. $2az = ak + 4 + (x + ay + b)^2$. 15. $az^2 = (x + b)^2 + 4y^2$.
16. $4z = (x + a)^2 + (y + a)^2 + (x - y)[(x - y)^2/2 - a^2]^{\frac{1}{2}}$
    $- \sqrt{2}\, a^2 \log\{x - y + [(x - y)^2 - 2a^2]^{\frac{1}{2}}\} + b$. 17. $z = (x \pm y)^2/8$.
18. Eliminate a from $f(x,y) = (x + a)[y + \varphi(a)]$, $(x + a)\varphi'(a) + y + \varphi(a) = 0$.

Pages 249, 250
1. $z = cx/a + f(bx - ay)$. 2. $z = y^2/x + f(y/x)$.
3. $f(x^2 + y^2 + z^2, lx + my + nz) = 0$.
4. $z = [x + yf(x^2 + y^2)]/[y - xf(x^2 + y^2)]$. 5. $e^x F(x - y)$.
6. $\mu = xf(x^2y)$. 7. $f(y/x, z/x) = 0$. 8. $f(z, y/x) = 0$.
10. $z = x^2/y$. 11. $2z^2 = x^2 + y^2 - 1$. 12. $e^x F(x - y, x - z)$.
13. $(u - z)^3(x + y + z + u) = f[(u - x)/(y - z), (x - y)/(y - z)]$.
14. $f = x^n F(y/x, z/x)$. 15. $f = F[(y - x)/xy, (z - x)/xz]$.

Chapter 11

Page 252
1. $z = A_0(x) + A_1(x)y + \cdots + A_{n-1}(x)y^{n-1} + B_0(y) + B_1(y)x + \cdots + B_{m-1}(y)x^{m-1}$. 2. $z = e^y A(x + y) + B(y)$.
3. $z = y(x - 1) + A(x) + B(y)e^{-x}$. 4. $z = xy^3/3 + A(x)y^2 + B(x)$.

5. $3z_{xx} + z_{xy} - 2z_{yy} = 0$.   **6.** $z_{xy} - z_x - z_y + z = 0$.
7. $2z_{xx} - 3z_{xy} + z_{yy} - z_y - 2z = 0$.   **8.** $z_{xx} - z_{yy} + z_x - z_y = 0$.
9. $z = A(x)e^{-fy} + B(y)e^{-gx}$.

Pages 256, 257
1. $z = A(y + ix) + B(y - ix)$.   **2.** $u = A(t + ax) + B(t - ax)$.
3. $z = A(ay - bx) + B(y)$.

Page 259
3. $z = A(y) + B(y + x) + C(2y - x)$.
4. $z = A(y + x) + B(y - x) + C(y + ix) + D(y - ix)$.
5. $z = f(y + 3x) + x^2 - f(4x)$.

Page 265
4. $z = -(x^2 + y + 3) + \Sigma A_k e^{\theta_k}$, where $\theta_k = m_k x + (m_k^2 - 1)y/(2m_k - 1)$.
5. $z = f(y + 2x) + e^{3x}g(y - 2x) - [x^2y/6 + (x^3 + 2x^2 + xy)/9 - 2y/27 + xe^{x+y}]$.
6. $z = \Sigma A_k e^{m_k x + (m_k^2 + 2m_k)y}$.
7. $z = e^{-ax}[f(y - mx)\cos bx + g(y - mx)\sin bx]$.
9. $z = \Sigma A_k x^{m_k} y^{n_k}$, $2n_k = 2m_k + 1 \pm (8m_k + 1)^{\frac{1}{2}}$.
10. $z = (Ae^{ax} + Be^{-ax})(C \sin ay + D \cos ay)$,
$z = (A \sin ax + B \cos ax)(Ce^{ay} + De^{-ay})$, $z = (Ax + B)(Cy + D)$.
11. $z = (ax^2 + Ax + B) + (-ay^2 + Cy + D)$.
12. In Exercise 10 replace x and y by $x + y$ and $x - y$, respectively.

Pages 268, 269
1. $T = 2l/na$; $x = l/n, 2l/n, \cdots$.   **2.** $y = h \sin x \cos at + k \sin 2x \cos 2at$.
4. Yes.   **8.** $y = \Sigma(ld_n/na\pi) \sin n\pi x/l \sin na\pi t/l$.
9. $2ay = he^{-x^2-a^2t^2} \sinh 2axt$.
10. $f(x)$ even with respect to the origin. $y = h \cos \pi x/2l \cos \pi at/2l$.

Page 272
1. $z = \Sigma \sin m\pi x/a \sin n\pi y/b \, [c_{mn} \cos c\pi k t + (d_{mn}/c\pi k) \sin c\pi k t]$,
where $k = (m^2/a^2 + n^2/b^2)^{\frac{1}{2}}$, and $z_t(x,y,0) = \Sigma d_{mn} \sin m\pi x/a \sin n\pi y/b$.
2. $T = 2/c(m^2/a^2 + n^2/b^2)^{\frac{1}{2}}$; $x = a/m, 2a/m, \cdots, y = b/n, 2b/n, \cdots$.
3. $T = 2\pi a/c\alpha_n$; $r = a\alpha_1/\alpha_n, a\alpha_2/\alpha_n, \cdots$.
4. $T = 2a/c\sqrt{5}$; $x + y = a$; $y = x$; $A \cos \pi y/a + B \cos \pi x/a = 0$.

Pages 274, 275
2. $u = Ax + B$.
3. $u = (1/4a^2\pi t)\int_{-\infty}^{\infty}\int_{-\infty}^{\infty} f(\lambda,\mu)e^{-[(x-\lambda)^2+(y-\mu)^2]/4a^2 t}\, d\lambda\, d\mu$.
6. $u = u_0(1 - e^{-ht}) + (e^{-ht}/2a\sqrt{t\pi})\int_{-\infty}^{\infty} f(\lambda)e^{-(x-\lambda)^2/4a^2 t}\, d\lambda$.
7. $u = u_0 + Ae^{\sqrt{h}x/a} + Be^{-\sqrt{h}x/a}$.

Page 278
1. $u = (100/\sinh \pi) \sin \pi x/a \sinh \pi(1 - y/a)$.

INDEX

Abel's result, 64
Algebra of operators, 197
Alteration, of the function, 117
   of initial conditions, 118
Analytic function, 150
Analytic solutions, 150–152
Approximations, method of successive, 109–123
Arbitrary constant, 5
Auxiliary equation, 66, 90

Backward differences, 223
Beats, 82
Ber and bei functions, 173
Bernoulli's equation, 36
Bessel's equation, 165–173
Bessel's functions, 166
   expansions in, 172, 272
   integral properties, 171, 172
   recurrence formulas, 173
   roots of, 168–170

Catenary, 132
Cauchy problem, 239, 241
Cauchy-Lipschitz method, 220
Cauchy-Riemann equations, 253
Central forces, 145
Characteristic curve, 232, 245
Characteristic point, 233
Characteristic strip, 235–242, 245
   equations of, 236–239
Charpit's method, 242–244
Checking, 214
Clairaut's equation, 16, 245
Classical equations, 150–173
Coefficients, undetermined, 70, 91, 264
Complete equation, 60, 70–75, 91–99
Complete solution, 233–235, 242–244

Completely linear equation, 257, 259–264
Conical element, 231, 235
Constant coefficients, equations with,
   complete, 70–73, 91–96, 101
   linear partial, 257–259, 262–264
   reduced, 65–69, 89–91
   systems of, 129, 141–143
Cord, flexible, 131

D, 93, 197
D^{-1}, 200
Damping, 80
∇, 223
Δ, 177, 197
Δ^{-1}, 140, 200
Difference tables, 178
Differences, finite, 177–206
   of polynomial, 177
Differential equations, Bernoulli's, 36
   Bessel's, 165–173
   Clairaut's, 16, 245
   Euler's ordinary, 76
   Euler's partial, 253–257
   exact, 49–52, 77
   first-order, 33–59
   homogeneous, 40–42
      linear partial, 257–259
   hypergeometric, 155–157
   Legendre's, 158–165
   linear (see Linear differential equations)
   linear fractional, 45–49
   nth order, 133
   numerical solution (see Numerical solution of differential equations)
   partial, first-order, 225–250
      second-order, 251–278
   Pfaffian, 136–140
   Riccati's, 38–40
   systems of ordinary, 120–149

DIFFERENTIAL EQUATIONS

Differential equations, systems of ordinary, of higher order, 135, 141–143
total, 135–140
Differentiation, numerical, 194–196, 199
Direction field, 9, 125
Dirichlet problem, 277
Drum, 271

E, 197
Electric current, 38
Envelopes, 17, 231–233
Equipotential curves, 23
Error, in interpolation, 182–185
in numerical differentiation, 195
in numerical integration, 188, 191, 194
in Simpson's rule, 188
Euler's equation, ordinary, 76
partial, 253–257
Euler-Maclaurin series, 203
Exact equations, 49–52, 77

$F(\alpha, \beta, \gamma, x)$, 156
Falling raindrop, 29
Field, direction, 9, 125
Finite differences, 177–206
First-order equations, 33–59
partial, 225–250
Flexible cord, 131
Flow, of heat, 272–275
of water from orifice, 32
Forced vibrations, 81
Fourier series, 268, 270, 276
Fractional equation, linear, 45–49

Gauss's formula, 205
Geometrical applications, 20
Gramians, 88
Green's theorem, 79
Guessing, aids to, 212–214

Heat conduction, 272–275
Homogeneous equations, 40–42
linear partial, 257–259
Hypergeometric equation, 155–157
Hypergeometric series, 156

Indicial equation, 153
Initial conditions, change of, 118

Integrable Pfaffian, 137–140
Integral curve, 9, 125, 136
Integral surface, 231
through a curve, 239–242
enveloping a surface, 240
Integrating factors, 52–57, 78, 247
Interpolation, 174–206
error in (see Error in interpolation)
parabolic, 175
Interval, change of, 204

$J_n(x)$, $J_o(x)$, 166

Kepler's laws, 148
Kutta, 221

Lagrange's formula, 176
Laplace transforms, 99–108
references, 106
table of, 107
Laplace's equation, 253, 261, 275–278
Legendre polynomials, 159–165
expansion in series of, 163
integral properties, 160–162
recurrence formulas, 165
roots of, 163
use in numerical integration, 190
Legendre's equation, 158–165
Line integrals, 51, 140
Lineal element, 9, 125
Linear dependence, 62, 85–89
Linear differential equations, with constant coefficients (see Constant coefficients)
first-order, 33–36
nth-order, 84–99
partial, first-order, 245–250
second-order, 253–278
second-order, 60–83
systems of, 122, 141–143
Linear fractional equation, 45–49
Linear interpolation, 184
Lipschitz condition, 112, 120, 134
Lubbock's formula, 206

Mayer's method, 139
Membrane, vibrating, 269–272

INDEX

Motion, equations of, 27, 144
  of particle, 26–30, 144
  of planet, 146–148
  simple harmonic, 80

n-parameter families, 19
Newton's formula, 179
  for backward interpolation, 205
Numerical differentiation, 194–196, 199
Numerical integration, 185–194, 200–205
Numerical solution of differential equations, 207–224
  finite difference methods, 223
  Laplace's equation, 276–278
  references, 207

Oblique trajectories, 23
One-parameter families, 14
Operators, 197–206
  algebra of, 197
  fundamental equations, 198
Order of differential equation, 1
  reduction of, 74, 97–99
Ordinary point, 152
Orthogonal curves, 22, 126
Osgood's theorem, 116

$P_n(x)$, 160
Parabolic interpolation, 175
Parameters, variation of, 72, 96
Parasite problem, 132
Partial differential equations, 225–278
Partial fractions, 94, 103
Particle, motion of, 26–30, 144
Particular solution, 5
Pendulum, 28
Pfaffian, 136–140
Picard, 109
Planar element, 231
Planetary motion, 146–148
Poisson's integral, 262
Potential, 261, 276

Radiocarbon dating, 31
Raindrop, falling, 29
Reduced equation, 61, 65–70, 84–91
Reduction of order, 74, 97–99
Regular singular point, 152–154
Remainder (*see* Error)
Riccati's equation, 38–40

Rodrigues's formula, 160
Rules, making of, 55
Runge-Kutta formulas, 221–223

Separation of variables, 4
Series, of Bessel functions, 172, 272
  Fourier, 268, 270, 276
  hypergeometric, 156
  $J_n(x)$, $J_o(x)$, 166
  of Legendre polynomials, 163
  symbolic, 94, 198–205, 264
  Taylor's, 12, 150
Simple harmonic motion, 80
Simpson's rule, 186–189, 203
Singular points, regular, 152–154
Singular solutions, 17, 234
Star, 151
Start of the solution, 208
Steepest descent, 23
Stirling's formula, 205
String, vibrating, 265–268
Strip, characteristic, 235–242, **245**
Sturm's theorem, 169
Subsidiary equation, 102
Successive approximations, method of, 109–123
Summation, 200, 206
Surface, integral (*see* Integral surface)
Suspension-bridge problem, 132
Symbolic methods, 92–96, 196–206
Symbolic series, 94, 198–205, 264
Systems of equations, 120–149, **217**

Tables, difference, 178
Taylor's series, 12, 150
Three-eighths rule, 203
Total differential equations, 135–140
Trajectories, 22

Undetermined coefficients, 70, 91, 264

Variables, separable, 4
Variation of parameters, 72, 96
Velocity, 6
Vibrating membrane, 269–272
Vibrating string, 265–268
Vibrations, 79–82

Weddle's rule, 204
Woolhouse's formula, **206**
Wronskians, 62–65, 85–88